Fuzzy Systems Design

Studies in Fuzziness and Soft Computing

Editor-in-chief
Prof. Janusz Kacprzyk
Systems Research Institute
Polish Academy of Sciences
ul. Newelska 6
01-447 Warsaw, Poland
E-mail: kacprzyk@ibspan.waw.pl

Vol. 3. A. Geyer-Schulz
*Fuzzy Rule-Based Expert Systems and
Genetic Machine Learning, 2nd ed. 1996*
ISBN 3-7908-0964-0

Vol. 4. T. Onisawa and J. Kacprzyk (Eds.)
*Reliability and Safety Analyses under
Fuzziness, 1995*
ISBN 3-7908-0837-7

Vol. 5. P. Bosc and J. Kacprzyk (Eds.)
*Fuzziness in Database Management
Systems, 1995*
ISBN 3-7908-0858-X

Vol. 6. E.S. Lee and Q. Zhu
Fuzzy and Evidence Reasoning, 1995
ISBN 3-7908-0880-6

Vol. 7. B.A. Juliano and W. Bandler
Tracing Chains-of-Thought, 1996
ISBN 3-7908-0922-5

Vol. 8. F. Herrera and J.L. Verdegay (Eds.)
*Genetic Algorithms and Soft Computing,
1996*
ISBN 3-7908-0956-X

Vol. 9. M. Sato et al.
*Fuzzy Clustering Models and Applications,
1997*
ISBN 3-7908-1026-6

Vol. 10. L.C. Jain (Ed.)
*Soft Computing Techniques in Knowledge-
based Intelligent Engineering Systems,
1997*
ISBN 3-7908-1035-5

Vol. 11. W. Mielczarski (Ed.)
*Fuzzy Logic Techniques in Power Systems,
1998*
ISBN 3-7908-1044-4

Vol. 12. B. Bouchon-Meunier (Ed.)
*Aggregation and Fusion of Imperfect
Information, 1998*
ISBN 3-7908-1048-7

Vol. 13. E. Orłowska (Ed.)
*Incomplete Information:
Rough Set Analysis, 1998*
ISBN 3-7908-1049-5

Vol. 14. E. Hisdal
*Logical Structures for Representation of
Knowledge and Uncertainty, 1998*
ISBN 3-7908-1056-8

Vol. 15. G.J. Klir and M.J. Wierman
Uncertainty-Based Information, 1998
ISBN 3-7908-1073-8

Vol. 16. D. Driankov and R. Palm (Eds.)
Advances in Fuzzy Control, 1998
ISBN 3-7908-1090-8

Leonid Reznik · Vladimir Dimitrov ·
Janusz Kacprzyk (Eds.)

Fuzzy
Systems Design

Social and Engineering Applications

With 95 Figures
and 21 Tables

Springer-Verlag Berlin Heidelberg GmbH

Prof. Leonid Reznik
Department of Electrical and Electronic Engineering
Victoria University of Technology
P.O. Box 14428 MCMC
Melbourne, VIC 8001, Australia

Prof. Vladimir Dimitrov
Centre for Research in Healthy Futures
University of Western Sydney
Bourke St.
Richmond, NSW 2753, Australia

Prof. Janusz Kacprzyk
Systems Research Institute
Polish Academy of Sciences
ul. Newelska 6
01-447 Warsaw, Poland

ISBN 978-3-662-11811-5   ISBN 978-3-7908-1885-7 (eBook)
DOI 10.1007/978-3-7908-1885-7

Library of Congress Cataloging-in-Publication Data
Die Deutsche Bibliothek – CIP-Einheitsaufnahme
Fuzzy systems design: social and engineering applications; with 21 tables / Leonid Reznik
... (eds.). – Heidelberg; New York: Physica-Verl., 1998
   (Studies in fuzziness and soft computing; Vol. 17)

© Springer-Verlag Berlin Heidelberg 1998
Originally published by Physica-Verlag Heidelberg in 1998.
Softcover reprint of the hardcover 1st edition 1998

Hardcover Design: Erich Kirchner, Heidelberg

SPIN 10679039   88/2202-5 4 3 2 1 0 – Printed on acid-free paper

Preface

This aim of this book is to construct bridges and filling gaps. However, one should not think that it is about civil engineering. The gaps considered are not real although they do exist in our reality - in our perception and thoughts, in our life and relations to each other.

The first gap which we would like to overcome is one between a theory and practice, between academia on the one hand, and a community of practitioners on the other hand. Unfortunately, this gap becomes more and more obvious and now it shows up even at the level of consideration. That is why we concentrate our discussion on design problems with a design being the top of a research and development process, its ultimate point and a point of departure for implementation, where all methods applied and decisions made at the preceding stages are actually tested. On the other hand, a lion's share of this book is occupied by the sections devoted to the design of fuzzy systems for solving practical problems.

This book considers applications in both social and engineering fields and this is another bridge which it is meant to build between these two application areas and corresponding communities, which can use it to learn from each other. Engineers, researchers and students working in fuzzy engineering may learn about social impact of fuzzy system applications and acquire some knowledge about an area, which without any doubt has an unlimited potential for future developments. On the other hand, social scientists and students may be given a hint how to apply real working methods and tools of fuzzy engineering in solving their problems.

Fuzzy sets and systems theory was introduced by Professor Lotfi A. Zadeh in 1965. Not at the least degree attempting to decrease his merit in the invention, development and promotion of a new approach, we would like to mention that this overture was prepared by a previous history of science and, first of all, logic. And one of the forces initiating the birth of fuzzy logic was some famous ancient paradoxes - the problems which could not be solved by classical logic. The history of fuzzy logic is filled with some paradoxes as well. One of the main paradoxes is that despite the founders of this theory initially expected main applications to appear in social sciences and large organisational systems, most of real applications until now have been developed in engineering fields, notably in fuzzy control. Nowadays the problem how to share the experience and combine the developed methodologies of engineering design with the great potential of social applications has emerged.

The volume consists of three parts. The first one includes chapters on philosophical and mathematical basics of fuzzy system design. Providing solid theoretical fundamentals it serves as a bridge between the two other parts considering social and engineering applications. One should note, however, that some authors are very successful in constructing bridges in their papers which include some theoretical developments and application problem solutions together.

The book starts with a discussion of the philosophical foundations of fuzzy logic given in the chapter "Fuzzy Leadership: Dancing with Organisational Reality" by P. Duignan. The paper proposes to apply a "fuzzy logic frame" as an attempt to gain deeper understanding of social reality, especially the nature of leadership. Different aspects of natural and social systems and applications of fuzzy systems for leadership and management are discussed based on the concept of a "cosmic dance of energy". In order to, understand why and how the individuals join together and share in a group or organisation, the metaphor of a "free-form dance" is examined. The problems of language description and interpretation are investigated. This paper should be interesting for any reader with either "social" or "engineering" background.

The chapter "Multi-Criteria Optimisation: an Important Foundation of Fuzzy System Design" by H.T. Nguyen and V. Kreinovich considers a problem which is quite common in different research areas, and typical in design. Namely: how to design a control device with both a good steady-state response and excellent transient characteristics? How to increase budget revenues and decrease taxes? The traditional way of trying to achieve several goals simultaneously is their combination into one. Besides the obvious fact that one achieves neither objective, it is not often clear which combination to apply. As the authors claim, fuzzy logic provides a natural tool how to handle multi-criteria optimisation problems. The paper proposes a mathematical foundation for a number of fuzzy logic based optimisation methods developed until now and proves that fuzzy logic application is really a "natural" way of design.

Another way of decision making in a fuzzy environment is proposed by V. Korotkih in the chapter "A Mathematical Framework for Human Decision Making as an Integrated part of the Whole". This paper considers a model of human collective decision making. The web of integer relations and structural complexity is proposed as a mathematical framework to model a problem situation in which each individual decision is considered in its relation to others or as a part of the whole process. The process of decision making in this case is supposed to be a multi-step and adaptive one. This web is later used to study properties of strategies of the optimal decision search and formulation of the optimal rules. As one may see this model can be definitely considered as a mathematical basis for the design of both social and engineering systems.

The study of decision making continues in the next chapter "Fuzzy Decision Making in Design on the Basis of the Habituality Situation Application" by A.E. Gorodetsky. Although this paper considers a mathematical formalism, it has a practical goal: to improve an efficiency of expert systems. The way how to reach this goal is quite obvious - one has to teach an expert system how to apply some tricks used by a human expert. One of those tricks is the decision making based on the so-called habituality situation application. If an environment situation at the decision moment is similar to the situation observed previously for which a good decision is known, there is no reason to waste time and other resources by looking for another decision. The chapter describes how to determine the habituality situation and to decide if the situation is similar enough to apply the well-known decision.

The same idea of creating intelligent artificial tools by including some features previously specific to the human designers is exploited in the next chapter "Fuzzy Logic Applications in Computer Aided Design" by B. Pham. As one can see from its title it is associated with the computer aided design. However, the author's main idea is rather non-standard for contemporary CAD systems. In order to improve the quality of a CAD system the paper proposes to include into the CAD systems some methods featuring a human-like design procedure. These methods are based on the application of aesthetic factors and their fuzzy models. It will allow to expand CAD applications from traditional limited engineering and architectural areas to other types of design such as sculpting or industrial design, and probably further to social system design, where aesthetic factors play a more important role. And definitely these CAD systems will be introduced with some characteristics, not typical for the computers, like creativity and artistic abilities.

J. Kacprzyk in his paper "Including Socioeconomic Aspects in a Fuzzy Multistage Decision Making Models of Regional Development Planning" considers the planning of socioeconomic regional development over a planning horizon. The development is considered from economic, social, infrastructural, etc. points of view. The development is chracterised by a life quality index consisting of 7 life quality indicators concerning: economic, environmental, housing, health service, infrastructure, work opportunity, and leisure time qualities. Both objective and subjective aspects are accounted for, the latter mainly in terms of resulting social satisfactions. The problem is formulated in terms of Bellman and Zadeh's (1970) multistage decision making (control) under fuzzy constraints on investments applied and fuzzy goals on the life quality attained and their resulting social satisfaction, seeking optimal investments over a planning horizon. Some stability requirement to reflect a human preference to operate under possibly stable conditions is also included. The model may be viewed as an example of a *soft* approach to the modelling of a complex socioeconomic problem.

In this book a number of articles of Part 2 illustrate various social applications of fuzzy logic. An original framework for fuzzy system design related to modern participatory management based on fuzzy logic is developed; it is also shown how fuzzy system design helps to unlock some destructive cycles continuously formed in complexity of human and organisational ecology.

R. Woog, V. Dimitrov and L. Kuhn-White in their paper "Fuzzy Logic as an Evocative Framework for Studying Social Systems" share their experience of fuzzy logic applications in social system research. This chapter answers a question why fuzzy logic is useful in social research. The authors emphasise a crucial role for fuzzy logic in social inquiry and such of its fields as a conversation mapping, a heuristic pattern formation, an emergence of meaning, a multi-layered interpretation, a study of temporality and non-foundations thinking. One of the major contributing strengths of a fuzzy logic approach to social inquiry is its ability to cope with multiple constructions of reality, as well as other forms of circumstantial and constructed complexity. In the application to research/inquiry methodology, fuzzy logic helps to avoid absolute statements and deny the legitimacy of all dichotomies, to embrace ambiguity and ambivalence, to seek freedom from a myopia of hyper-determined research projects and formulations, and to be aware of the tentativeness of causal explanations.

The discussion of management problems is continued in the paper "Fuzzy Logic and the Management of Social Complexity" by V. Dimitrov and K. Kopra. The presentation is concentrated on an application of fuzzy logic to management of paradoxes inherent in social systems and their effects as well as the chaos of social life. The role of fuzzy logic in understanding and healing some socio-political maladies of contemporary societies is paid a special attention to. Among the maladies considered are: a false face of representative politics, a misconception of liberty and equality as mutually exclusive, a destructive competition based on short-sighted economic rationalism, a disconnection of economy from society, a no-alternatives strategy, a false concept of personal identity. One can see how important these problems are for a modern society. Regardless of one's political and philosophical points of view, the chapter provides a reader with a research tool for a social system understanding and analysis.

The next chapter "Using Fuzzy Sets to Extend Holland's Theory of Occupational Interests" by M. Smithson and B. Hesketh considers a fuzzy system application to enhance a typological theory of careers and career choice introduced by Holland which is famous in psychology. This application has several advantages, such as a possible use of different types of analysis which both sharpen and extend the meaning of consistency and congruency, the focus on a relation of an individual to different groups, and a better handling of information contributed by measuring avoidance reactions. A large part of the paper is devoted to the report of empirical studies in which the fuzzy sets and systems methodology were applied, and to the analysis of the results obtained.

Another psychological application is considered in the next chapter "A Model for Fuzzy Personal Construct Psychology" by A. Anderson. The author attempts to fuzzify another well-known psychological theory - the personal construct psychology by G. Kelly. Personal constructs are a means of discrimination between items (usually called elements) in terms of similarity and contrast.

Clearly, this research subject could be important for specialists in different areas. The paper presents a mathematical model which combines the essence of the personal construct psychology with fuzzy logic. The proposed methodology is incorporated into a software package FUZZYGRID and a case study is considered.

The next chapter "On the Topology of Uncertainty" by D. McNeil and V. Dimitrov investigates the role of uncertainty in real life and how fuzzy logic may be applied to model it. The social framework of fuzzy logic presents a perspective which makes it possible to understand, how our current world order works. Despite ecological disasters, violence, injustice, poverty and wars, there is some kind of order that keeps humanity resisting the forces of distraction. Paradoxically, the driving mechanism of this order is uncertainty, embedded in our individual and social life. It is the endavour for dealing with this uncertainty and trying to grasp and manage its multifarious aspects that keeps people looking for consensus and togetherness, although it seems difficult sometimes. Fuzzy logic is a way of thinking that is responsive to human zeal to unveil uncertainty and deal with social paradoxes emerging from it.

Different fuzzy logic applications in engineering have become well known. Various software and hardware design tools have been developed and applied successfully. So Part 3, the engineering part of the book, contains a number of papers, devoted to the description of fuzzy engineering design methodologies. In order to share the experience gained we have selected papers describing not the application results only but the way how this results have been obtained, that is explaining the design procedures. The papers chosen consider mainly not the theory of fuzzy system design but practical aspects of design.

The chapter, which opens up a collection of the papers devoted to the problems of fuzzy control system design and applications, "Fuzzy Controller Design for Different Applications: Evolution, Methods, and Practical Recommendations" by L. Reznik briefly describes a short history of fuzzy controller design, underlining an interaction of different approaches in fuzzy system design. This competition between different methods has become a driving force for further improvements and developments of new synergetic design methodologies. Not attempting to propose a comprehensive mathematical theory of design, the paper aims at providing some advice on which way of design to apply, and how to choose the structure and parameters of a fuzzy controller. These recommendations could be useful for designers of different areas in fuzzy systems.

L.T. Koczy and D. Tikk in their paper "Approximation of Transfer Functions by Various Fuzzy Controllers" are trying to expand a very popular in control engineering as well as a simple mechanism of transfer functions to a domain of any fuzzy system. Or, at least, most widely used fuzzy systems with a reasonably low computational complexity can be included. If proved, this method will allow to construct a bridge to an application of very well developed analytical and design methodologies to artificial intelligence systems.

Next chapter "Fuzzy Performance Indicators for Manufacturing Processes" by L. Berrah, G. Mauris, L. Foulloy, and A. Haurat discusses the control of manufacturing processes. In addition to the accounting measures and the financial reports, supervisors usually require performance indicators for the "physical" evaluation of the manufacturing process evolution, leading to a more reactive control. According to pre-set objectives, the indicators evaluate the results obtained, and measure the extent of any detected drift or malfunctioning. These indicators can be formed on the base of both technical and human generated data. When the measures are effected by physical sensors, they are numerical and eventually pervaded by errors, often described by a Gaussian distribution, while the measures related to subjective estimates are often acquired by human operators who describe them as linguistic variables. Here an engineering problem becomes an organisational problem. Can fuzzy technology help solve both of them together?

The chapter "Advanced Neuro-Fuzzy Engineering for Building Intelligent Adaptive Information Systems" by N.K Kasabov is a brilliant example of an application of engineering methodologies to a new domain of information systems. The paper introduces a new architecture of fuzzy neural networks (FNNs). These FNNs have been suggested and applied by several authors for learning and tuning fuzzy rules and solving classification, prediction and control problems while applied mainly in an engineering area. In this paper the author proposes a new architecture of FNN, called a FuNN, and develops a general methodology for building modular FuNN-based adaptive intelligent systems. Interesting enough, possible applications considered in the chapter include adaptive intelligent speech interfaces to databases and an adaptive prediction of chaotic time-series with financial applications. The results obtained demonstrate how strong the "engineering factor" has become in business and information management applications.

The authors of the next chapter "Linguistic Variables in Surveys of Intentions to Purchase Energy Efficient Equipment", W. Mielczarski, G. Michalik, M.E. Khan are engineers. However, the topic of their research and their results can cause a strong interest of "social" scientists, businessmen, market researchers, and even politicians. They are applying linguistic variables and scales to analyse customer's purchase intentions. By the way, the same method seems to be suitable for forecasting the electorate's voting and for any type of forecasting based on customers' surveys. The approach proposed can be a tool for the analysis of linguistic terms, assigned subjective probabilities and sensitivity studies.

Although the application considered in the next chapter "A Universal Approach to Adaptive Fuzzy Logic Controller Design With an Application to a

Power Generator Excitation Control" by O. Ghanayem and L. Reznik is a traditional power engineering area of generator control, the proposed design methodology has a universal nature. The paper introduces a new method for on-line tuning of fuzzy controller scaling factors. An importance of a right choice of the scaling factors is crucial due to their influence on the stability and performance of a fuzzy control system. This is why the problem has been investigated for some time and some results have been obtained. Main features of this paper are the adaptation of both input and output factors simultaneously and on-line tuning. These features make it possible to apply this method in the design of different fuzzy systems including control systems for the plant with unknown models.

The next chapter "Fuzzy Logic Applications in Diagnosing Mechatronic Systems" by T. Rauma and M. Kurki is dealing with a tolerance of fuzzy logic models to uncertainty. The proposed methodology includes constructing of fuzzy models, modifying them, and applying them in solving different problems. It makes it possible to apply fuzzy models in solving complex problems of fault detection and diagnosis and building an interface with an operator. A few examples of real life applications in diagnosis of mechatronic systems are presented.

The last chapter "Application of Fuzzy Logic to Handoff Control in Cellular Mobile Communication Networks" by P.S.K. Leung considers an application of fuzzy system design in the area which has been marked by phenomenal growth and rapid technological advances during last two decades. In cellular engineering the problem addressed in the paper is quite typical. There are two conflicting criteria to be minimised: a number of unnecessary handoffs between base stations and a number of lost calls. As a solution a fuzzy system is designed to improve the handoff characteristics of a future generation of micro- and pico-cellular wireless communication systems. The paper investigates such design problems as a choice of inputs, rules, membership functions and inference methods.

As one can see different researchers from various countries and continents have made their generous contribution into this volume. The editors would like to express the great appreciation of their work and research results and thank all those who have made this publication possible.

Melbourne, Sydney, Warsaw   L. Reznik
March 1998   V. Dimitrov
                                                       J. Kacprzyk

Table of Contents

Part 1

Philosophical and Mathematical Fundamentals of Fuzzy System Design

FUZZY LEADERSHIP: DANCING WITH ORGANISATIONAL REALITY

Patrick A. Duignan
Foundation Professor of Educational Leadership
Australian Catholic University
Castle Hill, NSW 2154
Australia
E-mail: P.Duignan@mary.acu.edu.au

Abstract. In our attempts to understand the complexities of social systems and the dynamics of organisational relationships, we cannot be sure that our picture or frame is the only one that can explain our observations. The 'fuzzy frame' presented in this chapter, enshrined within the metaphor of the dance, offers a particular way of framing organisational reality which the author claims offers useful insights into the nature of organisational dynamics (especially leadership and relationships) that may be appropriate in the turbulent and uncertain organisational environments at the end of this century.

The 'cosmic dance of energy', as it relates to the dynamics of natural and social systems, is explained and analysed. The metaphor of 'free form dance' is used to try to better understand the 'emergent reality' when individuals join together and share in a group or organisation. The metaphor of the dance is also used to explore the concept of 'the edge of chaos' where order and chaos intertwine in a complex, ever-changing dance of relationships and structures. The dance metaphor is again employed to explore our never-ending dance with the language of description and interpretation which we use to help frame and reframe our observations of 'reality'.

All of this is to provoke us into a 'dance' with ideas that can shed new light on the nature and the dynamics of the organisations in which we spend most of our working lives and on the leadership and management responses appropriate in such dynamic systems.

> At the still point of the turning world. Neither flesh nor fleshless;
> Neither from nor towards; at the still point, there the dance is,
> But neither arrest nor movement. And do not call it fixity,
> Where past and future are gathered. Neither movementfrom nor towards,
> Neither ascent nor decline. Except for the point, the still point,
> There would be no dance, and there is only the dance.

> (From T. S. Eliot, 1963, "Burnt Norton", *Four Quartets*)

Eliot provides a major insight into the nature of reality by employing the dance imagery to implicitly reject the either/or logic, and its accompanying mental models, which has dominated Western thought since the time of Aristotle. The 'still point of the turning world' - dynamic balance - speaks to a reality that embraces the 'Fuzzy Logic' concepts of 'more or less', 'A and not-A', and the Bhuddist perspective that something can be simultaneously 'thing' and 'non-thing'. In this 'more or less' frame for viewing reality, there is no 'fixity'.

The fuzzy logic frame of reference provides a perspective on reality that embraces an idea or family of ideas which includes "shades of grey, blurred boundary, grey area, balanced opposites, both true and false, contradiction, reasonable not logical . . ." (Kosko 1994: 67). Fuzzy logic recognises that the world is full of uncertainty, ambiguity, contradictiony, vaguesness   Kosko reminds us that the fuzzy view of reality proposes that "almost all truth is grey truth, partial truth, fractional truth, fuzzy truth." (P. 80)

Eliot also rejects traditional Aristotelian logic when he reminds us of the great limitations of the either/or view (A or not-A or 'law of excluded middle') even in dealing with time and other binary classifications or dichotomies which are in common use:where the past and the future become as one and where phenomena are neither flesh nor fleshless; neither from nor towards; neither arrest nor movement, neither ascent nor decline. At the still point of the turning world there is only the dance.

The great Bertrand Russell also rejects Aristorle's law of excluded middle:

All traditional logic assumes that precise symbols are being employed. It is therefore not applicable to this terrestial life, but only to an imagined celestial one. The law of excluded middle [A or not-A] is true when precise symbols are employed but it is not true when symbols are vague [fuzzy], as. in fact, all symbols are ." (Bertrand Russell, quoted in Kosko, p. 92).

Lofti Zadeh (a seminal figure in the development of fuzzy logic), in presenting his 'principle of incompatibility', has argued that

As the complexity of a system increases, our ability to make precise and specific statements about its behavior diminishes until a threshold is reached beyond which precision and significance (or relevance) become almost mutually exclusive characteristics . . . . A corollary principle may be stated succinctly as, "The closer one looks at a real-world problem, the fuzzier becomes its solution" (Lofti Zadeh cited in Kosko p. 148).

It is not my intention to describe and discuss the details of fuzzy logic in this chapter. Discussions on the nature and the characteristics of fuzzy thinking and of fuzzy logic are presented in other chapters. In this chapter, I wish to explore T. S. Eliot's metaphor of the dance to highlight what I take to be the essential features of fuzzification and of fuzzy thinking, especially as they relate to our understanding of the nature and dynamics of social systems. I propose that the dance imagery captures the very essence of fuzzy thinking in a way that will be meaningful for leaders and managers of social systems and organisations.

THE DANCE

In our attempts to describe and understand the complexities of social systems and the dynamics of organisational relationships, we cannot be sure that our picture or frame is the only one that can explain our observations. The 'fuzzy frame' presented in this chapter, enshrined within the metaphor of the dance, offers a particular way of framing and reframing organisational reality which, I believe, offers useful insights into the nature and characteristics of social systems and organisational dynamics (especially leadership and relationships) that may be appropriate in the turbulent and uncertain organisational environments at the end of this century.

The metaphor of the dance is not only employed to explain actual phenomena which have no 'fixity' (eg. the dynamics of social systems, leadership, relationships) but is also employed as an approach to understanding, or a process of learning about, these phenomena in order to develop more complete pictures of the 'real world'.

In his book "The Dancing Wu Li Masters", Gary Zukav points out that the Chinese word for physics is "Wu Li" which essentially means "patterns of organic energy". Masters are those who teach the essence of an idea or concept and who allow others to get the feel of the idea before attempting to get them to understand it through analysis. As Zukav (1979: 7-8) points out (excuse sexist language):

> The Wu Li Master does not speak of gravity until the student stands in wonder at the flower petal falling to the ground. He does not speak of laws until the student, of his own, says, "How strange! I drop two stones simultaneously, one heavy and one light, and both of them reach the earth at the same moment!" He does not speak of mathematics until the student says, "There must be a way to express this more simply.

In this way, as Zukav says, "the Wu Li Master dances with his student."

I am not a Wu Li Master, but Zukav provides me with ways of thinking, learning about, and challenging my understanding of reality - especially the reality of the nature and dynamics of natural or social systems and of my attempts to understand and explain the nature of leadership and relationships in our systems and organisations. He points out that even Einstein recognised the difficulties of explaining the world of which we are such an integral part. Quoting Einstein's (1938) familiar lines, he reminds us of the fact that the best we can hope to do in explaining reality, the world, is to 'dance' with it.

Physical concepts are free creations of the human mind, and are not, however it may seem, uniquely determined by the external world. In our endeavour to understand reality we are somewhat like a man trying to understand the mechanism of a closed watch. He sees the face and the moving hands, even hears it ticking, but he has no way of opening the case. If he is ingenious he may form some picture of a mechanism which could be responsible for all the things he observes, but he may never be quite sure his picture is the only one which could explain his observations. He will never be able to compare his picture with the real mechanism and he cannot even imagine the possibility of the meaning of such a comparison.

Zukav tells of a Chinese physicist he regarded as a Wu Li Master and who described his methods of teaching thus: "Every lesson is the first lesson, every time we dance, we do it for the first time." He is not saying that he does not build on previous knowledge, or that he is constantly reinventing the wheel, but that each lesson is "a source of unlimited energy", a feeling of enthusiasm as if doing it for the first time or as Zukav says, "every dance that he dances, he dances for the first time." He is, therefore, continuously attempting to develop new understandings, to come up with different explanations, in other words, to reframe his view of reality.

In employing the 'fuzzy logic frame' as an attempt to gain a deeper understanding of social reality, especially the nature of leadership, I feel that I am involved in the dance for the first time. However, the real excitement is in trying to uncover, a little bit at a time, the reality that surrounds us and of which we are such a vital part. In Einstein's words we need to maintain "a holy curiosity" so that we can comprehend a little more of the great mysteries of reality every day. The main argument in this chapter is that we are all 'dancing together' with this reality and with our sometimes feeble attempts to make some sense of it. We try to interpret the past and understand the present in order to gain some insights into the future. Referring to physicists' attempts to develop more complete pictures of the real world and its possible futures, Zukav reminds us that:

> The Wu Li Masters move in the midst of all this, now dancing this way, now that, sometimes with a heavy beat, sometimes with a lightness and grace, ever flowing freely. Now they become the dance, now the dance becomes them. This is the message of the Wu LI Masters: not to confuse the type of dance that they are doing with the fact that they are dancing.

The concepts of 'dance' and 'dancing' will be used to illuminate the discussion in this chapter on different aspects of natural and social systems and on implications for leadership and management, assuming the appropriateness of these metaphors. We have already been introduced, in T. S. Eliot's poem to the concept of dance "as the still point of the turning world", where there is no "fixity". We have also used the concepts of the dance and dancing in relation to the Wu Li Dancing Masters to both their methods of teaching and their ways of developing more complete pictures of the real world.

In the upcoming sections, we will discuss the 'cosmic dance of energy' as it relates to the dynamics of natural and social systems. We will also examine the metaphor of 'free-form dance' to try to better understand the 'emergent reality' when individuals join together and share in a group or organisation. We will also use the metaphor of the dance to explore the concept of 'the edge of chaos' (the world of paradox and tension) where order and chaos intertwine in a complex, ever-changing dance of relationships and structures. Also, the dance metaphor will be used to try to better understand our never-ending struggle with the language of description and interpretation which we use to help frame and reframe our observations of 'reality'.

All of this is to provoke us into 'a dance' with ideas that, I believe, can shed new light on the nature and the dynamics of the organisations in which we spend most of our working lives and on the leadership and management responses appropriate in such dynamic systems.

I am not proposing here any grand theory of organisation and leadership based on chaos theory, fuzzy logic and/or quantum realities. That would be far too presumptuous. I wish, however, to examine new ways of thinking about organisation and new ways of approaching leadership and management in the light of this thinking.

First, I will discuss the nature and characteristics of natural and social systems using the metaphor of Free-Form Dance. Second, I will use the idea of "The Dance of the Starlings" to query some current linear, mechanistic and hierarchical organisational structures and processes that seem to constrain individual creativity and initiative and try to form (pressure) organisational members into a 'flock.' Third, I will examine the context of paradox and confusion in which we have to operate as leaders and managers, with a view to suggesting some ways of managing those seeming contradictions. Fourth, I will explore the nature of metaphor and its use in our language of description and interpretation of the dance of reality. Fifth, I want to argue that unless we 'dance' with the passionate side of organisational life, we are avoiding the Life/Death/Life nature and are, therefore, operating without the heart and the soul and are likely to present to others an 'emotionally crippled self. Finally, I will explore the implications of fuzzy thinking for leaders and leadership in social systems and organisations.

I invite you in this chapter to "Come, join the dance!"

DANCING WITH ORGANISATION FREE-FORM DANCE

Drawing on the work of the eminent physicist, David Bohm, who compared the movements of electrons in the laboratory to those of ballet artists responding to a musical score, Zohar and Marshall (1994: 1) use the metaphor of the "free-form dance company" to illustrate the dynamic unity of quantum laws and principles and to depict the dynamic interrelationships in organisations or social systems with "each member a soloist in his or her own right but moving creatively in harmony with others". These authors propose a 'quantum society' which is " firmly rooted in nature ...drawing its laws and principles, its self images and its metaphors from the same laws and principles underlying all else that is in the universe".

They argue that there is a need to change our social perceptions and values and to look at organisation and relationships in new ways. They propose the restructuring of the customs and institutions of society through "the power of this new vision". Essentially, the main challenge of their book, they state, is "learning to dance together" (p. 1).

Zohar and Marshall's approach to developing fuller understandings of social systems, while drawing heavily on the new philosophy of science, especially insights derived from quantum mechanics, includes many of the essential features of fuzzy thinking. They emphasise an interrelated, holistic and relational approach to explaining and understanding the nature, structures and operations of a system - natural or social. They reject thinking based on Aristotle's logic of the excluded middle (either/ or) and on linear and mechanistic perspectives and propose a frame for social systems that has the qualities of 'undivided wholeness' and 'social relational whole' (explained elsewhere in this chapter). Their view is strongly supported by a number of new and influential theories derived, especially, from the natural sciences. (Capra, 1982 and 1992; Davies, 1983; Prigogine and Stengers, 1984; Gleick, 1987). They propose a new confluence of thought and meaning between the natural and social sciences — referred to by Capra as "a new systems view" — which casts serious doubt on earlier systems views of organisations based on a Cartesian mechanistic, linear and causal view of the world. Capra (1982: 70) proposes a quantum theory perspective in which:

...nature does not show us any basic building blocks, but rather appears as a complicated web of relations between the various parts of a unified whole.

This "cosmic web" is intrinsically dynamic and is in a "continuous dancing and vibrating motion" whose stability is not based on equilibrium but on dynamic balance (pp 78-79). Zohar and Marshall (1994) refer to this quantum reality as "an unbroken web of overlapping, or correlated internal relationships" with a quality of "undivided wholeness".

Fundamental to this new systems view at the quantum level are the concepts of interrelatedness and interdependency (Capra, 1982: 285). Interrelationships and interactions between the parts of the whole are more important than the parts themselves. As Capra so eloquently states:

There is motion but there are, ultimately, no moving objects; there is activity but there are no actors; there are no dancers, there is only the dance (pp 83).

We may well ask how these characteristics of the quantum world of subatomic particles and waves help us understand our social world of systems and organisations. Zohar and Marshall (1994) provide us with an interesting perspective on this link. They argue that there is an uncanny and intriguing similarity between the way that quantum systems relate and behave and so much that we are now beginning to understand or hope for about human social relations.

They see the link between individuals, on the one hand, and social and physical reality, on the other hand, centring on the human brain which they argue is ". . . the natural link between our perceptions and values and the 'cosmic dance' of physical [and social] reality." Questioning traditional mechanistic models of the brain's operations and developing on the work of Roger Penrose, they postulate quantum correlations and structures for the brain which give it special insights into the quantum realm and which, in turn, may be infused into the social structures - systems and organisations - that our collaborative minds create (Zohar and Marshall, 1994: 41).

Using the concepts of 'consciousness' and 'self' which act to create unity and a meaningful whole from multiple and diverse bits of sensory information, they propose for the brain an ". . . emergent, holistic quantum substructure" and a structure that ". . . is greater than the sum of its parts [having] some identity over and beyond the identities of those parts" (pp 48). What they are really proposing is a 'holographic model' of mind in which:

> . . . each part of the holographic image contains information spread across the whole pattern. That is, if we break the hologram apart into small pieces, we will still be able to get the whole picture from each of the pieces - just as each cell in the human body contains the genetic code for the whole body (pp 48).

Holograms, they contend, are "the products of quantum processes [which] arise from information 'written' on a laser beam" (pp 49). A laser beam — a quantum structure known as a Bose-Einstein condensate — is an example of "the most highly ordered and highly unified structure possible in nature". However, the quantum properties of the structure give it special capabilities which ". . . allow both a 'fluid' (ever-changing) order and a high degree of unity" (pp 51). In effect, the structure (condensate) as a whole acts holistically as "one single particle" and has a very high level of coherence. Zohar and Marshall (1994: 61) propose that consciousness may arise "...from a coherent quantum field over large regions of the brain, a field in which neural information is evenly distributed throughout . . . ."

These authors conclude that just as traditional mechanistic models of the mind cannot account for how a thinking subject - and one with a sense of self and a unity of consciousness - could result from ". . . the inherently separate functioning of neural structure", similar mechanistic models of organisation and society "...have the same problem accounting for how any social cohesion could emerge from the coming together of a myriad separate individuals or separate . . . groups" (pp 63).

"Socially, they argue, the unity of consciousness or the unity of self have their counterpart in the unity of society, in the "we" of community relationship.

They reject the 'atomised' liberal democratic view of society and organisation where the collective ". . . dissolves into the sum of its parts" and where the "we" of community ". . . is just a phantom or a myth for what appears to exist when these individuals co-operate, or when their separate characteristics are added together" (pp 67). No deep and meaningful sharing of values or experience can occur in such circumstances.

Likewise, collectivist models of organisations and society are rejected because individual diversity, and individual creativity suffers for the cause of "collective identity". As the authors put it, "the errant individual must often be sacrificed for the 'greater good' of the collective identity".

Referring back to the dance, we are reminded that its essence cannot be captured by either the individualistic or collectivist perspectives. The dancers are not isolated atoms (particles) floating in the void, nor is their dance an indivisible whole without individual parts (waves).

Rather, they argue (1994: 77) the dance is:

. . . a repository of movements and characteristics not possessed by any one member of the dance company. It is a shared style - a shared rhythm and timing in which all the dancers participate as the 'background conditions' of their individual movements. Similarly, a society (or a culture) is a repository of skills, knowledge and potential - even of will - not possessed by any one of its members. Like the dance, it has a shared style, (shared purpose, shared values).

The dancers need the dance to fully express their creativeness and the dance means nothing without its dancers. Likewise, creative individuals need the organisation or society to realise their potential and the group or community cannot exist without individuals. A group or organisation is what Zohar and Marshall (1994: 83) refer to as "a social relational whole", in which its characteristics and behaviour ". . . emerge from the pool of latent (indeterminate) characteristics of its individual members, but until these latent characteristics are correlated they have only a phantom reality". A mob, for example, "is a social 'relational whole' of the most visible sort." It is in the relationship of the individuals (with their pool of latent, indeterminate, characteristics) that the shared emergent reality of the mob, in which individuals seemingly willingly participate, is evoked (p. 83). Communities that work together and share - learning communities - are 'social relational wholes'.

Through commitment to open communications, interactions and dialogue (essentially shared learning) they argue, we can focus on 'relationships' and discover the reality of living and working together in organisations. However, "as the world becomes more complex and interdependent, the ability to think systemically, to analyse fields of forces and understand their joint causal effects on each other, and to abandon simple linear causal logic in favour of complex mental models" (Schein: 1992: 372) will become critical if learning is to occur.

This shift in thinking about social groups and organisation is characterised by perspectives that emphasise complexity in organisational life rather than simplicity. Complex systems are said to be 'on the edge of chaos' where ". . . new ideas and innovative genotypes are forever nibbling away at the edges of the status quo", and where there is a ". . . constantly shifting battle zone between stagnation and anarchy, the one place where a complex system can be spontaneous, adaptive, and alive (Waldrop, 1992: 12). This is in T. S. Eliot's terms "at the still point of the turning world." In the words of Estes it is "a dance with Death; Death as a dancer, with Life as its dance partner" (to be discussed later). For Zohar and Marshall its characteristics are best captured by the images of the "Free-Form Dance".

Waldrop (1992), reporting the work of Christopher Langton at the Santa Fe Institute (a centre for the study of complexity) on dynamical systems at the edge of chaos (the point of balance between order and chaos) states that at that point in their transition "order and chaos intertwine in a complex, ever-changing dance . . . and nothing ever really settles down - the dynamics that took forever to settle down, the intricate dance of structures that grew and split and recombined with eternally surprising complexity".

Dynamic systems are, therefore, in a state of constant change - the dance. Stability means eventual decay and death. Dynamic systems, if they are to grow and flourish, need to adapt themselves to "a condition of perpetual novelty, at the edge of chaos" (Waldrop, p. 356). There are many implications, as a result, for the way we conduct ourselves as leaders and managers in our organisations. We will discuss some of these later.

THE DANCE OF THE STARLINGS: INDIVIDUAL-ORGANISATIONAL RELATIONSHIPS

David Whyte (1994) tells the story that on 26th November 1799, the great poet Samuel Taylor Coleridge observed a flock of starlings from his carriage window. He was in awe of what he observed. It puzzled him for the rest of his life as he searched for meaning in it and as he tried to analyse where he himself fitted into in the scheme of things (into the flock).

"The starlings," he wrote:

drove along like smoke... misty... without volition - now a circular area inclined in an arc - now a globe, now... a complete orb into an ellipse... and still it expands and condenses, some moments glimmering and shivering, dim and shadowy, now thickening, deepening, blackening! (cited in White, D. p.215-216)

He had observed 'the dance of the starlings'. In it he perceived some kind of order, grace harmony, style - a community of birds involved in their dance with each other and with nature. They had definite direction and purpose but also there was evidence of chaos. There seemed to be a complex relationship between the individual starlings and the flock. How did the 'flocking' occur? What rules governed such spectacular flying habits? The starlings seemed to know where they were going, they did it with harmony and synchronisation but they were unpredictable in terms of constancy of direction and shape of the flock. However, they performed the dance acting as a 'social relational whole'.

How could it be explained? Coleridge seems to have puzzled over its meaning for most of a lifetime. What do you make of it? Are we any better off today with our post-modern, scientific understanding of such concepts and theories as quantum mechanics, chaos theory and fuzzy logic in determining what makes the individual starlings 'fly' (dance) in such magnificent communal formations?

In 1987, Craig Reynolds, a scientist, developed a computer simulation program called "Boids" to help understand this 'flocking' phenomenon (reported in Waldrop, 1992, pp. 241-2).

He gave each 'Boid' three simple rules;

1. maintain a minimum distance from all objects and other birds;
2. try to match the velocity of your neighbours; and
3. try to move toward the perceived centre of the mass of birds in your neighbourhood.

What happened?

When he ran his programme, the 'Boids' quickly flocked every time and they flew smoothly around objects and didn't touch each other. Remember, Reynolds did not give them a rule stating 'form a flock.' When he did do so, they flocked but were cumbersome and jerky, like the movements of "animated dinosaurs". Certainly, they lost the rhythm of the dance!

Now I can guess what you are thinking, "People are not starlings". But - - -- !

Can the 'flocking' concept enlighten us in any way about life in communities and organisations? Well, organisations, or their leaders, in so many ways, say to their members, "Form a flock" - eg. develop a corporate culture based on shared values. Top-down management approaches often try to mandate 'flocking' behaviour. Policies and procedures in organisations are often blueprints for conformity and group think. Creativity and individual initiative are sometimes victims of cloning processes.

Change in organisational settings is often described and analysed in terms of 'top-down' and 'bottom-up' structures, inputs and processes without any real understanding of what the dynamic balance between these forms and forces should be. As hierarchical structures and processes mostly characterise organisational dynamics, then the direction of change is usually top-down. "A or not-A' thinking too frequently drives management thinking and practice. For many in management positions, power is very much an 'either/or' concept -

either you have it or you don't. Fuzzy thinking is regarded as typical of those lower down in the hierarchy who are not in touch with the 'real world' of management.

Change is too often approached through restructuring or 'fiddling' with external indicators. During restructuring or downsizing the shape of the flock is often mandated from above but the individual members aren't consulted and/or don't understand (or perhaps agree with) the new rules for flocking. No real attempts are made to understand what is making individual members 'fly' or to appreciate their inner states of being. Individuals are seldom asked about their views on the meaning of 'flocking' and on how the creative spirit of the individual can be protected, even encouraged, while still preserving the positive aspects of 'flocking'. The result is, of course, that the whole flock may perish through misadventure or, more likely, individual members ignore the new rules, no real learning takes place, and nothing actually changes at all.

One of the most challenging and puzzling questions one can ask about complex, modern organisations is, "Where does real change come from?" Is it as a result of strategic corporate planning and the dictates of top management? Perhaps, but if chaos theory has any currency in social systems, then change is more likely to come from the periphery, from the edge of chaos, from the still point of the turning organisation. Could it be that some "Boids" form a little cluster of their own and through the impact of positive feedback (a concept popular with chaos theorists) cause the flock to change direction and head for greener pastures, clearer air or cleaner water? Well it is worth some thought at the very least!

How can the story of the starlings then inform us as leaders and managers? Well, just like Coleridge I will reflect on it to see what messages, if any, are in it for me. I suggest that leaders and managers of social systems and organisations need to better understand such concepts as dynamic balance at the still point of organisations, and develop a greater appreciation of the balance between individual initiative, creativity and central organisational forces for conformity. We will pick up these themes again in the section on fuzzy leadership.

DANCING WITH CONFUSION AND PARADOX: MANAGING THE DOUBLE-HEADED ARROW

In a complex, uncertain, and turbulent world of constant change (some might say interminable change), we are frequently, as leaders, faced with tensions, dilemmas and paradoxes that are "inevitable, endemic and perpetual" (Handy, 1994). It is easy to become disoriented, confused, frustrated, even angry in such trying conditions. In response, our inclination may be to become defensive, withdrawn or even disengaged. Handy argues that paradoxes confuse us because we are asked to live with contradictions and with simultaneous opposites . . . To live with simultaneous opposites is, at first glance, a recipe for indecision at best, schizophrenia at worst.

Life is full of paradoxes. We face them every day as we make decisions about our careers and lives. We have to plan for the long term without neglecting our needs today. Handy claims that, parents may have to be "simultaneously tough and strict on their children *and* tender and relaxed. Organisations may have to be both tight *and* loose in their structures, big *and* small in their operations, strategic *and* tactical in their plans". Notice that Handy employs fuzzy thinking insofar as he treats paradox in terms of both/and thinking and he rejects either/or approaches.

Handy (1994: 48) also argues that the key to managing paradoxes is to understand them for what they are - part of the stuff of life. Managing paradox is, he says, like learning to ride a see-saw - "if you know how the process works, and if the person at the other end also knows, then the ride can be exhilarating". He cautions, however, "that if your opposite number does not understand, or wilfully upsets the pattern, you can receive a very uncomfortable and unexpected shock".

We must, according to Handy, learn to live with the ups and the downs, "knowing that the opposites are necessary to each other" (p. 48). This is a point well recognised in fuzzy logic. He advocates that we learn to frame the confusion and find pathways through the paradoxes by understanding what is happening and by learning to be different (p. 3). We need to learn to 'dance' with turbulent and confusing situations that frequently seem to present themselves as seeming contradictions and dilemmas. We must break the bonds imposed on us by the either/ or mindset.

Fuzzy Logic provides us with the frame and the tools to deal with such confusing and paradoxical situations. Kosko (1994: 102) points out that from a fuzzy logic perspective, paradox is a misnomer for most of the seeming contradictions we face in our modern organisations. He states:

The term *paradox* suggests exception. A fuzzy analysis shows the reverse. Paradoxes are the rule and not the exception. Pure black-and-white outcomes are the exceptions . . . . There are two Aristotelian extremes of black and white, 0 and 1, and infinitely many shades of grey between them. A grey shade means 'A and not-A' holds to some degree.

English (1995), too, proposes a useful and insightful approach to the understanding of, and managing and leading in, paradoxical situations that fits with Kosko's fuzzy perspective. He recommends that we should analyse the tensions inherent in complex organisations, not in terms of contradiction, polarity, and either/or frames but in terms of a *relationship* that encompasses both competition and complementarity (a *both/and* and *A and not-A* philosophy). We should, he says, determine, as best we can, the qualities and conditions of the relationship in each situation. In this way we can better understand and manage a change situation (usually characterised by uncertainty and confusion) by building a profile of the tensions - in Handy's terms we are framing the confusion.

A tension situation, says English, can then best be described as a double-headed arrow. Seeming polar opposites are actually in a complex relationship and influences are rarely one-way. Instead of being mutually exclusive, most seeming opposites are in a tension situation which is characterised partly by competition and partly by complementarity - as the song suggests, "You can't have one without the other".

By emphasising the relationship and complementarity instead of the seeming contradictions and opposites, we have a better chance of influencing the direction and intensity of the positive side of the tension. Otherwise, we may opt for the either/or approach, perhaps believing that seeming opposing forces are mutually exclusive and incompatible, and immediately we create a win-lose situation. In other words we have fallen for the either/or, *A or not-A* dichotomy which fuzzy logic rejects.

I recognise, however, that change is very complex and often more emotive than rational. I also acknowledge that some 'tension situations' are likely to explode no matter what we do. Some conflict situations may, indeed, be immune to any of our change management interventions. We have to do the best we can with the tools and understandings at our disposal. However, I believe that fuzzy logic and fuzzy thinking can assist leaders and managers to avoid over analysing situations into their component parts, or seeing situations in terms of mutually exclusive, either/ or, win/ loose strategies. It doesn't have to be zero or one. It is likely that most solutions fall somewhere between, at the still point of the turning world.

What is too often missing are attempts by managers to recognise the complementarity, the potential for harmony, the possibility of the dance. What happens in the free-form dance if individuals ignore the other dancers and try 'to steal the show'? What if some individuals try to deliberately sabotage the dance? What if the rules for the steps are so rigid that there is no room for individual improvisation? In the case of the starlings, if the birds simply 'did their own thing' or if they were ordered to 'flock', Coleridge would never have had to puzzle over their magnificence.

DANCING WITH LANGUAGE: METAPHORICAL ANALYSIS

Terry (1993: 160) argues that metaphors "open windows into reality. They identify the known from the unknown, the novel from the familiar. They link the well established with the less well understood". Drawing on the work of Bolan (1985: 353), Terry suggests that a metaphor simultaneously asserts that "A is B" and "A is not B". Had he been familiar with fuzzy logic he would have included, I have no doubt, that it also asserts "A and not A". Similar to the argument of the tension implied in the double-headed arrow (English: 1995), Terry quotes from Paul Riceour to argue that:

> such a process [metaphorical analysis] creates a tension between the two words employed. Resemblance plays a role but nonetheless the tension, the contradiction, between the words demands a dual interpretation of the words, one literal and one figurative .... A metaphor, in short, tells something new about reality.

When one reflects upon some of the metaphors employed within the frame that is being presented within this chapter and this book (eg dance, still point of the turning world, edge of chaos, fuzzy reality) one can appreciate the simultaneous resemblance and contradiction in our attempts to describe organisational reality. Part of the problem for us is that, as Terry (p. 160) points out, "a metaphor anchors a direct experience to something that goes well beyond it". In the analysis being attempted in this chapter, we are dealing with phenomena that not alone go well beyond the power of our language to describe and analyse but even, most probably, beyond our comprehension, at least at this point in human development.

Yet the metaphors we choose to describe our observations can greatly influence our perception of reality. Terry (P. 160 - 161) draws on the work of Lakoff and Johnson (1980) to demonstrate how two differing metaphors - *argument as war* and *argument as dance* - can colour our view of the world. The metaphor "argument as war" is very common in our speech even though we may not be directly aware of it. He states:

> We see the person we are arguing with as an opponent.... We attack his positions and we defend our own. We gain and lose ground. We plan and use strategies. If we find a position indefensible, we can abandon it and take a new line of attack.

This metaphor presents an either/or frame for managing or dealing with interpersonal relationships. On the other hand, using the metaphor "argument as dance" the parties involved "are seen as performers and the goal is to perform in a balanced and aesthetically pleasing way". Using this frame participants in an argument would "view arguments differently, carry them out differently, and talk about them differently". This metaphor fits quite well within the frame employed in fuzzy thinking.

Different metaphors, therefore, reveal different realities, "resulting in different perceptions of what is going on". The point to be reinforced here is that language and its use is part of the dance with reality. We need to be aware of the language of description and analysis we employ and the meanings inherent in it (particularly the metaphors, as they constitute powerful tools to communicate meanings).

Terry argues that to expect people with diverse backgrounds and values to accept a rigid interpretation of reality or predicted future reality (eg. in strategic plans, visions and shared mission statements) is expecting the impossible. We should accept and encourage multiple perspectives as a prerequisite for creative change and innovation. As leaders in complex organisations in uncertain and rapidly changing environments, we would do well to "engage in metaphorical analysis from many points of view" (p. 162). Terry promotes six such metaphors two of which are directly relevant to the argument in this chapter - 'ups versus downs' and 'life is a journey' - and they will be discussed briefly here.

He points to the fact that our common speech contains numerous up-down metaphors, each emphasising that to be 'up' is superior to being 'down'. Examples include (Terry: 171-172):

The *heights* of ecstasy, the *depths* of despair.
Lower than a snake's belly.
I'm *under* the weather.
I'm *down* at the mouth
I'm *downhearted.*
I'm not *up* to doing that.
They are *beneath* contempt.
Morale is *up.*
The computer is being *upgraded.*

This very powerful metaphor when applied to relationships either within or without our organisations creates a we/they tension or worse a superior/inferior syndrome. As Terry asserts, this metaphor has difficulty "envisioning people standing side by side. From a fuzzy logic perspective, it reinforces the either/or and the 'A or not-A' view of reality.

On the other hand, the metaphor 'life is a journey' "directs our attention inward and outward, highlighting the connection of self to community, nature, and God, while stressing our creative capacity to chart our own courses" (pp. 174-175). It contains a spiritual dimension as it reminds us of meaning beyond ourselves and "justifies and energises side-by-side partnerships" (p. 185). We are indeed inseparable from the significant, perhaps spiritual, relationships that give meaning to our lives. We cannot define 'self' outside such relationships. We are simultaneously 'A is B' and 'A is not B', *and* 'A and not-A' - the part of self that is in relationship with B. As Kosko (1994) points out, fuzziness acknowledges, even encourages, the world of beauty, morality and spirituality, as well as the world of fact and truth.

In summary, the metaphors leaders and managers select to frame organisational realities will themselves provide powerful orienting lenses for their descriptions and interpretations of the dance. Indeed, they will find that language proves to be a dynamic, dancing force in their descriptive and interpretive attempts. They must reflect on the nature of the language they choose and it may be wise for them to employ multiple metaphorical analysis in order to better appreciate the complexity, the multi-faceted nature, the fuzziness, of organisational reality.

DANCING WITH EMOTIONS: PASSION IN ORGANISATIONAL LIFE

Exhilaration, celebration, despair, love, jealousy, envy, hate, outrage, desire, apathy, growth, decay, grief, mourning, are all part of living and being in organisations. All contribute to the organisation's vitality. To think that the work environment is (or can be) free from life's passions is to delude oneself. Success and failure, the ups and downs, the calm and the storm are all essential to the growth cycle (life/death/life cycle) of organisations, as they are to our own growth. They are part of the undivided wholeness of organisational reality.

There is a rather poignant discussion of the Life/Death/Life cycle in the intriguing book by Clarissa Pinkola Estes (1992) entitled "Women Who Run With The Wolves". She argues that it is through close communion (or love) between two beings that we can communicate with the soul-world and "participate in fate as a dance with life and death" (p. 131, akin to T. S. Eliot's "At the still point of the turning world"). Using a hunting story about love from the tribes of the circumpolar regions called "The Skeleton Woman", and drawing on her rich sources of stories from many tribes and cultures on the 'dance of life and death' (eg Iceland, Wichita, Micmac, Celtic Ireland), Estes points out that women, especially, understand physically, emotionally, and spiritually that zeniths fade and expire, and what is left is reborn in unexpected ways and by inspired means, only to fall back to nothing, and yet be reconceived again in full glory.

We don't hear many direct references to 'love' and 'spirituality' in our modern organisations, yet they are central to true relationship. Love is, frequently, seen as being 'too soft' for male dominated organisations. We cannot, however, experience the full range of emotions in our organisations if love is barred. Indeed we cannot hope to appreciate the Life/Death/Life dance in our organisational lives if essential ingredients - love and spirituality - are ignored or neglected in our communion with others. Estes reminds us that:

Love in its fullest form is a series of deaths and rebirths. We let go of one phase, one aspect of love, and enter another. Passion dies and is brought back. Pain is chased away and surfaces another time. To love means to embrace and at the same time to withstand many many endings, and many many beginnings - all in the same relationship . . . . Throughout the world, though it is called by different names, many see this nature [Life/Death/Life] as a dance with Death; Death as a dancer, with Life as its dance partner (p. 162).

Estes argues that many of the emotional states such as energy, feeling, closeness, solitude, desire, ennui, all "rise and fall in relatively closely packed cycles" The acceptance of the Life/Death/Life nature "not only teaches us to dance these, but teaches that the solution for malaise is always the opposite; so new action is the cure for boredom, closeness is the cure for loneliness, solitude is the cure for feeling cramped". This is indeed fuzzification in its most potent expression.

Estes, however, cautions that without the knowledge of this dance, a person is inclined, during various still-water times, to act foolishly or without caution, making reckless choices (especially either/ or choices) thereby suffering the consequences (p. 163).

Life in our organisations embraces the full range of human passions. While we don't expect to perish in organisational life/death'life cycles, we can be severely buffeted by the passionate 'fire' of emotional outbursts and angry confrontations or by being alienated and excluded (frozen out) by the 'icy' stares and behaviours of others. We need to recognise that if we are to be successful, we will have to cope with and manage the passion and emotion of both the 'fire' and the 'ice' in life and in work.

The poet Robert Frost recognised the possibilities of both sets of passions in his poem "Fire and Ice".

Some say the world will end in fire
Some say in ice.
From what I've tasted of desire
I hold with those who favour fire.
But if it had to perish twice,
I think I know enough of hate
To say that for destruction ice
Is also great
And would suffice.

(Cited in Whyte, D. p.84)

FUZZY LEADERSHIP

The world is mostly grey but, all too often, people in leadership and management positions try to lead and manage as if it were black and white. However, a fuzzy logic frame suggests that most natural and social phenomena (events, ideas, things) have vague boundaries with their opposites (Kosko 1994: 24). As leaders and managers, where do we draw the line between justice and injustice, fairness and unfairness, right and wrong, and even truth and untruth?

Many leaders and managers are driven by a bivalence perspective - true or false, A or not-A. However, as fuzziness reigns supreme in the social world how are leaders and managers to decide where the boundary line should be drawn?. They would do well to remember that the fuzziness of reality (an undivided wholeness) cannot be easily dissected or compartmentalised by the application of Aristotelian logic (either/or) and linear reasoning.

The context for leadership in modern organisations is turbulent, uncertain, unpredictable,and full of paradox and tension. Leadership and management strategies and techniques derived from linear, mechanistic and traditional Western logic, which may have served our needs in the past, are now proving to be inadequate.

Wheatley (1992, proposes a "new scientific management paradigm" to assist leaders and managers frame the confusion of modern organisational life. In this new frame, Wheatley (1992: 142-143) argues that as leaders and managers we need to:

> become truly good scientists of our craft [and] seek out surprises, relishing the unpredictable when it finally decided to reveal itself to us. Surprise is the only route to discovery, the only path we can take if we're to search out the important principles that can govern our work. The dance of this universe extends to all the relationships we have. Knowing the steps ahead of time is not important; being willing to engage with the music and move freely onto the dance floor is what's key.

She states that of all the concepts from the new science (especially chaos theory) that have major implications for social systems and their leaders, especially in the area of change, none are more significant than the concept of 'self reference'. This is the capacity of natural systems to renew themselves within certain boundaries; each part of the system must remain consistent with itself, with all other parts of the system, and with the system as a whole, as it changes (p. 146). It underpins the belief that even with a large degree of local autonomy, order is maintained throughout change. Wheatley (p. 147) states that " we need to be able to trust that something as simple as a clear core of values and vision, kept in motion through continuous dialogue, can lead to order."

As leaders, we need to understand the dynamic balance (the still point) in our systems, and give primary value to the relationships that exist among seemingly discrete parts (Wheatley 1992) and appreciate the 'greyness' inherent in the problems and paradoxes in our lives,. Meadows (cited in Wheatley, 1992: 9), quoting from ancient Sufi wisdom, captures the very essence of fuzzy thinking as it applies to leadership when he states:

You think because you understand *one* you must understand *two*, because one and one makes two. But you must also understand *and*.

More and more students and practitioners of leadership are emphasising its relational aspects. This focus on the interconnectedness of phenomena and events within a wholistic framework, has led to a search for meaning in relationship and a search for ethical, moral and spiritual guidelines for behaviour. A fuzzy logic approach provides a most useful frame for shaping these relationships and behaviours.

Leaders in this brave new world need, above all, to be learners, to maintain 'a holy curiosity' and to be in a constant state of strategic readiness for perpetual novelty. They need to be prepared for the ever-changing dance of reality where nothing ever really settles down. Leaders who yearn for stability and tranquility are involved in wishful thinking.

A challenge for leaders is to try to determine the point of dynamic balance between individuals and organisation. Perhaps some lessons derived feom the metaphors of the free-form-dance and the flocking of the starlings, discussed earlier, could be instructive here

Leaders also need to be aware of the potency of the language in use in their organisations, especially the predominant metaphors. They would be wise to accept, appreciate, even encourage multiple perspectives and use metaphorical analysis to help interpret and appreciate these.

Leaders delude themselves if they believe that passion and emotion can be banished from organisational life. These are part of the dynamics and relationships of 'real' organisations and wishing them away will not make them go away. They are inherent in human relationships and they will find expression covertly if they are denied overt outlets. Leaders are challenged to cope with and manage both the fire and the ice in organisations.

Fuzzy logic, I believe, provides a framework for leaders in messy situations or conditions which are characterised by uncertainty, unpredictability, confusion, paradox and tension. It can assist leaders avoid the many pitfalls inherent in traditional either/or management approaches while providing guidance on how to deal with the greyness, the fuzziness, of life in our modern organisations.

CONCLUDING REMARKS

"Fuzzy Leadership: Dancing With Organisational Reality!" What does it actually mean? For me, it means that we need to be in tune with the perpetual dynamism inherent in vibrant and evolving systems. If chaos theory has any currency, then successful organisations are those that operate at 'the edge of chaos' or 'at the still point of the turning world.' We need to be aware that while we have to be future oriented we cannot predict the future and we cannot regret what might have been.

As leaders and managers, we need to engage in the dance, to be aware of, and perhaps encourage, what is happening at the still point of the organisation. We need to pay closer attention to the language we use (especially the metaphors and their underlying assumptions) to describe and interpret organisational reality- the dance of relationships. We, also, need to be alive to the dance of the Life/Death/Life cycle and appreciate that life is constantly balancing at the edge of chaos, "always in danger of falling off into too much order on the one side, and too much chaos on the other (Waldrop citing Langton, P. 235). Terry's (P. 40) words seem to speak directly to the dance that is leadership:

Leadership lives between paradigms, taking people from the comfortable to the less comfortable, from the familiar to the unfamiliar. As the new reality emerges, leadership opens our eyes to the new world that is manifesting itself.

And this new world may well be a better place - a rose garden. I will finish this chapter as I began it with a quotation from T. S. Eliot. For me it presents a powerful case for a view of leadership that derives its insights from a fuzzy logic interpretation of reality:

> What might have been is an abstraction
> Remaining a perpetual possibility
> Only in a world of speculation.
> What might have been and what has been
> Point to one end, which is always present.
> Footfalls echo in the memory
> Down the passage which we did not take
> Toward the door we never opened
> Into the rose garden.

REFERENCES

Capra, F (1992), *Belonging to the Universe; New Thinking About God and Nature*, London: Penguin.

Capra, F. (1982), *The Turning Point*, New York: Simon and Schuster.

Davies, P. (1983), *God and the New Physics* London: J. M. Dent (Now in Penguin).

Eliot, T. S. (1963), *Collected Poems 1909 - 1962*, London: Faber & Faber.

English, A. (1995), *The Double-Headed Arrow*, Ph D. Thesis, University of New England, unpublished.

Estes, C. P. (1992), *Women Who Run with the Wolves*, London: Rider.

Gleick, J. (1987), *Chaos*, London: Sphere Books.

Handy, C. (1994), *The Empty Raincoat: Making Sense of the Future*, London: Random House.

Kosko, B (1994) *Fuzzy Thinking: The New Science of Fuzzy Logic*, London: Flamingo.

Prigogine, I. and Stengers, I. (1984), *Order Out of Chaos*, London: Heinemann.

Schein, E. H. (1992 - 2nd ed.), *Organisational Culture and Leadership*, San Francisco: Jossey-Bass.

Terry, R. W. (1993), *Authentic Leadership: Courage in Action*, San Francisco: Jossey Bass.

Waldrop, M. M. (1992), *Complexity: The Emerging Science at the Edge of Order and Chaos*, Middlesex: Penguin Books.

Wheatley, M. J. (1992), *Leadership and the New Science: Learning About Organisations from an Orderly Universe*, San Francisco: Berrett-Koehler.

Whyte, D. (1994), *The Heart Aroused: Poetry and the Preservation of Soul in Corporate America*, New York: Doubleday.

Zohar, D and Marshall, I. (1994), *The Quantum Society: Mind, Physics and a New Social Vision*, London: Flamingo.

Zukav, G (1979), *The Dancing Wu Li Masters: An Overview of the New Physics*, Sydney: Bantam.

MULTI-CRITERIA OPTIMIZATION: AN IMPORTANT FOUNDATION OF FUZZY SYSTEM DESIGN

Hung T. Nguyen[1] and Vladik Kreinovich[2]

[1]Department of Mathematical Sciences
New Mexico State University
Las Cruces, NM 88003, USA
Email: hunguyen@nmsu.edu

[2]Department of Computer Science
University of Texas at El Paso
El Paso, TX 79968, USA
Email: vladik@cs.utep.edu

Abstract

In many real-life design situations, there are several different criteria that we want to optimize, and these criteria are often in conflict with each other. Traditionally, such *multi-criteria* optimization situations are handled in an *ad hoc* manner, when different conflicting criteria are artificially combined into a single combination objective that is then optimized. The use of unnatural *ad hoc* tools is clearly not the best way of describing a very natural aspect of human reasoning. Fuzzy logic describes a much more natural way of handling multi-criterion optimization problems: when we cannot maximize each of the original conflicting criteria 100%, we optimize each to a certain extent.

Several methods have been proposed in fuzzy logic to handle multi-criteria optimization. These methods, however, still use some *ad hoc* ideas. In this paper, we show that some approaches to multi-objective optimization can be justified based on the fuzzy logic only and do not require any extra *ad hoc* tools.

Keywords: multi-criteria optimization, fuzzy system design, fuzzy sets, fuzzy logic.

1 Introduction

In some real-life situations, when we design a complicated system, we know exactly what we want to to optimize. For example, when we design a race car, our goal is to maximize its speed. In such situations, the problem of finding the best design becomes a clearly defined mathematical problem. Let X denote the set of all possible designs. Then, the problem can be formulated as follows:

GIVEN:

- a (crisp) *objective* function $f : X \to R$, and

- a (crisp) set $C \subseteq X$ (of all designs that satisfy certain *a priori* criteria)

TO FIND $x \in X$ for which

$$f(x) \to \max_{x \in C}.$$

There are several methods of formalizing and solving the maximization problem for more realistic cases in which the conditions on x are formulated in uncertain terms, and are, therefore, described by a *fuzzy* set $\mathbf{C} \subseteq X$ (see, e.g., survey [3] and references therein).

In the majority of real-life situations, however, the objectives of the designed system are not easy to formulate in precise terms. Usually, there are many different criteria $f_1(x), \ldots, f_n(x)$ that we want to optimize, and these criteria are often in conflict with each other. For example, the optimal design for an atomic power station must be both maximally safe and maximally money-saving. If we simply formulate these two maximalities in crisp terms, we will get an inconsistent criterion, because the design that is 100% safe will make the station hundreds of time more expensive, and the cheapest design is clearly not safe. Such situations are called "multi-criteria optimization".

Traditionally, such situations are handled in a somewhat *ad hoc* manner, when different conflicting criteria $f_1(x), \ldots, f_n(x)$ are (rather artificially) combined into a single combination objective $f(x)$ that is then optimized. This combination is usually performed by using an aggregation function $h(y_1, \ldots, y_n)$: $f(x) = h(f_1(x), \ldots, f_n(x))$. The simplest (and most frequently used) aggregation function is a linear function $h(y_1, \ldots, y_n) = w_1 \cdot y_1 + \ldots + w_n \cdot y_n$.

The use of (not very natural) *ad hoc* tools is clearly not the best way of describing a very natural aspect of human reasoning. Fortunately, fuzzy logic describes a much more natural way of handling multi-criterion optimization problems: when we cannot maximize each of the original conflicting criteria 100%, we optimize each *to a certain extent*.

Several methods have been proposed in fuzzy logic to handle multi-criteria optimization; see, e.g., Hwang and Yoon [7], Chen and Hwang [4], Klir and Yuan [8], and references therein. These methods are much more natural than the crisp ones, because they are based on the laws of fuzzy logic that reflect the properties of human reasoning; however, the descriptions of all existing methods use, in additional to fuzzy logic, some ad hoc assumptions and formulas: most of these methods use an aggregation function to combine different criteria $f_i(x)$. In this paper, we show that some of these approaches can be justifies based on the fuzzy logic only and, thus, do not require any extra *ad hoc* tools.

This result builds on the theorems proved in our 1996 paper on fuzzy optimization [3].

2 Fuzzy multi-criteria optimization problems are difficult to formalize

Informally, the multi-criterion optimization problem can be described as follows:

GIVEN:

- a positive integer n;

- n (crisp) functions $f_1, \ldots, f_n : X \to R$, and

- a (fuzzy) set $\mathbf{C} \subseteq X$.

TO FIND $x \in X$ for which

$$f_i(x) \to \max_{x \in \mathbf{C}}$$

for all $i = 1, \ldots, n$.

What is given can be easily formalized:

Definition 1. *By a* multi-criterion maximization problem under fuzzy constraints, *we mean a tuple* $(f_1, \ldots, f_n, \mathbf{C})$, *where* $f_1, \ldots, f_n : X \to R$ *are (crisp) functions from a set X into the set R of all real numbers, and $\mathbf{C} \subseteq X$ is a fuzzy subset of X.*

However, *what we want* is not immediately clear. Since the condition on the desired element x is formulated in fuzzy terms, this problem is difficult to formalize even for a single objective function (see discussion in [3]). For several objective functions, it is even more difficult to formulate. To overcome this double difficulty, it is reasonable to try to handle the two difficulties one by one; in other words:

- first, we will try to formulate the multi-objective optimization problem for the case of crisp constraints (i.e., for the case when $\mathbf{C}$ is a *crisp* set);

- and then, we will try to use general fuzzy techniques to extend this formulation to the general case of fuzzy constraints (i.e., to the case when $\mathbf{C}$ is a fuzzy set).

Let us start with the first part. There are two main ways of representing crisp knowledge for a computer (i.e., in a computer-accessible form):

- as a more mathematics-oriented, *non-procedural* knowledge; usually, in terms of first order logic (or one of its modifications), and

- as a more computer-oriented, *procedural* knowledge; usually, in terms of if-then rules.

In the following sections, we will:

- formalize the crisp version of the multi-objective optimization problem in both languages;

- apply standard fuzzy extension methods (see, e.g., [5, 6]) to extend these descriptions to fuzzy case; and then

- analyze and compare the resulting definitions.

3 Multi-criteria optimization in terms of logic

Let us describe the statement "functions $f_1, \ldots, f_n$ attain their maxima on a set C at x" (denoted hereafter as $S(x)$) in terms of (classical) logic, and then translate it into fuzzy logic.

This statement means that:

- x belongs to C, and

- if y belongs to C, then $f_i(x) \geq f_i(y)$ for $i = 1, \ldots, n$.

Formally,

$$S(x) \leftrightarrow x \in C \,\&\, \forall y (y \in C \to \forall i (f_i(y) \leq f_i(x))).$$

We want to extend this expression to fuzzy logic. How to do that? Let's start with atomic formulas $x \in C$ and $f_i(y) \leq f_i(x)$:

- For a fuzzy set $\mathbf{C}$, the formula $x \in \mathbf{C}$ is described by a membership function $\mu_C(x)$.

- The system of inequalities $\forall i (f_i(y) \leq f_i(x))$ is a crisp statement, so it can be represented by its truth value $t[\forall i (f_i(y) \leq f_i(x))]$ ($t[A] = 1$ if A is true, and $t[A] = 0$ if A is false).

To combine these atomic formulas, we must choose fuzzy operations $f_\&$, $f_\vee$, and $f_\to$ that correspond to $\&$, $\forall$, and $\to$. Then, as a result, we will get the desired membership function for S:

$$\mu_S(x) = f_\& (\mu_C(x), f_\vee (f_\to (\mu_C(x), t[\forall i (f_i(y) \leq f_i(x))]))).$$

How do we choose these fuzzy operations? We consider $\&$ and $\forall$ together, because $\forall$ is nothing else but many "and"s. If we have a finite set X with elements $x_1, \ldots, x_n$, then $\forall x A(x)$ means $A(x_1) \,\&\, A(x_2) \,\&\, \ldots \,\&\, A(x_n)$. If we have an infinite set $X = \{x_1, x_2, \ldots, x_n, \ldots\}$, then we can consider $\forall x A(x)$ as an infinite "and" $A(x_1) \,\&\, A(x_2) \,\&\, \ldots \,\&\, A(x_n) \,\&\, \ldots$, and interpret it as a limit (in some reasonable sense) of finitely many "and"s.

So, it is sufficient to choose a fuzzy analogue of "and"; then, a fuzzy analogue of $\forall$ will be automatically known.

It turns out that the fact that we need to apply $\&$ infinitely many times, and still get a meaningful number, drastically restricts the choice of an

&$-$operation. Namely, we must combine the values $f_\rightarrow(\mu_C(x), t[\forall i(f_i(y) \leq f_i(x))])$ that correspond to all possible y's. If we take $y_1, y_2, \ldots, y_n, \ldots$ all close to each other, then the aggregated degrees of certainty will also be close. The closer y_i to each other, the closer the aggregated values to each other. In the limit, we get the following problem: to combine infinitely many identical values a.

If we take $f_\& = \min$, then we get

$$f_\&(a, \ldots, a, \ldots) = \lim_{n \to \infty} f_\&(a, \ldots, a) \ (n \text{ times}) = \min(a, \ldots, a) = a.$$

For $f_\&(a, b) = a \cdot b$, we get

$$f_\&(a, \ldots, a, \ldots) = \lim_{n \to \infty} f_\&(a, \ldots, a) \ (n \text{ times}) =$$

$$\lim_{n \to \infty} a \cdot a \cdot \ldots \cdot a \ (n \text{ times}) = \lim a^n = 0$$

for all $a < 1$. So, for $f_\&(a, b) = a \cdot b$, we get a meaningless result $\mu_S(x) = 0$ for all x. It turns out that we get the same meaningless result for all &$-$operations different from min. Let's formulate this result in precise terms.

Definition 2. [15, 8, 13] *An* &$-$*operation* (t$-$norm) *is a continuous, symmetric, associative, monotonic operation* $f_\& : [0,1] \times [0,1] \to [0,1]$ *for which* $f_\&(1, x) = x$.

Usually, three types of &$-$operations are used: min, strict operations, and Archimedean operations:

Definition 3. *An* &$-$*operation is called Archimedean if* $f_\&(x, x) < x$ *for all* $x \in (0, 1)$, *and* strict *if it is strictly increasing, as the function of each of the variables.*

The following result is known:

PROPOSITION 1. [3] *If* $f_\&$ *is an Archimedean or a strict operation, then for every* $a \in (0, 1)$,

$$\lim_{n \to \infty} f_\&(a, \ldots, a) \ (n \text{ times}) = 0.$$

Due to this result, it is reasonable to choose $\& = \min$ and, correspondingly, $\forall = \inf$. Hence, we arrive at the following definition:

Definition 4. *Let* $(f_1, \ldots, f_n, \mathbf{C})$ *be a multi-criteria maximization problem under fuzzy constraints, and let* $f_\rightarrow : [0.1] \times [0, 1] \to [0, 1]$ *be a function. We will call* $f_\rightarrow$ *an* implication operation. *By a* solution *corresponding to* $f_\rightarrow$, *we mean*

$$\mu_S(x) = f_\&(\mu_C(x), \inf_y(f_\rightarrow(\mu_C(x), t[\forall i(f_i(y) \leq f_i(x))]))).$$

How can we choose $f_\rightarrow$? There exist many fuzzy analogues of $\rightarrow$ (see, e.g., [12, 14, 8, 13]. For our purposes, however, the choice is not so big, because in our formula, we only have crisp conclusions. Let's analyze how different implication operations behave in this case. We will consider the simplest implication operations first, and then we will discuss the general case.

3.1 Kleene-Dienes operation

This operation (see, e.g., [2]) is based on the well-known expression from classical logic: $(a \rightarrow b) \leftrightarrow (\neg a \lor b)$. To use this formula, we must know $\neg$ and $\lor$.

Definition 5. *By a $\neg-$operation, we mean a strictly decreasing continuous function $f_\neg : [0, 1] \rightarrow [0, 1]$ such that $f_\neg(0) = 1$ and $f_\neg(f_\neg(a)) = a$.*

Definition 6. *[15, 8, 13] An $\lor-$operation (t$-$conorm) is a continuous, symmetric, associative, monotonic operation $f_\lor : [0, 1] \times [0, 1] \rightarrow [0, 1]$ for which $f_\lor(0, x) = x$.*

Definition 7. *Assume that $f_\lor$ and $f_\neg$ are $\lor-$ and $\neg-$operations. The function $f_\rightarrow(a, b) = f_\lor(f_\neg(a), b)$ will be called a Kleene-Dienes implication.*

PROPOSITION 2. *Let $(f_1, \ldots, f_n, C)$ be a multi-criteria maximization problem with fuzzy constraints. Then, the solution corresponding to Kleene-Dienes implication has the form*

$$\mu_{KD}(x) = \min(\mu_C(x), f_\neg(\sup_{y:\, \exists i(f_i(y)>f_i(x))} \mu_C(y))).$$

Proof. Since $b \in \{0, 1\}$, we can eliminate $f_\lor$: indeed, $f_\lor(0, x) = x$, and $f_\lor(x, 1) = 1$ for an arbitrary $\lor-$operation. Q.E.D.

Comment. In particular, for $f_\neg(z) = 1 - z$, we get the expression

$$\mu_{KD}^*(x) = \min(\mu_C(x), 1 - \sup_{y:\, \exists i(f_i(y)>f_i(x))} \mu_C(y)).$$

3.2 Zadeh's operator

This implication operator is based on another formula from classical logic: $(a \rightarrow b) \leftrightarrow (\neg a \lor (a\&b))$. Since we already know that $\& = \min$, we arrive at the following definition:

Definition 8. *Assume that $f_\lor$ and $f_\neg$ are $\lor-$ and $\neg-$operations. The function $f_\rightarrow(a, b) = f_\lor(f_\neg(a), \min(a, b))$ will be called a Zadeh implication.*

PROPOSITION 3. *Let $(f_1, \ldots, f_n, C)$ be a multi-criteria maximization problem with fuzzy constraints. Then, the solution corresponding to Zadeh's implication has the form*

$$\mu_Z(x) =$$

$$\min(\mu_C(x), f_\neg(\sup_{y:\ \exists i(f_i(y) > f_i(x))} \mu_C(y)), \sup_{y:\ \forall i(f_i(y) \le f_i(x))} f_\vee(\mu_C(y), f_\neg(\mu_C(y)))).$$

Proof easily follows from considering the cases $b = 0$ and $b = 1$. Q.E.D.

3.3 Other implication operations

It is easy to check that for crisp b, all other known implication operations either turn into one of these two, or lead to a crisp formula. For example, let's consider the most frequently used operations listed in [14]:

- Lukasiewicz's $\min(1, 1 - a + b)$ turns into $1 - a$ if $b = 1$ and 1 if $b = 0$ (same as Kleene-Dienes).

- Gödel's [11] 1 if $a \le b$, b else, gets only crisp values if b is crisp.

- Gaines's [11] 1 if $a \le b$ and b/a else leads to 1 if $b = 1$ and to 0 if $b = 0$, i.e., also, only to crisp values.

- Kleene-Dienes-Lukasiewicz [2] $1 - a + a \cdot b$ for crisp b coincides with Kleene-Dienes's.

- Willmott's $\min(\max(1 - a, b), \max(a, 1 - a), \max(b, 1 - b))$ reduces for crisp b to Zadeh's formula [16].

This "lack of choice" can be partially explained (see also [3]) by the fact that usually, two methods of describing an $\rightarrow$ −operation are used:

- We can describe $\rightarrow$ directly in terms of &, $\vee$, and $\neg$. We have already considered these methods.

- We can also describe $a \rightarrow b$ indirectly: as a statement that, being added to a, implies b (i.e., as a kind of a "solution" of the equation $f_\&(a, a \rightarrow b) = b$). If this equation has several solutions, we can choose, e.g., the largest one, or more generally, the largest c for which $f_\&(a, c) \le b$. Since b is crisp, we get a degenerate solution:

  - If $b = 1$, then $f_\&(a, c) \le 1$ is always true, so $c = 1$.
  - If $b = 0$, then $f_\&(a, c) = 0$ is usually only true for $c = 0$.

So, for crisp b, this definition leads to an operation with crisp values only.

4 Multi-criteria optimization in terms of if-then rules

Let's describe the (crisp) conditional multi-criteria optimization problem in terms of if-then rules. Computational algorithms that compute the maximum are usually iterative, so it is difficult to find if-then rules that would directly select the desired solution. However, it is very easy to describe rules that will delete everything but the desired solution:

- If x does not satisfy the condition, then x is not the desired solution.

- If for some x and for some i, there exists another element y that satisfies the constraint C and for which $f_i(y) > f_i(x)$, then x is not the desired solution.

In logical terms, these rules take the following form:

$$\neg C(x) \rightarrow \neg S(x),$$

$$C(y) \,\&\, (f_i(y) > f_i(x)) \rightarrow \neg S(y).$$

To generalize these rules to the case when the constraint set $\mathbf{C}$ is fuzzy, we will use the standard (Mamdani's) methodology from fuzzy control (see, e.g., [10, 9, 3]). According to this methodology, if we have a set of rules, then for a certain conclusion to be true, it is necessary and sufficient that for one of the rules that lead to this conclusion, all the conditions are satisfied. In logical terms,

$$\neg S(x) \leftrightarrow \neg C(x) \vee (C(y_1)\,\&\,(f_1(y_1) > f_1(x))) \vee \ldots \vee (C(y_1)\,\&\,(f_n(y_1) > f_n(x))) \vee$$

$$(C(y_2)\,\&\,(f_1(y_2) > f_1(x))) \vee \ldots \vee (C(y_2)\,\&\,(f_n(y_2) > f_n(x))) \vee \ldots$$

Here, $\vee$ is applied to statements that correspond to all possible values of y.

Next, in fuzzy control, we substitute degrees of membership instead of atomic statements, and use $\&-$, $\vee-$, and $\neg-$operations instead of $\&$, $\vee$, and $\neg$. As a result, we get the following formula:

$$\mu_{\neg S}(x) = f_\vee[\mu_{\neg C}(x),$$

$$f_\&(\mu_C(y_1), t[f_1(y_1) > f_1(x)]), \ldots, f_\&(\mu_C(y_1), t[f_n(y_1) > f_n(x)]),$$

$$f_\&(\mu_C(y_2), t[f_1(y_2) > f_1(x)]), \ldots, f_\&(\mu_C(y_2), t[f_n(y_2) > f_n(x)]), \ldots].$$

Here, the $\vee-$operation combines infinitely many terms, so (similarly to what we have shown in Section 3), we can conclude that the only way to avoid the meaningless situation in which $\mu_{\neg S}(x) = 1$ for all x is to use $f_\vee = \max$. So, we arrive at the following definition.

Definition 9. *Let* $(f_1, \ldots, f_n, \mathbf{C})$ *be a maximization problem under fuzzy constraints, and let* $f_\neg$ *be a* $\neg-operation$. *A fuzzy set* $\mathbf{S}_R$ *will be called a rule-based solution to this problem if its membership function* $\mu_R(x)$ *satisfies the following equality:*

$$f_\neg(\mu_R(x)) = \max[f_\neg(\mu_C(x)), \sup_{y,i} f_\&(\mu_C(y), t[f_i(y) > f_i(x)])].$$

It turns out that this solution coincides with the one described in the Section 3.1:

PROPOSITION 4. *For every multi-criteria maximization problem under fuzzy constraints, the rule-based solution $\mu_R(x)$ coincides with the solution $\mu_{KD}(x)$ corresponding to Kleene-Dienes implication.*

Proof. The predicate $f_i(y) > f_i(x)$ is crisp, so its truth value is either 0, or 1. By definition of an &−operation, $f_\&(a, 0) = 0$, and $f_\&(a, 1) = a$. Therefore:

- If $f_i(y) \leq f_i(x)$, we have $t[f_i(y) > f_i(x)] = 0$, and $f_\&(\mu_C(y), 0) = 0$.

- If $f_i(y) > f_i(x)$, then $t[f_i(y) > f_i(x)] = 1$, and $f_\&(\mu_C(y), 1) = \mu_C(y)$.

When computing sup of a set of non-negative numbers, we can neglect 0's, and thus consider only y for which $\exists i(f_i(y) > f_i(x))$. So, we get

$$f_\neg(\mu_R(x)) = \max[f_\neg(\mu_C(x)), \sup_{y:\, \exists i(f_i(y) > f_i(x))} \mu_C(y)].$$

Now, because of the properties of an ¬−operation, we have $\mu_R(x) = f_\neg(f_\neg(\mu_R(x)))$, so

$$\mu_R(x) = f_\neg(\max[f_\neg(\mu_C(x)), \sup_{y:\, \exists i(f_i(y) > f_i(x))} \mu_C(y)]).$$

Since $f_\neg$ is decreasing, $f_\neg(\max(a, b)) = \min(f_\neg(a), f_\neg(b))$, so

$$\mu_{DR}(x) = \min[\mu_C(x), f_\neg(\sup_{y:\, \exists i(f_i(y) > f_i(x))} \mu_C(y))].$$

This is exactly the expression μ_{KD} for the Kleene-Dienes implication. Q.E.D.

5 Scale-invariance: an important property of the proposed solution

We will see that the proposed solution has the following important property: the resulting solution $\mu_S(x)$ does not depend on the choice of units in which we measure the values of the objective functions $f_i(x)$, or on any other re-scaling of these functions.

This "scale-invariance" is an important requirement: e.g., if we want to design the largest spherical oil reservoir, which is possible under the current technology, then we can formulate this problem as $f_1(x) = R \to \max$, where R is the radius of the reservoir x, or as $f_1'(x) = V \to \max$, where $V = (4/3)\pi \cdot R^3$ is the volume of this reservoir. From the user's viewpoint, maximizing the radius is exactly the same problem as maximizing the volume, so it is desirable that the solutions corresponding to f_1 and f_1' be the same.

- This, unfortunately, is not always the case for traditional *ad hoc* multi-criteria optimization methods: for example, if we maximize a linear combination of different criteria, then maximizing the combination $f(x) = w_1 \cdot f_1(x) + w_2 \cdot f_2(x) + \ldots$ can lead to different results than optimizing the combination $f'(x) = w_1 \cdot f_1'(x) + w_2 \cdot f_2(x) + \ldots$

- Let us show that this requirement of scale-invariance is satisfied for the proposed solution:

Definition 10. *Let a set X be fixed.*

- *By a re-scaling, we mean a strictly increasing function $g : R \to R$ from real numbers to real numbers.*

- *We say that a function $f_i' : X \to R$ is a re-scaling of the function $f_i : X \to R$ if $f_i'(x) = g_i(f_i(x))$ for some re-scaling function $g_i(y)$.*

- *We say that a multi-criteria maximization problem $\mathcal{P}' = (f_1', \ldots, f_n', \mathbf{C})$ under fuzzy constraints is a re-scaling of the maximization problem $\mathcal{P} = (f_1, \ldots, f_n, \mathbf{C})$ if each function f_i' is a re-scaling of the corresponding function f_i.*

PROPOSITION 5. *Let $\mathcal{P}$ be a multi-criteria optimization problem under fuzzy constraints, and let $\mathcal{P}'$ be a re-scaling of the problem $\mathcal{P}$. Then, for a given implication function, a function $\mu_S(x)$ is the solution to the problem $\mathcal{P}$ if and only if it is the solution to the re-scaled problem $\mathcal{P}'$.*

Proof. The proof easily follows from the fact that the definition of the solution (Definition 4) does not use the actual values of f_i, it only uses the relation $f_i(y) \leq f_i(x)$ which is preserved under any strictly increasing transformation $f_i(x) \to g_i(f_i(x))$. Therefore, the solutions that correspond to $\mathcal{P}$ and to $\mathcal{P}'$ indeed coincide. Q.E.D.

6 From fuzzy solution to crisp solution

In the above sections, we described the *fuzzy solution* $\mu_S(x)$ to the multi-criteria optimization problem. This fuzzy solution, in effect, supplies the user with a list of possible designs x, together with the degrees $\mu_S(x)$ to which each design x satisfies all the required criteria.

In some real-life situations, this is all the user wants, so that she can make the final decision herself. In some cases, however, the user would prefer a computer to choose the design for him. In these cases, it is natural to choose a design x^* with the largest degree of "requirements satisfaction" $\mu_S(x)$. The experience of fuzzy optimization shows that this is indeed a reasonable choice (see, e.g., Zimmermann [18]).

Definition 11. *Let* $(f_1, \ldots, f_n, \mathbf{C})$ *be a multi-criteria optimization problem under fuzzy constraints, and let* $\mu_S(x)$ *be the solution to this problem. We say that an element* $x^* \in X$ *is a* crisp solution *to this problem if*

$$\mu_S(x^*) = \max_{x \in X} \mu_S(x).$$

Comments.

- It is worth mentioning that in some cases, better results can be obtained by using more complicated defuzzification procedures [1, 17].

- From the fact the (fuzzy) solution $\mu_S(x)$ is invariant under an arbitrary re-scaling of the objective functions, we can now conclude that the crisp solution (defined as the maximum of $\mu_S(x)$) is also scale-invariant:

PROPOSITION 6. *Let* $\mathcal{P}$ *be a multi-criteria optimization problem under fuzzy constraints, and let* $\mathcal{P}'$ *be a re-scaling of the problem* $\mathcal{P}$. *Then, for a given implication function, an element* $x^* \in X$ *is a crisp solution to the problem* $\mathcal{P}$ *if and only if it is a crisp solution to the re-scaled problem* $\mathcal{P}'$.

Acknowledgments. This work was partially supported by NSF Grant No. EEC-9322370. The authors are thankful to Leonid Reznik, Vladimir Dimitrov, and Janusz Kacprzyk, the editors of this volume, for their help, and to Bernadette Bouchon-Meunier and Anatole Lokshin for valuable discussions.

References

[1] P. Angelov, "A generalized approach to fuzzy optimization", *International Journal of Intelligent Systems*, 1994, Vol. 9, pp. 261–268.

[2] W. Bandler and L. J. Kohout, "Fuzzy power sets and fuzzy implication operators", *Fuzzy Sets and Systems*, 1980, Vol. 4, pp. 13–30.

[3] B. Bouchon-Meunier, V. Kreinovich, A. Lokshin, and H. T. Nguyen, "On the formulation of optimization under elastic constraints (with control in mind)", *Fuzzy Sets and Systems*, 1996, Vol. 81, No. 1, pp. 5–29.

[4] S. J. Chen and C. L. Hwang, *Fuzzy multiple attribute decision making: methods and applications*, Springer-Verlag, N.Y., 1992.

[5] D. Dubois and H. Prade, "Measuring properties of fuzzy sets: a general technique and its use in fuzzy query evaluation", *Fuzzy Sets and Systems*, 1990, Vol. 38, pp. 137–152.

[6] D. Dubois and H. Prade, "Scalar evaluations of fuzzy sets: overview and applications", *Applied Mathematics Letters*, 1990, Vol. 3, No. 2, pp. 37–42.

[7] C. L. Hwang and K. Yoon, *Multiple attribute decision making: methods and applications*, Springer-Verlag, N.Y., 1981.

[8] G. Klir and B. Yuan, *Fuzzy sets and fuzzy logic: theory and applications*, Prentice Hall, Upper Saddle River, NJ, 1995.

[9] V. Kreinovich, C. Quintana, R. Lea, O. Fuentes, A. Lokshin, S. Kumar, I. Boricheva, and L. Reznik, "What non-linearity to choose? Mathematical foundations of fuzzy control", *Proceedings of the 1992 International Conference on Fuzzy Systems and Intelligent Control*, Louisville, KY, 1992, pp. 349–412.

[10] E. H. Mamdani, "Application of fuzzy logic to approximate reasoning using linguistic systems", *IEEE Transactions on Computing*, 1977, Vol. 26, pp. 1182–1191.

[11] M. Mizumoto and H. J. Zimmermann, "Comparison of fuzzy reasoning methods", *Fuzzy Sets and Systems*, 1982, Vol. 8, pp. 253–283.

[12] H. T. Nguyen, *Lecture Notes on Fuzzy Logic*, LIFE, Tokyo Institute of Technology, Tokyo, Japan, 1993.

[13] H. T. Nguyen and E. A. Walker, *A first course in fuzzy logic*, CRC Press, Boca Raton, Florida, 1997.

[14] P. Ruan and E. E. Kerre, "Fuzzy implication operators and generalized fuzzy method of cases", *Fuzzy Sets and Systems*, 1993, Vol. 54, pp. 23–37.

[15] B. Schweizer and A. Sklar. *Probabilistic metric spaces*, North Holland, Amsterdam, 1983.

[16] R. Willmott, "Two fuzzy implication operators in the theory of fuzzy power sets", *Fuzzy Sets and Systems*, 1980, Vol. 4, pp. 31–36.

[17] R. R. Yager and D. P. Filev, *Essentials of fuzzy modeling and control*, J. Wiley, N.Y., 1994.

[18] H. J. Zimmermann, *Fuzzy set theory and its applications*, Kluwer, Boston, 1991.

A MATHEMATICAL FRAMEWORK FOR HUMAN DECISION MAKING AS AN INTEGRATED PART OF THE WHOLE

Victor Korotkih

Department of Mathematics and Computing
Central Queensland University
Rockhampton, QLD 4702
Australia

E-mail: v.korotkich@cqu.edu.au

Summary. A mathematical framework to model people's thinking, formulated in terms of fuzzy logic, is proposed. It captures individual's strategies when they are integral parts of the situation in which they participate. The mathematical framework is a web of relations between integers which is developed using a concept called structural complexity. The approach used embodies the idea that natural phenomena cannot be seen as isolated entities but as integrated parts of a whole. The web of integer relations is used to study properties of strategies connected with optimality and a mechanism which approximately constructs optimal rules is presented.

Key words: human decision making, fuzzy rules, complexity, optimization, integers

1. Introduction

The standard approach to develop a theory of a natural phenomenon is to represent the system under study and its environment as independent and separable objects. This approach is appropriate in certain limited contexts, as in classical physics, but involves difficulties in other contexts, as in quantum theory and social science. In particular, this approach cannot be used when a problem involving human decision makers is considered and their thinking is an integral part of the situation in which they participate.

This problem, arising in many areas [1, 2], has the following generic form. There is a group of decision makers who seek to understand and use in a proper way the situation in which they participate. Their understanding serves as the basis for decisions they have to make in succesive steps. These decisions at each step create an event which provides a result of their actions. Expectations of some of them are validated by the result, others are not. Then each decision maker independently employs the event they created together to understand the situation. At each step, the decision maker updates his or her understanding of the situation and incorporates this into the next decision. A simple example of this type of situation is illustrated in the El Farol problem [2] which is discussed in section 2. Decision makers' thinking through their decisions affects the course of events and the course of events affects their thinking. The thinking and the course of events cannot be considered separately and independently. Consequently, the standard approach cannot be used to describe the problem and what seems to be called for is the development of a new mathematical framework to investigate this kind of problem.

The paper proposes a mathematical framework to address this kind of problem and uses the framework to study their properties connected with optimality. The starting point is an approach which considers the world as a dynamic web of relations [3] and embodies the idea that natural phenomena cannot be seen as isolated entities but as integrated parts of a whole [4], [5]. An attempt to translate this approach into an adequate mathematical form is made in [6]. The heart of the translation is the idea to use a web of relations between integers as a model for the whole. The web of integer relations provides a formal setting for structures to be seen not as independent and separable objects, but as parts of the whole. The paper aims to describe the decision making process, consisting of human decision makers and the course of events, as an integrated part of the whole.

In section 2 we consider the decision making process in terms of the El Farol problem. This problem is a typical representative of the decision making processes and is illuminating to our intuitions. Since humans employ mostly words in thinking, we formulate the problem in terms of fuzzy logic [7] and propose a model of fuzzy decision making. To characterize the performance of the decision making procedure two measures of different types, local and global, are considered. The local measure is based on Cartesian space and comes from the fact that sequences, which are the main object of description in the model, can be determined locally step by step. The model is built specifically in a way to eliminate the difference between decision making procedures in terms of the local measure and to focus on their characterization in terms of the global one.

In section 3 we begin to explore a new way to characterize the performance of the decision making procedure by using the global measure. The key to this measure is the idea that sequences must have a global nature in addition to their local one. To describe the global measure it is necessary to go rather deeply into the meaning of structural space and a concept of structural complexity [8]. Structural space attempts to provide a proper global description of sequences, just as Cartesian space provides a local description. By analogy with the local measure derived from Cartesian space, we use a measure of proximity in structural space to characterize the performance of the decision making procedure. This measure deals with the procedure's ability to determine global properties of sequences.

Structural complexity is an abstract concept and it would be difficult to use it without giving a mental picture. An illustrative example is given in section 4 to help assimilate the concept. The example provides for the most part a visual access to the web of integer relations which we propose as a mathematical framework to investigate the problem. Since there is something essential in human understanding, we hope the visualization from the the example may provide a means of seeing answers to questions, which are not possible to obtain in a formal way.

In section 5 a formal definition of the web of integer relations is given. The word "web" is used to informally capture the structure of the set. The intuitive motivation for the name is provided by the example, which shows that a subset of the web has a tree structure.

In section 6 with the mathematical framework at hand we study properties of the decision making procedures connected with optimality. In particular, we use the global measure defined in section 3 to characterize the performance of the decision making procedure. We define optimality of the decision making procedure in terms of structural complexity.

In section 7 under certain assumptions we present a mechanism which constructs rules of an approximation to the optimal decision making procedure. It represents our initial idea, to understand the decision making process as an integrated part of a whole, in algorithmic terms.

The optimal decision making procedure presented in section 7 reinforces our idea to use the web of relations between integers and structural complexity as a new mathematical framework for human decision making.

2. A Model of Fuzzy Decision Making

For definiteness, it is worthwhile to consider the decision making process in terms of the El Farol problem, suggested by W. Brian Arthur [2]. This problem is a typical representative of the kind of decision making processes. It is hopefully illuminating to our intuitions. Since humans employ mostly words in thinking, we formulate the problem in terms of fuzzy logic.

Assume there is a group of people, decision makers, who enjoy going to El Farol, a small bar, to listen to the music on Thursday evenings. However, none of them wants to visit the bar if it is going to be crowded. For example, one person may consider the bar to be crowded when more than half of the people are present whilst another one may consider the bar to be crowded when it is full to capacity. As a result each person seeks to understand and use in a proper way the problem situation in which he or she participates. In particular, on Thursday each person, before going to the bar, employs a decision making procedure to predict if it will be crowded at the bar in the evening. Each person decides independently to go to the bar only if his or her prediction is that it will be not crowded. The situation at the bar is described by a linguistic variable attendance whose linguistic values [9], for simplicity, are crowded and not crowded. It is assumed that each person only has the information of how many people were in the bar over all past weeks. This information is used by each person to figure out whether the bar was crowded for him or not over all past weeks. Because different people have different decision making procedures and different meanings of whether "the bar is crowded", some will turn up at the bar, while others expect it will be crowded and do not go there. While the number of people going to the bar is uncertain, there is no uncertainty in each person's decision, he or she takes an action: go or not to go to the bar. These decisions create a result, the number of people at the bar on Thursday evening.

As the new attendance number is become known on the next day, expectations of some of them are validated by the result, others are not. For example, someone may have gone to the bar and found it not crowded, while another person may have stayed at home even though the bar was not crowded. Then each person updates the accuracy of their decision making procedure for the next Thursday. Therefore, each person's decision making through their actions affects the problem situation, and the problem situation affects their decision making. This means that they cannot be considered separately and independently. Consequently, a standard mathematical approach cannot be used to describe the El Farol problem and what seems to be called for is the development of a new mathematical framework to investigate it.

To clarify what we are dealing with we use another way of looking at the problem. The group of people in the El Farol problem exhibits collective behaviour that emerges from their individual decision making procedures and the problem itself contains all the components that characterize what is meant by a complex, adaptive system [2]. A key issue in the area is to propose the right type of a mathematical framework leading to a theory of complex, adaptive sytems. In particular, the theory has to explain how a useful collective behaviour can be designed from individual decision making procedures. For example, it could be desirable to have a collective behaviour with the attendance number approximately equal the number of seats in the bar and roughly the same number of visits per a person in one year. It is important that this collective behavior, if emerged, would be a result of self-organization of the group but not of central control.

To propose a mathematical structure for the problem we formulate a model of fuzzy decision making. In the model a person is characterized by a decision making procedure which is supposed to predict or infer an answer to the query if it will be crowded at the bar on Thursday evenings or not. This prediction is used by the person as the basis for the decision to go or not to go to the bar. Therefore, the decision making procedure takes only linguistic values crowded and not crowded. For simplicity we use symbols -1 and $+1$ respectively for denotation of the values. Namely, if the procedure predicts that it will be crowded in the bar then the value can be denoted by -1. This means that the person does not go to the bar. If it predicts that it will not be crowded in the bar then the value can be denoted by +1. This means that the person does not stay at home and goes to the bar. Accordingly, the result, due to the attendance number, can be encoded by symbols -1 and $+1$. If the attendance number is interpreted by the person that it was crowded in the bar then the result can be denoted by -1 and $+1$ otherwise. Thus, in what follows symbols -1 and $+1$ are labels of a granule attendance and will stand for its linguistic values of crowded and not crowded respectively.

Let $B_n = \{s : s = s_1, \ldots, s_n, s_i \in \{-1, +1\}, i = 1, \ldots, n\}$ be the set of all sequences of length n, where for simplicity we drop all commas separating symbols in the notation. The length of a sequence s, denoted $|s|$, is the number of symbols in the sequence. If $|s| = 0$, then the sequence s is the empty sequence $\emptyset$. A decision making procedure A for a period of n weeks $n = 2, 3, \ldots$ is a set $\{\chi_k, k = 1, \ldots, n\}$ of fuzzy rules. Each rule $\chi_k, k = 1, \ldots, n$, in general, is a function of all information given by the past $k - 1$ weeks or steps and takes $s_k \in \{-1, +1\}$ as the prediction of the result $s'_k \in \{-1, +1\}$ at the kth step. This information can be encoded as sequences $s(k-1) = s_1, \ldots, s_{k-1}, s'(k-1) = s'_1, \ldots, s'_{k-1}, k = 2, \ldots, n$, where $s_i, s'_i \in \{-1, +1\}$, $i = 1, \ldots, k - 1$ are the prediction and result at the ith step respectively. We have empty sequences $s(0) = s_0 = \emptyset, s'(0) = s'_0 = \emptyset$ when $k = 1$. Hence, for a rule $\chi_k, k = 1, \ldots, n$ we have

$$s_k = \chi_k(s(k-1), s'(k-1)). \tag{2.1}$$

In the model it is assumed that there is no statistics underlying $s'_k, k = 1, \ldots, n$. To put this another way, the particular value of s'_k is totally indeterminate before the prediction, i.e., symbols -1 and $+1$ appear with the same probability. After n steps a decision making procedure A for a sequence $s' = s'_1, \ldots, s'_n \in B_n$ of results produces, due to (1), a sequence $s = s_1, \ldots, s_n \in B_n$ of predictions and this condition will be formally denoted herein by $s = A(s')$.

Given sequences $s, s' \in B_n$ such that $s = A(s')$, a natural question is how to characterize the performance of a procedure A. Intuitively, the performance of a procedure A must be connected with its ability to determine the sequence s'. This determination is resulted in the procedure's generation of the sequence s and a measure of proximity between the sequences is required to formally capture it. We consider two measures which are of different types, namely local and global. The global measure will lead us to the mathematical structure in quest.

The local measure is based on Cartesian space and comes from the fact that a sequence can be determined locally step by step. In particular, each sequence $s' = s'_1, \ldots, s'_n \in B_n$ is specified locally as s'_k in a particular moment of time $k, k = 1, \ldots, n$, i.e., at the kth step. Locally a procedure A predicts correctly the sequence at the kth step $k = 1, \ldots, n$ if $s_k = s'_k$ and does not predict it correctly if $s_k = -s'_k$, where s_k is given by (1). From practical considerations it arises naturally to have a procedure that locally makes a correct prediction as many times as possible. This dictates to characterize the performance of decision making procedures by their numbers of correct predictions locally done.

Namely, let $N(s, s')$ be a measure that counts the number of components in which sequences $s, s' \in B_n$ are equal. Evidently, for sequences $s, s' \in B_n$ such that $s = A(s')$ the measure $N(s, s')$ gives the number of times a procedure A locally makes a correct prediction.

Let $\mathbf{A}$ be the set of all decision making procedures. Two procedures $A_1 = \{\chi_{1,k}, k = 1, \ldots, n\}, A_2 = \{\chi_{2,k}, k = 1, \ldots, n\} \in \mathbf{A}$ are considered not to be equal if there exists a sequence $s' = s'_1, \ldots, s'_n \in B_n$ and an index k, $k = 1, \ldots, n$ such that their predictions are not equal

$$\chi_{1,k}(s(k-1), s'(k-1)) \neq \chi_{2,k}(s(k-1), s'(k-1)),$$

where for $k = 2, \ldots, n$ we have $s_i = \chi_{1,i}(s(i-1), s'(i-1))$, $s_i = \chi_{2,i}(s(i-1), s'(i-1))$, $i = 1, \ldots, k-1$.

When one has a measure of performance, two types of analysis to define optimality with respect to it are of importance - the worst case and average case. A decision making procedure A^* (with respect to the local measure) is said to be optimal in the worst case, if

$$\min_{s' \in B_n, s = A^*(s')} N(s, s') = \max_{A \in \mathbf{A}} \min_{s' \in B_n, s = A(s')} N(s, s')$$

and optimal in the average case, if

$$\sum_{s' \in B_n, s = A^*(s')} N(s, s') = \max_{A \in \mathbf{A}} \sum_{s' \in B_n, s = A(s')} N(s, s'),$$

where the sum is taken over all $s' \in B_n$. The following two results describe the situation about optimal decision making procedures defined with respect to the local measure [10]. Namely, let $A \in \mathbf{A}$ be an arbitrary decision making procedure then

$$\min_{s' \in B_n, s = A(s')} N(s, s') = 0,$$

which means that for each procedure A the worst case is that it is never correct. The second result is that, for any $A \in \mathbf{A}$,

$$\sum_{s' \in B_n, s = A(s')} N(s, s') = n2^{n-1},$$

which means that the average number of correct predictions is always the same, $n2^{n-1}$.

Consequently, all decison making procedures are equivalent when the local measure $N(s, s')$ is used. In order to distinguish between decision making procedures, we cannot rely on the local measure, we must develop a global one.

3. Structural Space and Complexity

In this section we begin to explore a new way to characterize the performance of the decision making procedure. The key to this measure is the idea that sequences must have a global nature in addition to their local one. To describe the global measure it is necessary to go rather deeply into the meaning of structural space and a concept of structural complexity. Structural space attempts to provide a proper global description of sequences, just as Cartesian space provides a local description. By analogy with the local measure derived from Cartesian space, we use a measure of proximity in structural space to characterize the performance of the decision making procedure. This measure deals with the procedure's ability to determine global properties of sequences.

Structural space can be formally introduced by a connection with Cartesian space [6]. In particular, every sequence $s = s_1...s_n \in B_n$ can be uniquely associated to a corresponding point $\vartheta(s)$ in structural space. This point $\vartheta(s)$ is specified by numbers $(\vartheta_1(s), \ldots, \vartheta_k(s), \ldots)$, which are its coordinates in structural space. These numbers, called structural numbers or values, are global properties of the sequence, because they are defined by the whole sequence, and given by the formula

$$\vartheta_k(s) = \sum_{i=0}^{k-1} \alpha_{ki}(n^i s_1 + \cdots + 1^i s_n), \quad k = 1, 2, \ldots , \tag{3.1}$$

where $\alpha_{ki} = \binom{k}{i}(-1)^{k-i-1}/k!, i = 0, \ldots, k-1$ and $\binom{k}{i}, i = 0, \ldots, k-1$ are the binomial coefficients. By definition $\vartheta_0(s) = \vartheta_0(s')$ for all $s, s' \in B_n$. As (2) suggests, a distinguishing characteristic of coordinates in structural space is that they are defined by using powers of integers $1, 2, \ldots, n$.

A characteristic feature of structural space is the fact that its points can be completely specified by a finite number of coordinates, which varies from one point to another. By comparison, a point $s = (s_1, \ldots, s_n) \in \Re^n$ is completely specified in the cartesian space $\Re^n$ if and only if all its cartesian coordinates $s_1, \ldots, s_n$ are known. Consequently, the number is the same for all points and there is little reason to consider it as a characteristic of points in $\Re^n$.

In contrast this feature of structural space makes it natural to define the concept of structural complexity of a sequence. Namely, the structural complexity of a sequence $s \in B_n$ with respect to a sequence $s' \in B_n, s \neq s'$, denoted $C(s, s')$, is defined as the integer $k, k = 1, 2, \ldots$ such that $\vartheta_0(s) = \vartheta_0(s'), \ldots, \vartheta_{k-1}(s) = \vartheta_{k-1}(s')$, but $\vartheta_k(s) \neq \vartheta_k(s')$ [8]. This means that by using the first k coordinates of the point $\vartheta(s)$ it is possible to distinguish it from the point $\vartheta(s')$ and thus to distinguish the sequence s from the sequence s'.

The structural complexity of a sequence $s \in B_n$, denoted $C(s, B_n)$, and $C(s, s)$ are defined by

$$C(s, B_n) = \max_{s' \in B_n, s' \neq s} C(s, s'), \quad C(s, s) = C(s, B_n) + 1.$$

It is conjectured that as s varies over B_n then $C(s, B_n)$ is varied through the range [10]

$$1 \leq C(s, B_n) \leq \lfloor \log_2(n) \rfloor + 1,$$

where $\lfloor a \rfloor$ is the integer part of a real a. For example, when $s = +1 \ldots +1 \in B_n$ then it is easy to show that $C(s, B_n) = 1$ and when $s \in B_n$ is the initial segment of length n of the celebrated Prouhet-Thue-Morse (PTM) sequence [11]-[13] then it is conjectured that $C(s, B_n) = \lfloor \log_2(n) \rfloor + 1$.

It is important for us that structural complexity $C(s, s')$ also is a measure of proximity in structural space. The structural complexity $C(s, s')$ of sequences $s, s' \in B_n$ keeps account of their successive equal structural numbers. Specifically, if $C(s, s') = k, k \geq 1$ then $\vartheta_0(s) = \vartheta_0(s'), \ldots, \vartheta_{k-1}(s) = \vartheta_{k-1}(s')$ and $\vartheta_k(s) \neq \vartheta_k(s')$. Note that the structural complexity $C(s, s')$ of two sequences $s, s' \in B_n$ is the first structural number for which they differ. For example, if $\vartheta_0(s) = \vartheta_0(s'), \vartheta_1(s) = \vartheta_1(s')$ but $\vartheta_2(s) \neq \vartheta_2(s')$ then $C(s, s') = 2$.

The idea that underlies the definition of structural complexity is that the coordinates $(\vartheta_1(s), \ldots, \vartheta_k(s)), k = C(s, B_n)$ are a complete specification of a point $\vartheta(s)$ in structural space and consequently the structural numbers are a complete specification of the sequence s in B_n, by which it can be recognized. Since the structural numbers completely determine the whole sequence they determine all of its properties and therefore it is possible, in principle at least, to read off all the properties of the sequence from the structural numbers.

Given a procedure A and sequences $s, s' \in B_n$ such that $s = A(s')$, all these facts motivate us to characterize the performance of the decision making procedure A by comparing the successive structural numbers of the sequence s' with those of the sequence s. Clearly, the mathematical form of this characterization is exactly the structural complexity $C(s, s')$.

Up to this point structural complexity is an abstract concept and it would be difficult to use it without giving a mental picture. But a sense of what it entails can be gained by the following illustrative example. It helps assimilate the concept and demonstrates a set of images connected with structural complexity, i.e., an interesting phenomenon of the formation of geometrical patterns and integer relations. But it provides for the most part a visual access to the web of integer relations which we propose as the mathematical framework to investigate the problem. Since there is something essential in human understanding, we hope the visualization from the example may provide a means of seeing answers to questions, which are not possible to obtain in a formal way.

4. An Illustration of Structural Complexity

Consider a sequence $\varsigma \in B_n$ connected with the Fibonacci numbers, tiling and quasicrystals [14]. The sequence ς is formed by the substitution rules $1 \to 1-1, -1 \to 1$. Starting with $+1$, successive substitutions produce $+1 - 1, +1 - 1 + 1, +1 - 1 + 1 + 1 - 1, \ldots$. Let $s = +1 - 1 + 1 + 1 - 1 + 1 - 1 + 1 + 1 - 1 + 1 + 1 - 1 + 1 - 1$ be the initial segment of length 15 of ς. There is a sequence $s' = -1 + 1 + 1 + 1 + 1 - 1 - 1 + 1 + 1 + 1 - 1 + 1 - 1 - 1 + 1$ such that $C(s, s') = 4$. Let us show how the structural complexity $C(s, s') = 4$ and its derivation can be made more explicit and geometrical. This helps appreciate that structural complexity is not an abstract notion but in fact it has a clear relationship with geometry, symmetry and integers.

We deal with the structural complexity $C(s, s') = 4$ and use a geometrical interpretation of structural numbers [8]. This geometrical interpretation can be given as follows. Finite sequences admit a natural representation in terms of piecewise constant functions. Namely, let $\Delta_\hbar[t_0, t_n], n = 1, 2, \ldots$ denote the partition of an interval $[t_0, t_n] \subset \Re^1$ with $n + 1$ equally spaced points $t_k, k = 0, \ldots, n$ such that $t_0 = 0, t_{k+1} - t_k = \hbar, k = 0, \ldots, n-1, n = (t_n - t_0)/\hbar$. Let $W(\Delta_\hbar[t_0, t_n])$ be the set of all piecewise constant functions $f : [t_0, t_n] \to \Re^1$ such that each function f is constant on $[t_0, t_1], (t_k, t_{k+1}], k = 1, \ldots, n-1$. A code of a function $f \in W(\Delta_\hbar[t_0, t_n])$, denoted $c(f)$, is a sequence $s = s_1, \ldots, s_n$ such that s_k is the value of the function f on $(t_{k-1}, t_k], k = 1, \ldots, n$. Let a mapping $\rho_\hbar$ take each sequence $s = s_1, \ldots, s_n, s_i \in \Re^1, i = 1, \ldots, n$ to a function $f \in W(\Delta_\hbar[t_0, t_n])$, denoted $f = \rho_\hbar(s)$, such that $c(f) = s = s_1, \ldots, s_n$ and $f^{[i]}(t_0) = 0, i = 1, 2, \ldots$.

An integer code series for the value of the kth $k = 1, 2, \ldots$ integral $f^{[k]}$ of a function $f \in W(\Delta_\hbar[t_0, t_n]), f = \rho_\hbar(s), c(f) = s = s_1, \ldots, s_n, s_i \in \Re^1, i = 1, \ldots, n$ at the point t_n gives

$$f^{[k]}(t_n) = \sum_{i=0}^{k-1} \alpha_{ki}(n^i s_1 + \cdots + 1^i s_n)\hbar^k.$$

Therefore, for $\hbar = 1$ and $s = s_1, \ldots, s_n \in B_n$ we have

$$f^{[k]}(t_n) = \vartheta_k(s), \quad k = 1, 2, \ldots ,$$

where $f = \rho_1(s)$ and $\vartheta(s) = (\vartheta_1(s), \ldots, \vartheta_k(s), \ldots)$.

What this means is that the kth coordinate of the point $\vartheta(s)$ in structural space is the value of the kth integral of the function $f = \rho_1(s)$ at the point t_n. Actually, the definition of coordinates in structural space was chosen to ensure the validity of this condition.

Now, we may start to visualize the structural complexity $C(s, s') = 4$ and graph the first integrals $f^{[1]}, f'^{[1]}$ of functions $f = \rho_1(s), f' = \rho_1(s')$ as shown in Figure 1. When regarded separately the graphs exhibit no visible

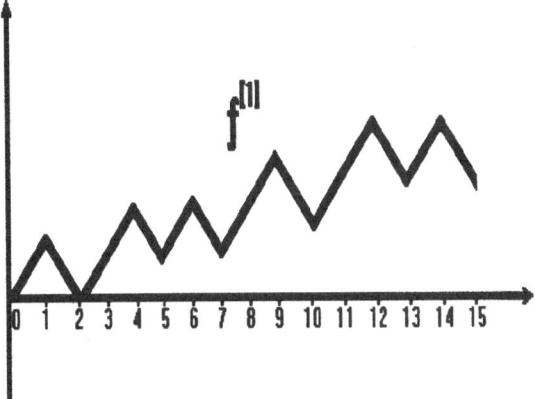

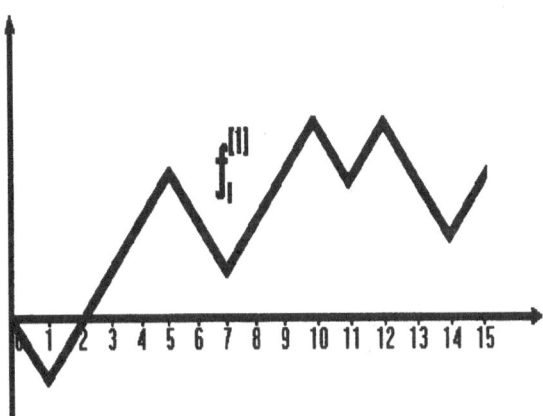

Fig. 4.1. Graphs of $f^{[1]}$ and $f'^{[1]}$ exhibit no visible properties when regarded separately

46

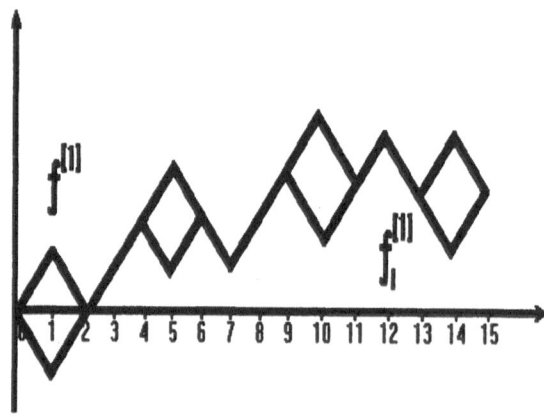

Fig. 4.2. Graphs of $f^{[1]}$ and $f'^{[1]}$ depicted together uncover visible properties

properties, but depicted together, properties emerge as in Figure 2. These properties are manifested explicitly in terms of the symmetry of the graph of a function $g^{[1]}$, where $g = \rho_1(s - s') = f - f'$ (see Figure 3). Graphs of functions $g, g^{[i]}, i = 1, 2, 3$ have visual symmetry and their patterns are brought out in Figures 3 and 4 by shading. From Figures 3 and 4 without computations and using a symmetrical and geometrical viewpoint, it follows easily $g^{[k]}(t_{15}) = 0$, $k = 1, 2, 3$, $g^{[k]}(t_{15}) > 0$, $k = 4, 5, \ldots$. Consequently $f^{[k]}(t_{15}) = f'^{[k]}(t_{15})$, $k = 1, 2, 3$, $f^{[k]}(t_{15}) > f'^{[k]}(t_{15})$, $k = 4, 5, \ldots$ or by definition $\vartheta_k(s) = \vartheta_k(s')$, $k = 1, 2, 3$, $\vartheta_k(s) > \vartheta_k(s'), k = 4, 5, \ldots$. But this is exactly the complexity $C(s, s') = 4$ and the reasoning gives a clear access to it.

Thus, the structural complexity $C(s, s') = 4$ describes a curious phenomenon, connected with pattern formation, that is amenable to explanation in terms of visible things. Namely, Figures 3 and 4 may be seen as though they were experimental results successively representing levels of the patten formation. It is interesting to observe that as we integrate, it appears as if the patterns merge together in pairs to form new patterns belonging to the next level. Since $g^{[k]}(t) > 0, t \in (t_0, t_{15}], k = 4, 5, \ldots$ then there are no patterns of the kth integral of the function g.

The levels combined together in a diagram in Figure 5 exhibit an interesting order involved in the formation of the patterns. It is worthy of note that the diagram exhibits the order without any need for further explanation. The eye at once sees the crucial image that yields the order. This order is of global, long-range character. It concerns the patterns at all levels simultane-

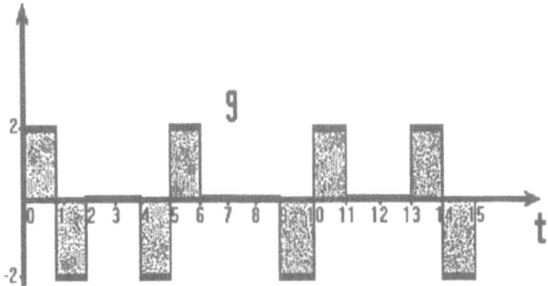

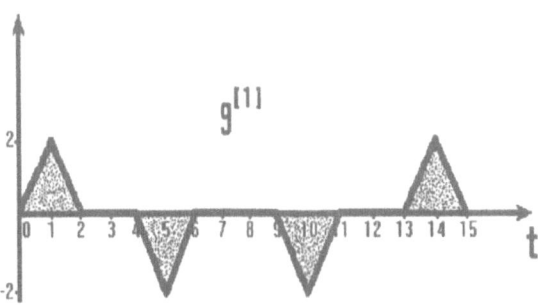

Fig. 4.3. Graphs of g and $g^{[1]}$. As we integrate, it appears as if the patterns merge together in pairs to form new patterns belonging to the next level

48

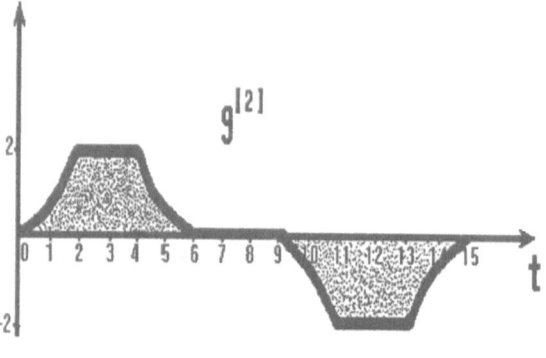

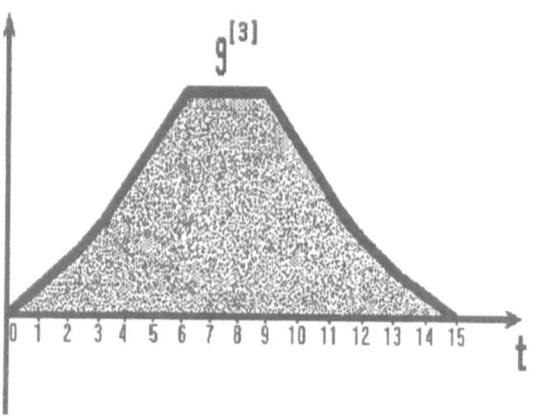

Fig. 4.4. Graphs of $g^{[2]}$ and $g^{[3]}$. The patterns may be seen as abstract entities corresponding to physical structures at different scale levels

ously and also extends across all levels in an enfolded manner. In particular, the order describes how patterns of a level have to make the exactly right connections between themselves in terms of their positions to grow from the level to the next one. One can envisage the patterns of each coming level as next in order of complexity to patterns of the previous level, since every pattern of the successive level is formed from patterns of the previous one. Therefore, the structural complexity $C(s, s') = 4$ gives an account of the phenomenon and admits an interpretation in terms of the maximum level or complexity of patterns that the sequences s, s' can produce. Figure 5 shows how the sequence $s - s' = +2 - 2\,0\,0 - 2 + 2\,0\,0\,0 - 2 + 2\,0\,0 + 2 - 2$ encodes the information about the phenomenon. This fact marked the motivation to consider $C(s, s')$ as a measure of complexity for structures and defined its name.

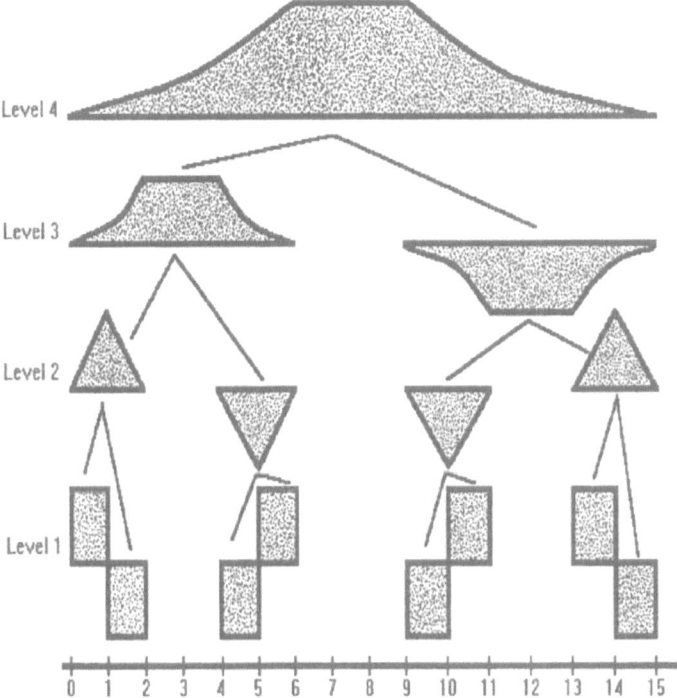

Fig. 4.5. This diagram exhibits an interesting order involved in the formation of patterns. It explicitly reveals symmetry in structural complexity and the connection between Cartesian space and structural space. Cartesian space is represented by a finite sequence (bottom row). The sequence through the integers $1, 2, \ldots, 15$ is connected with the process of pattern formation which is represented by a tree. The process is encoded by the structural complexity $C(s, s') = 4$

It is of prime importance that it is possible to translate these geometrical considerations into a very distinctive mathematical form. This is a place where relations between integers come into view. To be specific, relations between integers appear as we express the following conditions:

$$f^{[1]}(t_{15}) = f'^{[1]}(t_{15}), \quad f'^{[2]}(t_{15}) = f'^{[2]}(t_{15})$$

$$f^{[3]}(t_{15}) = f'^{[3]}(t_{15}), \quad f^{[4]}(t_{15}) > f'^{[4]}(t_{15})$$

consecutively in terms of (2) . Namely, since

$$f^{[1]}(t_{15}) = \alpha_{10}(15^0 s_1 + \cdots + 1^0 s_{15}), \quad f'^{[1]}(t_{15}) = \alpha_{10}(15^0 s'_1 + \cdots + 1^0 s'_{15})$$

then

$$15^0(s_1 - s'_1) + \cdots + 1^0(s_{15} - s'_{15}) = 0. \tag{4.1}$$

Next, we have

$$f^{[2]}(t_{15}) = \alpha_{21}(15^1 s_1 + \cdots + 1^1 s_{15}) + \alpha_{20}(15^0 s_1 + \cdots + 1^0 s_{15}),$$

$$f'^{[2]}(t_{15}) = \alpha_{21}(15^1 s'_1 + \cdots + 1^1 s'_{15}) + \alpha_{20}(15^0 s'_1 + \cdots + 1^0 s'_{15}).$$

By using (3) and canceling by α_{21}, we find

$$15^1(s_1 - s'_1) + \cdots + 1^1(s_{15} - s'_{15}) = 0. \tag{4.2}$$

Similarly, we arrive at

$$15^2(s_1 - s'_1) + \cdots + 1^2(s_{15} - s'_{15}) = 0. \tag{4.3}$$

and

$$15^3(s_1 - s'_1) + \cdots + 1^3(s_{15} - s'_{15}) > 0. \tag{4.4}$$

Combining (3), (4), (5) and substituting $s_i, s'_i, i = 1, \ldots, 15$ in the explicit form, we obtain a system, which rearranged for some elegance, gives a system of Prouhet's type identities [11]

$$1^0 + 6^0 + 11^0 + 14^0 = 2^0 + 5^0 + 10^0 + 15^0$$

$$1^1 + 6^1 + 11^1 + 14^1 = 2^1 + 5^1 + 10^1 + 15^1$$

$$1^2 + 6^2 + 11^2 + 14^2 = 2^2 + 5^2 + 10^2 + 15^2. \tag{4.5}$$

The system reveals relations between integers $1, 2, 5, 6, 10, 11, 14, 15$. Similarly, from (6) we get an inequality

$$1^3 + 6^3 + 11^3 + 14^3 < 2^3 + 5^3 + 10^3 + 15^3. \tag{4.6}$$

Thus the structural complexity $C(s, s') = 4$ is equivalent to the system of relations (7) and inequality (8) between integers. The identities (7) seem to formally capture "all relations" existing between sequences s, s' and inequality (8) suggests that these relations are exhausted. In a fundamental sense

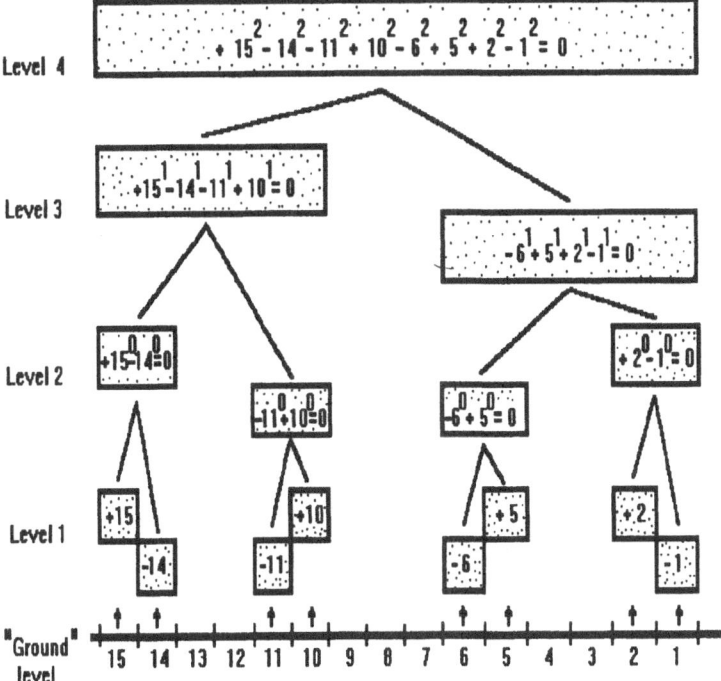

Fig. 4.6. Here we put integer relations into a tree. It corresponds to the tree of patterns in Figure 5 and is an integrated part of a web of integer relations

the equivalence indicates that structural complexity is not to be explained by a deeper underlying concept or principle, but by integers, which are self-consistent entities. At this stage the explanation of structural complexity must end unless we change the integers.

Next, it is important to observe that there is a correspondence between the patterns and the relations and that the formation of patterns corresponds to the formation of integer relations. In Figures 5 and 6 we witness a connection between two different modes of mathematical understanding, that is, geometry and integers, which comes to light in an interesting manner. In particular, it tells us that geometrical patterns are just relations between integers and the integers themselves may be considered as the ultimate building blocks from which processes of formation start. Consequently, the structural complexity $C(s, s') = 4$ has two faces, one is geometrical and the other is the algebraic. With this example in mind we continue with formal descriptions.

5. A Web of Relations Between Integers as a Model of the Whole

For sequences $s = s_1, \ldots, s_n, s' = s'_1, \ldots, s'_n \in B_n, s \neq s'$ the structural complexity $C(s, s') = k + 1, \ k = 1, \ldots, C(s, B_n) - 1$, due to (2), can be reduced, much as the structural complexity $C(s, s') = 4$ in the example, to a set $WR(s, s', B_n)$ of relations between integers

$$n^0(s_1 - s'_1) + \cdots + 1^0(s_n - s'_n) = 0$$

$$n^{k-1}(s_1 - s'_1) + \ldots + 1^{k-1}(s_n - s'_n) = 0, \tag{5.1}$$

provided the following inequality holds

$$n^k(s_1 - s'_1) + \ldots + 1^k(s_n - s'_n) \neq 0.$$

The system (9) considered by itself exhibits no evidence about the structure of the set $WR(s, s', B_n)$ and that it is connected with the formation of two-dimensional patterns. However, the example can help to give a visual access to the formal definition of a web of integer relations, so that we do not traverse it blindly. In particular, the web of relations between integers $1, 2, \ldots$, denoted WR, is defined as follows

$$WR(B_n) = \bigcup_{s \in B_n} \bigcup_{s' \in B_n} WR(s, s', B_n), \quad WR = \lim_{n \to \infty} WR(B_n).$$

Elements of this set are relations similar to (9) and the word "web" is used to informally capture the structure of the set. The intuitive motivation for the name is provided by the example, which shows that a subset of the web

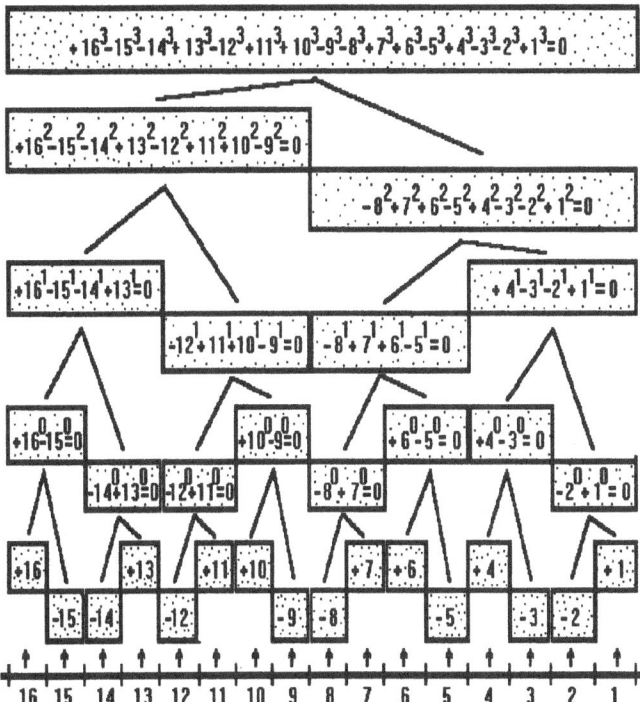

Fig. 5.1. A part of the web connected with the initial segment of length 16 of the Prouhet-Thue-Morse sequence. It is important that all relations and integers with their signs can be replaced by the corresponding patterns

has a tree structure (see also Figure 7 which gives a visual picture of a part of the web connected with the Prouhet-Thue-Morse sequence).

Consider a decision making procedure A and sequences $s, s' \in B_n, s \neq s'$ such that $s = A(s')$ with the structural complexity $C(s, s') = k$ and the related set of relations $WR(s, s', B_n)$. We can say that this set is an integrated part of the whole WR, since its elements are formed in the overall process of integer relation formation. Therefore, we propose the web as a mathematical framework to characterize the integration of human decision makers with their actions.

The web comes from the integers $1, 2, \ldots$ and is formed as one unbroken whole in the overall process of integer relation formation. It can be characterized as a logically isolated theory [15]. This means that the web is so rigid that there is no way to modify it by a small amount without the theory leading to logical absurdites. Namely, we cannot make even a small change in the integers and can only combine true integer relations of a single level to produce true integer relations of the next one. It is of prime importance that the web can be seen identically as a web of two-dimensional geometrical patterns. A pattern in the web is determined by the process of its formation and is a building block for the formation of other patterns where it is involved.

6. Structural Complexity as a Global Measure of Performance

With the mathematical structure at hand we should now be ready to study properties of the decision making procedures. It arises naturally to be interested in properties connected with optimality. As stated in section 2, the local measure $N(s, s')$ was not sufficient to distinguish between procedures. We now proceed with the idea to characterize the performance of a decision making procedure A by the structural complexity $C(s, s')$, where $s = A(s')$ and $s, s' \in B_n$. This characterization is consistent with our objective to describe the decision making process as an integrated part of a whole. The structural complexity $C(s, s')$ encodes the formation of the set $WR(s, s', B_n), s \neq s'$ which is an integrated part of the web of relations between integers. Figuratively, $C(s, s')$ is proposed to measure how the thinking s of human decision makers and the process s' they participate in "fit into the wholeness of all processes in the universe".

It then arises naturally to try to characterize a decision making procedure A as a whole by the average structural complexity or without loss of generality, the total structural complexity, it produces

$$\sum_{s' \in B_n, s = A(s')} C(s, s').$$

An important question is the principle governing the dynamics of the decision making process. The principle can be seen as an analogy to the principle of least action in physics. The emphasis on structural complexity motivates us to define an optimal decision making procedure A_c^*, which maximizes total structural complexity, i.e., A_c^* is a procedure which satisfies the following condition

$$\sum_{s' \in B_n, s = A_c^*(s')} C(s, s') = \max_{A \in \mathbf{A}} \sum_{s' \in B_n, s = A(s')} C(s, s'). \qquad (6.1)$$

It makes sense to compare the local measure $N(s, s')$ and the global one $C(s, s')$ with respect to how they reveal themselves in optimality definitions. The local measure is intuitively convincing and the practical consideration to have a procedure that locally gives a correct prediction as many times as possible can be viewed as a justification to introduce optimality that is based on it. The situation with the global measure is different. Despite the fact that we have provided a mental picture for the global measure nevertheless structural complexity is a measure in an abstract space and the idea to maximize it in the decision making process has no immediate support from practical considerations. To put it simply, it may be not easy to persuade a person not to go to the bar because it increases a measure in an abstract mathematical space. Therefore, we need a different means to justify the use of the global measure to define optimality.

What underlies the definition of optimality in terms of structural complexity is a hope to observe the optimal procedure in some form in nature. In particular, we should be quite satisfied if it could have a connection with self-organization in nature. This fact could indicate that in some sense nature has structural complexity as an optimization criterion and provide a justification to define optimality with respect to the global measure.

In view of

$$\max_{s \in B_n} C(s, s') = C(s', B_n),$$

it is evident that the optimal procedure A_c^* to have the extremum property (10) for each $s' \in B_n$ must strive to produce a sequence $s = A_c^*(s')$ such that $C(s, s') = C(s', B_n)$. Moreover, the greater the structural complexity of a sequence, the greater may be its contribution to the sum. It implies that the optimal procedure A_c^* must be inclined to produce a sequence $s = A_c^*(s')$, which satisfies the condition $C(s, s') = C(s', B_n)$, depending on the value of $C(s', B_n)$. Intuitively, it seems that this criterion may be opposite to the same tendency of the optimal procedure with respect to the local measure. Therefore, if this reasoning is true, then the greater $C(s', B_n)$ for a sequence $s' \in B_n$, the less $N(s, s')$ will be for the optimal procedure A_c^*, where $s = A_c^*(s')$. In general, there may be complementarity between the local and global measures, i.e., the closer one focuses on one measure, the more ambiguous the other becomes.

To illustrate the two measures $N(s, s,')$ and $C(s, s')$, let us consider the following example. Let $\tau(8) = +1 - 1 - 1 + 1 - 1 + 1 + 1 - 1$ and $\bar{\tau}(8) = -1+1+1-1+1-1-1+1$, where τ is the PTM sequence starting with $+1$ and $\bar{\tau}$ is the negation of τ. For those two sequences, the local measure indicates they are quite far apart in Cartesian space because they never agree,

$$N(\tau(8), \bar{\tau}(8)) = \min_{s,s' \in B_8} N(s, s') = 0, \quad \text{where} \quad \max_{s,s' \in B_8, s \neq s'} N(s, s') = 7,$$

while the global measure indicates the sequences are close together, only one step from each other in structural space,

$$C(\tau(8), \bar{\tau}(8)) = \max_{s \in B_8} C(s, B_8) = 4, \quad \text{where} \quad \min_{s,s' \in B_8} C(s, s') = 1.$$

The following approximation may be considered as the first step in the search for the optimal decision making procedure A_c^*. Given sequences $s, s' \in B_n$ such that $s = A(s')$, the main idea of the approximation is to focus on their symbols associated with incorrect steps of a decision making procedure A and consider them not as separate symbols but as whole sequences. Formally these sequences, denoted by $s \ominus s', s' \ominus s$ and called incorrect, are defined as follows.

If there exists an index $k_1, 1 < k_1 \leq n$ such that $s_j = s'_j, 1 \leq j \leq k_1 - 1, s_{k_1} = -s'_{k_1}$ and $s_1 = -s'_1$ for $k_1 = 1$, then $(s \ominus s')_1 = s_{k_1}$. Else, let $s \ominus s'$ be the empty sequence. Assume the first $i, 1 \leq i < n$ symbols of $s \ominus s'$ are defined, i.e., $(s \ominus s')(i) = (s \ominus s')_1, \ldots, (s \ominus s')_i$. If there exists an index $k_{i+1}, k_i + 1 < k_{i+1} \leq n$ such that $s_j = s'_j, k_i + 1 \leq j \leq k_{i+1} - 1, s_{k_{i+1}} = -s'_{k_{i+1}}$ and $s_{k_{i+1}} = -s'_{k_{i+1}}$ for $k_{i+1} = k_i + 1$, then $(s \ominus s')_{i+1} = s_{k_{i+1}}$. Else, let $s \ominus s' = (s \ominus s')(i)$. Set $s' \ominus s$ as the negation of $s \ominus s'$, i.e., $s \ominus s' = -(s' \ominus s), (s \ominus s')_i = -(s' \ominus s)_i, i = 1, \ldots, |(s \ominus s')|$. Clearly, the sequence $s \ominus s'$ is defined consecutively from incorrect steps with the ith symbol of $s \ominus s'$ corresponding to the ith $(i = 1, \ldots, |s \ominus s'|)$ incorrect step. For example, let $s = +1 + 1 - 1 + 1 + 1 - 1, s' = +1 - 1 + 1 + 1 - 1 + 1$ then $s \ominus s' = +1 - 1 + 1 - 1, s' \ominus s = -1 + 1 - 1 + 1$.

Next, a question arises. Given sequences $s \ominus s', s' \ominus s$ such that $s = A(s'), s, s' \in B_n$ how can one characterize a procedure A? Note, that despite the fact not a single symbol in $s \ominus s'$ is determined correctly by the procedure A, at the same time $C(s' \ominus s, s \ominus s') - 1$ structural numbers of $s' \ominus s$ are determined by it and the sequence $s' \ominus s$ itself can be uniquely specified by structural numbers $\vartheta_1(s' \ominus s), \ldots, \vartheta_k(s' \ominus s)$, where $k = C(s' \ominus s, B_l), l = |s' \ominus s|$. Consequently, this fact motivates us to try to characterize the procedure A by the number of successive structural values of $s' \ominus s$ determined by it. The procedure A makes this determination by producing the sequence $s \ominus s'$ whose structural numbers may coincide with the structural numbers of the sequence $s' \ominus s$. Clearly, the mathematical form of this characterization is exactly the structural complexity $C(s \ominus s', s' \ominus s)$. These remarks suggest to use the following quantity to characterize the decision making procedure A

$$\sum_{s' \in B_n, s=A(s')} C(s \ominus s', s' \ominus s),$$

where the sum through s' ranges over all B_n sequences. This quantity measures how many structural numbers are determined by the procedure A in all incorrect sequences. As an approximation to the optimal procedure A_c^* a procedure, denoted A_c, which maximizes

$$\sum_{s' \in B_n, s=A_c(s')} C(s \ominus s', s' \ominus s) = \max_{A \in \mathbf{A}} \sum_{s' \in B_n, s=A(s')} C(s \ominus s', s' \ominus s),$$

is sought.

7. Constructing an Optimal Procedure

A good theory tells us what is observable in nature. An ideal situation with the optimal procedure could be if it would be a mathematical description of a behaviour that is observable in nature. If this were the case, it would indicate that in some sense nature has the global measure, i.e., structural complexity, as an optimization criterion and provide a justification to define optimality with respect to it. Under certain assumptions it is possible to present a mechanism which constructs rules for the decision making procedure A_c[10]. It represents our initial idea, to understand the decision making process as an integrated part of a whole, in algorithmic terms. This allows us to correlate the idea with known algorithmic structures and provides supporting evidence. In particular, it is imporatnt that the mechanism turns out to be connected with models of self-organization in nature. This result strengthens the idea that structural complexity seems to play an important role as an optimization criterion in nature. Since structural complexity is nothing but a concept which describes the formation of integer relations in the web, it also reinforces our idea to use the web as a new mathematical framework for the problem.

The mechanism admits two different descriptions. The first description, called structural, is deterministic in character. It is expressed in terms of structural complexity and allows us to discover rules that are at work in the decision making procedure A_c explicitly. The second description, called natural, is based on using causal powers of natural phenomena. We start with the structural description of the mechanism.

7.1 Structural Description of the Mechanism

Within the context of this description the mechanism at each kth step $k = 1, \ldots, n$ constructs rules $\chi_{c,k}, k = 1, \ldots, n$ of the decision making procedure A_c based on the structural complexity $l = C(s(k) \ominus s'(k), s'(k) \ominus s(k))$, where $s(k) = A_c(s'(k))$ and the sign of

$$\vartheta_l(s(k) \ominus s'(k)) - \vartheta_l(s'(k) \ominus s(k)).$$

In this sense it may be said that the mechanism learns how to construct rules from mistakes. At the first step incorrect sequences are empty and the mechanism sets the rule as $s_1 = \chi_{c,1}(\emptyset) = -1$. At the $(k + 1)$th step $k = 1, \ldots, n - 1$ two cases are possible:

1. $s(k) \ominus s'(i) = \emptyset$, i.e., the incorrect sequence is empty,
2. $s(k) \ominus s'(k) \neq \emptyset$, i.e., the incorrect sequence is not empty.

In the first case the condition $s(k) \ominus s'(k) = \emptyset$ means that no incorrect steps have been made up to the kth step. Under these circumstances the mechanism continues to set $s_{k+1} = \chi_{c,k+1}(\emptyset) = -1$.

The second case is a place where structural complexity comes in and governs the rule's construction. In this case we have $s(k) \ominus s'(k) \neq s'(k) \ominus s(k)$. This means that there exists an integer

$$l = C(s(k) \ominus s'(k), s'(k) \ominus s(k)), \ 1 \leq l \leq C(s(k) \ominus s'(k), B_{l'}), \ l' = |s(k) \ominus s'(k)|$$

such that

$$\vartheta_0(s(k) \ominus s'(k)) = \vartheta_0(s'(k) \ominus s(k)), \ldots, \vartheta_{l-1}(s(k) \ominus s'(k)) = \vartheta_{l-1}(s'(k) \ominus s(k)),$$

$$\vartheta_l(s(k) \ominus s'(k)) \neq \vartheta_l(s'(k) \ominus s(k)).$$

Then two cases are possible:

1. $\vartheta_l(s(k) \ominus s'(k)) - \vartheta_l(s'(k) \ominus s(k)) > 0$, and
2. $\vartheta_l(s(k) \ominus s'(k)) - \vartheta_l(s'(k) \ominus s(k)) < 0$.

In this case the mechanism at the $(k + 1)$th step constructs the rule as follows. First, the structural complexity $C(s(k) \ominus s'(k), s'(k) \ominus s(k))$ is defined and second, s_{k+1} is set to minimize the difference between the lth structural values of the incorrect sequences $s(k + 1) \ominus s'(k + 1), s'(k + 1) \ominus s(k + 1)$. Namely, it sets $s_{k+1} = -1$, if

$$\vartheta_l(s(k) \ominus s'(k)) - \vartheta_l(s'(k) \ominus s(k)) > 0,$$

and sets $s_{k+1} = +1$, if

$$\vartheta_l(s(k) \ominus s'(k)) - \vartheta_l(s'(k) \ominus s(k)) < 0.$$

Thus, the construction of the $(k+1)$th rule $\chi_{c,k+1}$ is reduced to the finding of the structural complexity $l = C(s(k) \ominus s'(k), s'(k) \ominus s(k))$, which may vary from one construction to another, and the sign of $\vartheta_l(s(k) \ominus s'(k)) - \vartheta_l(s'(k) \ominus s(k))$. Clearly, if at the $(k+1)$th step $k = 1, \ldots, n-2$ it turns out that $s_{k+1} = s'_{k+1}$ then

$$s(k+1) \ominus s'(k+1) = s(k) \ominus s'(k), \quad s'(k+1) \ominus s(k+1) = s'(k) \ominus s(k).$$

Hence, the mechanism constructs the rule $\chi_{c,k+2}$ in the same way as the rule $\chi_{c,k+1}$ and in particular, $\chi_{c,k+2}$ has the same value as $\chi_{c,k+1}$. Thus, if at the kth $k = 1, \ldots, n-1$ step we have $s_k = \chi_{c,k}$ then $s_{k+i} = \chi_{c,k+i} = s_k = \chi_{c,k}, i = 1, \ldots, n-k$ until an incorrect step occurs. In other words the procedure A_c, having started with some prediction, continues this kind of prediction until it is unsuccessful.

The mechanism constructs rules in a specific manner, which has a remarkable property distinguishable by the construction of incorrect sequences. We start describing it by the following observation. In fact, at each kth step $k = 1, \ldots, n$ a rule χ_k cannot determine the next s'_k value. Nevertheless, a rule χ_k can determine correctly the next symbol of the incorrect sequence, if it appears at the kth step. Indeed, if the rule χ_k at the kth step sets $s_k = -1$, but $s'_k = +1$, then the next symbol of the incorrect sequence is $+1$ (the incorrect sequence does not change if $s_k = -1$) and if the rule χ_k sets $s_k = +1$, but $s'_k = -1$, then the next symbol of the incorrect sequence is -1 (the incorrect sequence does not change if $s_k = +1$). Therefore, each decision making procedure through the rules has its own logic of the incorrect sequence's construction. The important property of the mechanism is that it constructs rules for the procedure A_c in such a way that the kth symbol of its incorrect sequence $s \ominus s'$, as long as it is not the empty sequence, must be the kth symbol $k = 1, \ldots, |s \ominus s'|$ of the PTM sequence $\bar{\tau}$, where $s = A_c(s'), s' \in B_n$ [8].

The property provides another mechanism description. In fact, this description allows the mechanism to be possibly realized on a natural phenomena basis. In particular, suppose there exists a generator $G(\bar{\tau})$ of the PTM sequence $\bar{\tau}$, which can consequently produce one $\bar{\tau}$ symbol per iteration. The generator can be built on the basis of a dynamical system that has period doubling [16]. It is known that the system can produce the PTM sequence $\bar{\tau}$ and represents models of self-organization in nature [17]. In such a manner structural complexity becomes connected with self-organization. A decision making procedure which is optimal in terms of structural complexity seems to be a mathematical description of a process observable in nature. Since structural complexity is nothing but a concept to describe the formation of integer relations in the web, figuratively, we favour nothing more than the well-known view that "integers rule the world".

With the generator $G(\bar{\tau})$ being available the natural description of the mechanism can be represented in the following steps. This description equates the computational powers of computing devices with the causal powers of natural phenomena.

7.2 Natural Description of the Mechanism

Step 0. Set the number of steps $k = 1$ and the number of incorrect steps $m = 1$.

Step 1. Generate the mth value $\bar{\tau}_m$ of the PTM sequence $\bar{\tau}$ from the generator $G(\bar{\tau})$ and for the value of the rule $\chi_{c,k}$ set $\chi_{c,k} = \bar{\tau}_m$. Set $s_k = \chi_{c,k}$.

Step 2. If $s_k = s'_k$, then goto Step 3. Otherwise goto Step 4.

Step 3. If $k = n$, then goto Step 5. Otherwise set $\chi_{c,k+1} = \chi_{c,k}$ and $s_{k+1} = \chi_{c,k+1}$. Increment the number of steps $k = k + 1$ and goto Step 2.

Step 4. If $k = n$, then goto Step 5. Otherwise increment the number of steps $k = k + 1$ and the number of incorrect steps $m = m + 1$ and goto Step 1.

Step 5. Stop.

The procedure A_c can be formulated as a simple principle of behaviour. This principle of "win - stay, lose - consult generator $G(\bar{\tau})$" considers a generator $G(\bar{\tau})$ as a "source of universal wisdom", that can provide the answer to every question. The role played by the PTM sequence $\bar{\tau}$ in the principle is similiar to the meaning of the Chaitin's Omega Ω [18]. It is worthwhile to compare the description of the principle with known strategies Tit-for-Tat and Pavlov. These strategies are of a special interest in social science and arise in the study of the problem how humans can cooperate despite temptations to defect [19]–[21]. Tit-for-Tat starts with one decision and then always proceeds with a decision that repeats the previous result. The Pavlov rule says to stick to the former decision if it won but to switch to another if it failed.

8. Conclusion

A model of human decision making, formulated in terms of fuzzy logic, has been considered. We have suggested the web of integer relations and structural complexity as a mathematical framework to address problems when people's thinking is an integral part of the situation in which they participate. This mathematical framework pertains equally to complex, adaptive systems. It is significant that suggesting the web we have the following considerations in mind. If one proposes a mathematical framework one should expect to be asked for an explanation why it is sound. The explanation must be given in terms of other notions. In its turn the notions can be also asked for explanations. This generates a chain of explanations which stops when it comes to self-evident notions. In essence, the web of integer relations is proposed to describe human thinking and therefore must be of a deep character. We offer the explanation of the web in terms of integers which are regarded as fundamental features as they are irreducible to anything more basic.

We have used the web to define optimal decision making and its observability in some form in nature is claimed as verification of the definition. We have presented the mechanism which approximates rules' construction of the optimal decision making procedure. The mechanism is connected with models of self-organization in nature. The mechanism admits two different descriptions, structural and natural. The natural description is formulated as a simple principle. This principle is a new strategy of behaviour and it will hopefully be found useful in the current study of social systems [22].

Acknowledgements

The author would like to thank Zelda Zabinsky for helpful assistance in the final polishing of the paper.

References

1. G. Soros, *The Alchemy of Finance*, John Wiley & Sons, Inc., New York (1994).
2. J. Casti, *What if ...* , New Scientist, No. 2038, 36-40, 13 July (1996).
3. D. Bohm, *Wholeness and the Implicate Order*, Ark Paperbacks (1984).
4. R. Penrose, *Nonlocality in and Objectivity in Quantum State Reduction*, in Fundamental Aspects of Quantum Theory, edited by J. Anandan and J.L. Safko, World Scientific, Singapore (1994).
5. J.A. Wheeler, *Foundations of Quantum Mechanics*, in Complexity, Entropy and the Physics, edited by W.H. Zurek, Vol. VIII, Santa Fe Institute Studies in the Sciences of Complexity, Addison-Wesley, Redwood City, Califonia, 3-28 (1991).
6. V.V. Korotkih, *Symmetry in Structural Complexity and a Model of Formation*, in From Local Interactions to Global Phenomena, edited by R. Stocker, H. Jelinek, B. Durnota and T. Bossomaier, IOS Press, Amsterdam, 84-95 (1996).
7. L.A. Zadeh, *Fuzzy Logic = Computing with Words*, IEEE Transactions on Fuzzy Systems, Vol. 4, No. 2, 103-111 (1996).
8. V.V. Korotkih, *Integer Code Series with Some Applications in Dynamical Systems and Complexity*, Communications on Applied Mathematics, Russian Academy of Sciences, The Computing Center, Moscow (1993).

9. L.A. Zadeh, *Outline of a New Approach to the Analysis of Complex Systems and Decision Processes*, IEEE Trans. Syst., Man, Cybern., Vol. 3, 28-44 (1973).

10. V.V. Korotkih, *Multicriteria Analysis in Problem Solving and Structural Complexity*, in Advances in Multicriteria Analysis, edited by P.M. Pardalos, Y. Siskos and C. Zopounidis, Kluwer Academic Publishers, 81-90 (1995).

11. E. Prouhet, *Memoire sur Quelques Relations entre les Puissances des Nombers*, C. R. Acad. Sci. Paris, 33, 225 (1851).

12. A. Thue, *Über unendliche Zeichenreihen*, Norske vid. Selsk. Skr. I. Mat. Nat. Kl. Christiana, 7, 1 (1906), (Reprinted in Selected Mathematical Papers of Axel Thue, T. Nagell, editor, Universitetaforlaget, Oslo, 1977).

13. M. Morse, *Recurrent Geodesics on a Surface of Negative Curvature*, Trans. Amer. Math. Soc., 22, 84 (1921).

14. N.G. de Bruijn, *Algebraic Theory of Penrose's Nonperiodic Tilings of the Plane*, Nederl. Akad. Proc., Series A, 69 (1981).

15. S. Weinberg, *Dreams of a Final Theory*, Vintage (1993).

16. M.J. Feigenbaum, *Quantitative Universality for a Class of Nonlinear Transformations*, J. Stat. Phys., Vol. 19, 25-52 (1978).

17. J.-P. Allouche and M. Cosnard, *Sequences Generated by Automata and Dynamical Systems*, in Dynamical Systems and Cellular Automata, Academic Press. Inc., 17 (1985).

18. G.J. Chaitin, *Information, Randomness, and Incompleteness*, World Scientific (1987).

19. M.A. Nowak and R.M May, *Evolutionary Games and Spatial Chaos*, in Nature, Vol. 359, No. 6398, 826-829, October 29 (1992).

20. M.A. Nowak and K. Sigmund, *A Strategy of Win-Stay, Lose-Shift that Outperforms Tit-for-Tat in the Prisoner's Dilemma Game*, in Nature, Vol. 364, No. 6432, 56-58, July 1 (1993).

21. K. Sigmund, *Games of Life: Explorations in Ecology, Evolution, and Behaviour*, Penguin (1995).

22. V. Dimitrov and K. Kopra, *Fuzzy Logic and The Management of Complexity: The Social Perspective*, in Fuzzy Logic and The Management of Complexity, edited by J. Dimitrov and V. Dimitrov, Vol. 1, University of Technology Sydney, 13-17 (1996).

FUZZY DECISION MAKING IN DESIGN ON THE BASIS OF THE HABITUALITY SITUATION APPLICATION

A.E. Gorodetsky
IPMash - Russian Academy of Sciences
V.O. 61 Bolshoy pr.
199178 St. Petersburg
Russia
Email: gae@msa.ipmu.ru

Abstract. This paper considers a decision acceptance process made by artificial expert systems. In order to improve its efficiency it is proposed to accept a decision based on similarity between a current situation of a choice and one of the previous situations for which an efficient decision is known. The decision acceptance problem is formalised mathematically and the algorithm of its decision based on the so-called habituality situation is introduced.

Keywords: expert systems, habituality situation, decision acceptance

1. Description of the decision acceptance problem in design

Progress, made by now in the areas of fuzzy mathematics and soft computing systems has allowed to expand dramatically a scope of problems, which can be solved on the basis of methods employing a linguistic information and an application of an intelligent software, for example expert systems (ES). The problems of fuzzy design and fuzzy management are the most perspective among such problems, giving the greatest technological effect. Design process is characterised by a large number of different decisions including the structure and parameter choice. Usually the fuzzy description of a problem of the decision acceptance (DA) arises in the following cases [1]:

- Because of limited resources for the situation modelling and simulation, that does not allow to receive an existing precise information and forces system analysts to use an expert knowledge in a fuzzy linguistic form;
- Because of limitations of the numerical information being available, that does not allow to find the decision by formal methods within the existing restrictions on resources, when an expert finds the decision on the base of his/her experience, which can be described as a set of fuzzy rules, i.e. as system of the logic equations with casual logic variables;

- At early complex object design stages, while a number of alternative design versions are available, but it is not known precisely, which properties the object will have, when the resources to study comprehensively all versions are unavailable, and the experience of the designers is expressed qualitatively (in the linguistic form).

In the first and the third cases the DA problem appears to be "sinked" into a fuzzy environment, in a second one the DA problem is fuzzy according to its description. Thus, in all cases the designer is compelled to address the knowledge of the experts and to create an ES shell, using knowledge of the experts in DA process. The information illegibility in such ES is caused by a presence in the descriptions of DA problems of the concepts and relations with non-sharp borders and of the statements with a multiple-valued scale of the validity. The object can either belong to a class, described by the given concepts, relations or statements, or can not belong to it, but an intermediate degree of fitting is possible also. Concepts and the relations, describing such classes are fuzzy and one can formalise them using the theory of fuzzy sets.

2. Mathematical formulation of the DA problem

In general, the problem of the DA can be characterised mathematically as follows: $< M, Z, K, F, S, W, T>$. Here, the initial data are:

1. Set of alternatives, or the environment of a choice:

$$M = Bd \cup Bk \cup H$$

where:

$Bd = D \cup At$ and $At = A \cup B \cup C$;

H is a set of the influencing factors (effecting ES at DA moment), which can be considered as the set of disturbances $\{h\} = H$;

Bk - Set of rules (logic equations) of a knowledge base (KnB), applied by ES at DA moment;

Bd - Set of the given databases (DB), used by ES at DA;

D - Set of concepts (logic variables) of the DB, used by ES at DA;

At - Set of attributes of concepts of the DB, used by ES at DA;

A - Set of quantitative attributes, used by ES at DA;

B - Set of qualitative attributes, used by ES at DA;

C - set of graphic attributes, used by ES at DA.

2. Sets of outcomes Z of alternative possible results of ES operation, or set of the allowed decisions, received by the machine of a logic conclusion (MLC) of the ES from the logic equations system, used at DA.

3. Criterion of an outcome estimation T, which is a rule vector calculated by MLC of the ES from the attributes of concepts, included in allowable decisions.

4. Displays F of the set of attributes in the set of estimations T, stored in DB ES, or set by an operator.

5. The structure of preferences S of the expert, stored in DB ES, or assigned by the operator and applied in the decision efficiency E calculation.

6. The decision algorithm or a rule, or a way of actions W, allowing to produce the required action O over a set of alternatives I, ensuring a required decision (e.g. optimum search, ranging, etc.). Usually this algorithm is determined at the stage of the ES design and then it is supplied by providing its MLC, or it is searched for or it is specified during the ES learning stage.

3. Habituality situation and its role in the decision acceptance

The solution of a preference problem, which is the most usual problem in design, can be received theoretically by the application of fuzzy reasoning, probabilistic reasoning and mathematical logic. However, looking at the problem description given above one can conclude that this decision process is very labour and time consuming, so its frequent application required in design decreases the CAD effectiveness significantly.

The solution can be found in the application of the way which is often applied by a human designer in decision making and which is used in learning different systems. A human designer often makes a decision based on the previous experience or professional intuition, comparing the current situation with ones observed before or similar to them.

Based on this approach it is possible to specify a significant number of cases, when the decision of a fuzzy DA problem can be much accelerated. To achieve it the formalised description of a human behaviour as a purposeful system, the so-called habituality situation, was firstly introduced in [4].

Here the DA analysis is performed by ES by the way similar to one applied by a human decision maker. According to [4], a habituality situation is a situation, in which the possible results are classified into two non-intersecting complete classes Z1 and Z2, and a subject prefers Z1 rather than Z2, assuming it more valuable, while each way of action W has the identical efficiency E by any possible result. There should be entered probabilistic marks of an estimation of value of result Q, efficiency of a way of actions E and habituality degree L, allowing to analyse subject abilities to use a habituality situation.

Unlike [4], we assume a habituality situation as a situation, when ES at DA (at the time moment t) has an environment of a choice M (t_i) same or almost same, that happened earlier (in any moment of time t_{i-k}, k = 1,2,..., n), when ES received a result, which had a value higher than an allowable threshold, or in a more general case, required estimation To, i.e. Mi (t) α M (t$_{i-j}$) / E > K, where α is a two-local predicate on analysed sets, assigned, for example, by the mathematical logic or fuzzy reasoning. Obviously, the use of a habituality situation in ES will allow considerably to speed up the decision of a DA problem by the choice of the desired result by analogy or "intuition".

However, for this purpose ES should be given the mechanism of the habituality situation detection. In design this mechanism is required to solve two complex problems, first of all. Firstly, it is necessary to design criterion To for Z results estimation, and secondly, to determine the measures of the sets affinity, including non-metric and fuzzy, forming an environment for a choice ES at DA.

4. Mathematical formalisation of the ES operation

To begin with, we shall briefly describe what result of ES operation is, how it is represented and how it can be estimated. Actually an action of any ES consists of solving of the logic equations system and searching for the best one among the received allowed decisions. This search is performed by the attribute analysis of logic variables, included in allowed decisions, and by the calculation of the results estimates.

The result of an ES action for the time interval $t = t_2 - t_1$ is the solution of a logic equations system achieved by MLC during this interval of time. Let us denote it as an event W, meaning that with probability P > 0 for the time $t = t_2 - t_1$ one predicant x(t) will reach a result y (t), that is P {x (t) -> y (t)} > 0. It is obvious, that in order the i-th event to take place an ES should make i-th action Ak. The number of events, which can be reached by MLC for a certain interval of time can be applied to characterise the ES speed.

The result z of j-th ES operation for an interval of time $t = t_2 - t_1$ is the solution received by MLC as the decision of system of the logic equations for this interval of time. Therefore, on the reception of the result z (t) for a time interval $t = t_2 - t_1$ j-th ES performs a set of actions {Ak}, which is directed towards the decision of the logic equations system for achieving this result z. Such ES behaviour can be characterised mathematically by a way of actions:

$$W (t2 - t1) <=> \{Ak(I) /t = t2 - t1\} = > z\}$$

One could note, however, that in the same environment of a choice ES can choose different ways of actions. Probability of a choice of a way of actions W in the environment E at the k-th moment of time can be calculated as D (t(k)) = P {W/M, t(k)}. In most practical cases an ambiguity of ways of actions, especially in expert control systems, is undesirable. However, in human-machine systems it is possible to specify a large number of situations, when the presence of a plenty of alternative ways of actions, offered by ES in not completely certain situations can be quite useful, for example at the situation of a choice of the strategy of realisation of any operation with active counteraction. Besides that for the analysed ES correct ways of actions W, achieving best results Z are known, then the probability of a choice of this way of actions D calculated on the results of ES statistical tests will be a very good estimation to characterise the ES quality. Unfortunately, in many cases revealing of correct ways of actions is rather complex and labour-consuming process, especially, if it concerns the long-term forecast.

Let the probability, that during time interval t=t2 - t1 in the choice environment I the ES will choose an action way W, achieving result z to be called the efficiency E of a way of action W on the result z, i.e. $E = P \{W -> z(p) / M (t2-t1)\}$. This parameter allows to estimate the ES quality by doing statistical tests and calculations, but it can be applied only with those cases when it is known which results z are to be produced by the ES without preliminary revealing of correct ways of actions. As a possible way of actions W in the environment I we shall name a way of actions, which has a non-zero probability of a choice by some ES, i.e. $Wb = W / P \{W/M\} > 0$.

The set of all possible ways of ES actions $\{Wb\}$, determined as a result of its statistical tests, defines the ES action area or its subject area, which quantitatively can be estimated as a number of possible ways of action n. The possible way of actions W of the particular ES in the environment I is a way of actions, which has a nonzero probability to be chosen by the given ES.

The set of possible ways of actions $\{Wb\}$ in the environment I, is a subset of all possible ways of actions, i.e. $\{Mn\} \leq \{Mb\}$, and accordingly $n(n) < n(b)$. The capacity of the set of possible ways of actions of the particular ES or the affinity $n(n)$ to $n(b)$ characterises the ES universality or completeness of its knowledge base.

Any result, received by the j-th ES by the way of actions W, can be characterised with its value, i.e. the degree in which it is used in solving the given problem q. The value can be either positive or negative. In some cases value of the result is associated with the set of the type $\{\delta 1 < \delta 2 < ... < \delta n\}$, where $\delta 1, \delta 2, ..., \delta n$ are linear ordered linguistic estimates such as " very small value ", " small value ", " average value ", etc.. However, such estimations can be ranged and numerical measures can be received. Therefore, $Q \in < \{0,1\}, \{-1, 1\}, \{0, 100\}, \{1, n\} >$. In view of this variety the scaled value should be introduced according to $q*(i) = Q_i - q_{min}$, where q_{min} is the minimum value of the ES action, and

$$Q_i = \frac{q_i^*}{\sum_i q_i}$$

Obviously, $Q_i \in (\{0,1\}$ and $\sum_i Q_i = 1$. Frequently instead of specific value such estimation of the result correctness, as credibility $U_i \leq Q_i$ is applied. Usually the assumed probability that the ES will reach the correct k-th result by i-th way is assigned as the result credibility. Sometimes credibility of all k results is defined by the formula: $U = 1/Nz$, where Nz is the number of the answers, received by the ES in the environment I. The not less important parameter of the ES quality is the reliability V of the ES operation results.

Below, as the reliability of the k-th answer, received by the ES by a way of action W in the environment I, we shall assign a probability that the ES in the environment I during an interval of time $t = t2 - t1$ will choose a way of action W producing the result z, which has a nonzero calculated value or credibility ($Q > 0$):

$$Vi = (P \{M (t - t) -> W -> Zi) /((Qi > 0)V(U > 0))$$

In some cases Vi can be associated with a set like $\{-1,1\}$, $\{0,100\}$, $\{1, n\}$ and even $\{\delta_1 < \delta_2 < ... < \delta_n \}$. However, it is necessary to proceed to scaled values, in order to receive the set like $\{0,1\}$.

Any ES in the environment I during the time interval $t2-t1$ can choose at least one i-th way of action W, giving k-th result with the nonzero scaled value. This result refers to as the possible result $z_b \in Z_b$ ($Z_b = \{z_b\}$).

Any k-th possible result, which can be received by the ES for $\Delta t = t2-t1$ in the environment I by the action way W can be referred to as a potential result $z_n \in Z_n$ ($Zn = \{z_n\}$, if the probability of its choice by the ES for an interval of time $t = t2-t1$ is higher than zero:

$$z_n = Z_b/P\{W (t_2 - t_1) --> z_b\} > 0.$$

It is obvious, that $Z_n \subseteq Z_b$

As the final result of the ES action during a time interval $t = t_2 - t_1$ or its nearest desirable result in the choice environment I we shall name result z_u, that belongs to the complete set of mutually exclusive possible results Z_n and has the maximum scaled value during the current time interval: $z_u = z_n/Q_i = \max /\Delta t = t_2 - t_1$.

Thus, the total result is a reachable result, which the ES tends to reach most of all in the environment $I(t_2 - t_1)$ at the interval of time $t = t_2 - t_1$. A task, or an intermediate desirable result z_j of the ES for some environment set $\{M (t_2 - t_1)\}$ at the time interval $t = t_2 - t_1$ will be referred to as a last result $z_j = z_u$ from the set of allowed possible results $\{z_1, z_2,..., z_n\}$ for the ES during the time interval $t = t_2 - t_1$, sorted out according to their scaled values, so that $Q_1 < Q_2 . < ... < Q_n$, while after achieving any result z_k ($k < n$) from this set probability of achieving the next result z_{k+1} $\{z_1, z_2, ..., z_n\}$ during the time interval $\Delta t = t_{k+1} - t_1$ increases, i.e. $P\{z_{k+1}/\Delta t = t_{k+1} - t_1\} > P \{z_k/\Delta t = t_k - t_1\}$. So, the task is a result, which the ES tends to reach most of all in the environment $I(t_2 - t_1)$ at the interval of time $t = t_2 - t_1$.

A target, or a long-term desirable result z_y of the ES for the set of a choice environments $\{M(t_m - t_1)\}$ on an interval of time $\Delta t = t_m - t_1$, is the recent result $z_a = z_n$ from a set of potential results $\{z_1, z_2,..., z_n\}$, ordered so that $Q_1 < Q_2 . < ... < Q_n$

for j-th ES on an interval $\Delta t=t_n-t_1$, while the result z_n is not an accessible result for j-th ES in the given set of environments for an interval $\Delta t=t_n-t_1$

However, there exist the result $z= z_n$, $(1 < n < m)$, which is an aim for the j-th ES in this set of environments on an interval $\Delta t=t_m-t_1$ such, that at its achievement probability grows that j-th ES will receive result $z_y=z_m$ in a later moment of time t_n $(n>m)$, i.e.

$$P\{z_y=z_n/\Delta t=t_n - t_1\} > P \{z_y = z_m / \Delta t=t_n - t_1\}$$

Thus, the aim is a desirable result, unachievable for an examined interval of time, but accessible in the future, and it is possible to move closer to it for the given period.

An ideal or an ultimate desirable result z_a of the j-th ES for the complete set of mutually exclusive environments $\{M (t_n- t_1)\}$ and a rather long period of time $\Delta t=t_m-t_1$ is the most recent result $z_a=z_n$ from a set of results $\{z_1, z_2,...,z_n\}$, ordered so that $Q_1 < Q_2... < Q_n$ for j-th ES during $\Delta t=t_n-t_1$. Z_n is not a possible result for j-th ES in any environment from the set $\{M (t_n- t_1)\}$ during $\Delta t=t_n-t_1$, but any not yet received result from a set of probable Z_b for this j-th ES at the moment t_j $(1 < j < n)$ is for j-th ES either a total result, or a problem, or an aim at the moment t_j.

Thus, the ideal is a result, which can never be approached, but to which it is possible to move very closely. In spite of the fact that the aim is unreachable for a considered time, and the ideal is, in general, unreachable, the estimation of an approach degree and progress speed to them is meaningful and can speak about the ES quality.

As the reception of good results in the i-th situation by j-th ES can be very time consuming, that can be explained, first of all, by large volumes of calculation of such their quality estimations, as specific value Q, credibility U and reliability V, and on their base of the awareness efficiency [5]:

$$O = 1/ [(|U - V| +\delta r +\delta v) (n_d + n_r)],$$

where n_r is the quantity of rules in a KnB, used j-th ES for receiving the k-th result in the i-th situation of a choice; n_d is the quantity of data DB, used j-th ES for receiving the k-th result in the i-th situation of a choice; δr, δv are errors in the definition of U and V. Therefore, an application of a habituality situation in ES at fuzzy DA is very desirable.

5. Formalised choice problem and its solution

The fact that the choice environment is repeated, which shows a closeness of a habituality situation, can be discovered by comparison of the set $I(t_k)$, describing the choice environment of the j-th ES at the given k-th moment of time t_k, with sets $I (t_i)$, where $i =k-1, k-2,..., k-n$, describing j-th ES environment of a choice at the previous i-th moment of time, in which the ES received good results (optimum decisions) and these results (the decisions) are kept in the memory. Every set I (t_k) contains subsets of data $Bd(t_k)$, knowledge $Bk(t_k)$ and handicaps $F(t_k)$.

In the trivial case it is possible to consider that the habituality situation comes when the specified sets at the given time moment coincide with the similar sets at any previous moment, for which the following conditions are valid $Q > Q_{add}$, $U > U_{add}$, $V > V_{add}$ or $O > O_{add}$.

However, the probability of such situations occurrence in practice is negligibly small though, because the set of handicaps are the sets of casual sizes, or fuzzy sets. Nevertheless the not complete coincidence of the specified sets does not yet mean, that the accepted by analogy in this case decision will appear to be not optimal.

The simplest estimate is the affinity of sets that can be represented in terms of metrics, to which sets of quantitative attributes A naturally belong. If the sets are represented by the number sets, then their affinity can be estimated by calculating the metrics like:

$$\rho\left(a_i, a_{i-n}\right) = \sqrt{\sum_i \left(a_i - a_{i-n}\right)^2}$$

or

$$\rho\left(a_i, a_{i-n}\right) = \sup\left|a_i - a_{i-n}\right|$$

or

$$\rho\left(a_i, a_{i-n}\right) = \sum_i \left(a_i - a_{i-n}\right)$$

If the sets are implemented as the sets of functions, their affinity can be estimated by calculating the metrics like:

$$\rho\left(\varphi, \psi\right) = \sup_{Q \in A} \left|\varphi\left(a_i\right) - \psi\left(a_{i-n}\right)\right|$$

or

$$\rho\left(\varphi, \psi\right) = \int_a \left|\varphi\left(a_i\right) - \psi\left(a_{i-n}\right)\right| da$$

The affinity of the non-metric sets, for example the sets of qualitative attributes B can be estimated by verifying either each of them is situated in the binary relation with a similar set from the previous situations of a choice: B (t_i) ε B (t_{i-m}), where m = 1,..., n; i - n = 0 and ε is the two-berth predicate on the analysed sets, which is given, for example, by an indication of the formulas of the mathematical logic language and is stored in a special section of a database.

The most difficult case for estimating is one of the sets describing the knowledge base state, as during the operation ES is subjected to an evolution and changes. It is rather difficult to estimate a degree of discrepancy of the knowledge bases without the decision of the logic equations systems.

If the analysis of a choice environment at the given moment of time t_i discovers that there existed previously such a situation that the metrical or non-metrical difference between them is lower than the threshold specified, then this situation is therefore habitual to a situation in which the ES was at the previous moment of time t_{i-k}.

Thus, j-th ES does not solve the logic equations system and makes a decision, that the optimal result is the result received at t_{i-k}, i.e. $z_i = z_{i-k}$ with the specified value $Q_i = Q_{i-k}$ ($U_i = U_{i-k}$, $V_i = V_{i-k}$, $O_i = O_{i-k}$). The habituality situation can be defined simply, if the affinity of all analysed sets is estimated by one discreet metrics:

$$\rho_{ij} = \begin{cases} 1, M(t_i) = M(t_j) \\ 0, M(t_i) \neq M(t_j) \end{cases}$$

However, this definition of a habituality situation sharply reduces a possible number of approaches to the habituality situations, as a significant number of elements of the database sets DB have a random nature. The approach, assuming the DB and KnB design DB and KnB in such a manner that their describing sets are metrical is much more fruitful. In this case the approach of a habituality situation can be estimated by one variable, named a habituality degree of the ES in the i-th situation of a choice to the (i-k)-th situation of a choice and calculated according to the formula:

$$\Omega_{ik}(t_i, t_{i-k}) = \sqrt{\sum_i \omega_i^2}$$

where $\omega_i = \dfrac{\rho_{ij} - \rho_{min}}{\sum_i \rho_{ij}}$ is a specific metric between i and j metric sets, and ρ_{min} is the minimum metric. Then, if at any previous time moment t_{i-k}, at which $Q_{i-k} > Q_{add}$ ($U_{i-k} > U_{add}$, $V_{i-k} > V_{add}$, $O_{i-k} > O_{add}$), the calculated habituality degree $\Omega_i > \Omega_{add}$, where Ω_{add} is the maximum possible allowed value of a habituality degree, set by a user, then at a moment t_i a situation is the same, as it at the time moment t_{i-k}.

Thus, an application in ES of revealing habituality situation mechanisms is rather perspective and could be very useful, as it allows to raise dramatically a speed of acceptance of the decision in similar situations. The ES decision making process becomes similar to the behaviour of a human decision maker and the mechanism allows to model mathematically a decision process based on intuition.

References

1. Borisov A.N., Alekseev A.V., et.al. Fuzzy Information Processing in Decision-making Systems (in Russian) Moscow, Radio and Svjas', 1989
2. Zadeh L. A. Fuzzy Sets // Information and Control. - 1965, Vol. 8, N 3. - P. 338 - 353.
3. Saaty T. L. Measuring the Fuzziness of Sets // Journal of Cybernetics. - 1974. - Vol. 4, N 4. P. 53 - 61.
4. Ackoff R. L., Emery F. E. On Purposeful Systems / ALDINE, ATHERTON, CHICAGO AND NEW YORK - 1972.
5. Gorodetsky A.N. Physical Metrology: Theoretical and Application Problems (in Russian), St.Petersburg, Nauka, 1996.

FUZZY LOGIC APPLICATIONS IN COMPUTER AIDED DESIGN

Binh Pham

Research Centre for Intelligent Tele-Imaging
School of Information Technology & Mathematical Sciences
University of Ballarat
Ballarat VIC 3353
Australia
E-mail: binh@ballarat.edu.au

Abstract. Much research work has been devoted to developing techniques for improving the capabilities of computer aided design (CAD) systems for limited engineering and architectural purposes, where the domain of objects to be designed is somewhat restrictive and constrained in scope. Other types of design such as sculpting or industrial design, where aesthetic factors play a more important role, have not been adequately catered for. The main reason for this neglect is that aesthetic factors have fuzzy characteristics, are subjective and difficult to specify. This paper identifies the needs for fuzzy logic in the development of CAD systems, and in particular, discusses how fuzzy logic can be used to model aesthetic factors. We categorise aesthetic intents, analyse the requirements for their representations and present a systematic scheme to realise aesthetic intents using fuzzy logic. Finally, we show how this scheme can be applied to enrich the process of producing artistic work such as brush painting.

Keywords:: fuzzy logic, CAD, design, aesthetic intents, brush painting

1. INTRODUCTION

The purpose of drafting is to produce polished and professional artwork which have been designed by an artist or designer. Drafting does not require great drawing skills nor imagination. It requires instead the ability to pay attention to precise details and produce them meticulously once they have been defined. These precise characteristics of drafting match well with what are provided by current CAD systems in terms of their capabilities for geometric modelling and computerised facilities for editing. These systems are becoming increasingly more sophisticated and powerful. They provide tools for constructing and manipulating 2D and 3D objects in a variety of ways as well as for producing high quality and realistic colour images of these objects. Sophisticated modes of display and presentation (e.g. animation, fly-through) facilitate the communication between designers, users and manufacturers. Interactions between users and CAD systems have also been shifted from text-based to graphical user-interfaces with hierarchical menus. Furthermore, data can also be entered or selected interactively to avoid the tedium of constructing input files in advance. These efforts attempt to allow a designer to work in a more intuitive fashion.

However, there is an increasing demand to produce objects that are more artistically pleasing. People are not merely satisfied with the effects of photographic exactness, but demand the style and expressiveness produced by the artistic flair. Elaborate styling exercises have been made an integral part of marketing for some products, noticeably cars (e.g. [27]), household and office products. It is therefore increasingly more important to include aesthetic factors in any automatic design process. This is even more paramount for artistic applications such as sculpting. The sculptures produced by Ferguson [7] are excellent examples of what could be achieved by exploring the inter-relationship between mathematics and art. He managed to depart from the rigid and precise nature of mathematics to produce enigmatic and sensual objects that evoke the sensory experience.

In a previous paper, Pham [19] analysed the limitations of current CAD systems and evaluated the progress in a number of research areas most relevant to the creative design process. The bulk of progress so far has been concentrated on graphical techniques to provide more flexibility in shape representation and manipulation (e.g. [1,2, 10, 12, 25, 6, 21, 3]). However, there is an increasing interest in other paradigms for design such as parametric, feature-based, artificial intelligent and knowledge-based (e.g. [4, 5, 8, 22, 26, 28, 31]). In particular, there is a definite trend to incorporate more expert knowledge to allow CAD systems to assist a designer in a more intelligent and effective fashion. To date, the expert knowledge has been mainly concerned with engineering properties and experiences. Apart from a few attempts {e.g. [29, 11, 24]), the aesthetic intents of a designer have been left largely unexplored. The main reasons for this neglect is that the fuzzy characteristics of aesthetic factors have made them difficult to specify and implement.

Fuzzy logic which was introduced by Zadeh [30] in 1965, has been used extensively in many areas, especially in social sciences and engineering (e.g. [14,15]). Little attempt, if any, has been made to apply fuzzy logic to CAD systems. This paper identifies the areas where fuzzy logic can benefit the development of CAD systems, and discusses how fuzzy logic can be applied and combined with traditional approaches in order to model aesthetic aspects of 2D and 3D creative design activities. We first deal with a general problem of how to achieve a designer's aesthetic intents, then demonstrate how the paradigm can be applied to a concrete example. Section 2 discusses the differences between the approach used by a designer and by a CAD system and identifies the areas that could benefit significantly from the use of fuzzy logic. The limitations of existing object representations for the creative stage of design and the requirements for viable representations are explored. Section 3 presents a scheme for categorising aesthetic intents, while Section 4 discusses how these intents can be represented, specified and implemented using fuzzy logic. It is envisaged that this framework would provide an intuitive and effective automatic scheme to assist designers to achieve the desired effects for their products. Section 5 discusses a 2D artistic activity where the dynamism and expressiveness of brush painting can be modelled using this fuzzy scheme.

2. THE NEEDS FOR FUZZINESS IN CAD SYSTEMS

To understand the need to introduce fuzzy logic into the development of CAD systems, we have to look at the differences between the way designers and CAD systems work. Creativity in design occurs at the conceptualising level when ideas are vague, tentative. To start a new design, the common practice is for an industrial designer to draw thumbnail sketches of the object, its components or its views from different orientations. The sketches are constantly compared with the list of requirements and appropriate annotations for references are made. This list of requirements covers a broad range of factors, from aesthetic (e.g. rhythm, balance) to human factors (e.g. comfort, reach, clearance) and manufacturing feasibility. This information depends largely on his or her experience and often only exists in the designer's mind. Different design alternatives that arise during this process, may be variations of the same theme, or distinct products resulted from mental leaps. The designer's mind moves quickly back and forward between these alternatives, mentally evaluates their aesthetic values or suitability. Detailed information is not a concern at this stage, and is left to be worked out much later.

A sculptor explores alternative ideas by creating a rough prototype using clay or putty. Fine details are added gradually during the refinement process. The nature of design constraints in this case is more of an artistic than of a practical concern. In both industrial design and sculpting, a top-down approach is often used for planning, while detailed refinement is carried out iteratively in a bottom-up fashion.

Current CAD systems, on the other hand, construct an object using a bottom-up approach. The designer must have a clear idea of what each object component looks like before specifying geometric primitives and operations required for constructing the component. The number of ways for further shape manipulation are fixed and restrictive, and the systems do not support object representation at different abstract levels. Hence, it is not practical for designers to use the systems for exploring alternatives which differ significantly from each other. To bridge the gap between the requirements by a designer in the early stage of design and what a CAD system is able offer, it is important to find appropriate object representations that capture the fuzzy characteristics and allow interpretation at different abstract levels.

The most commonly-used object representations in CAD systems · are wireframes, polyhedra, CSG, boundary representation and volumetric. These representations are exact, precise and consistent. They offer no scope for fuzziness. The designer needs to know the precise details of each object before it can be constructed.

At the early stage, the most important feature that a designer wishes to capture is the global form of an object, hence a desirable object representation must be able to capture the qualitative characteristics which distinguish one form from another (e.g. linear, curved, round, elongated, twisted). It should also provide a mechanism for detailed information at a specified accuracy to be added to the object model, because although detailed information are not important at this stage, they may be required at later stages.

In addition, the designer must be able to control the degree of coarseness of the model at any stage. Thus, it is essential to look for a fuzzy representation for CAD objects to be used at the early stage of design, and a scheme to reduce this fuzziness as the design proceeds.

Design constraint specification is another aspect of CAD systems where fuzziness is essential. A number of attempts have been made to identify and formalise design constraints that relate to functionality, physical and manufacturing properties. A database of such constraints serves as guidelines for design, and as a mechanism for automatic checking the suitability of a design. The constraints are often specified in terms of a single numerical value, a range of values, or a small number of simple qualitative descriptors which can be mapped to appropriate numerical values. Zucker and Demaid [31] provided a critical analysis of such simple schemes for dealing with engineering properties and proposed an object-oriented representation scheme to take care of the agglomeration of properties and the interpretation of such properties within an appropriate context. No such work has been attempted to cater for the aesthetic aspect of design. The main reason for this neglect is that aesthetic factors often are very subjective and cannot be measured explicitly. We now present a formal framework to deal with this problem

3. CATEGORISING AESTHETIC INTENTS

In order to construct a theoretical framework for integrating aesthetics intents to CAD systems, we need to decompose aesthetic intents into categories that can be used for further analysis. We propose to categorise aesthetic intents into four main types:

- *forms*
- *physical attributes*
- *emotion-evoking attributes*
- *specific styles*

Forms relate to the shape of an object or its components. In simple cases, it concerns with general geometric shape characteristics (e.g. squarish, oblong), fairing or styling of curves and surfaces (e.g. car body shape). In more complex situations, form may concern with procedures to perform complicated transformations of the object shape (e.g. morphing, twisting, pinching, pulling, deforming of axes).

Physical attributes concern with aesthetic factors such as shading, light reflection, texture. These factors may depend on physical properties (e.g. material, paint) or geometric properties (e.g. the way light reflects on a surface depends on its shape). In the latter case, the last two types of intents are interdependent. Both of them may be either quantitative and qualitative.

Emotion-evoking attributes refer to those characteristics of the object appearance, that induce emotional responses. Typical of such attributes are soft, aggressive, expressive, sensual, rich and desolate. Although these attributes are hard to define exactly, one can identify a number of form or physical attributes that would provoke such emotional responses. For example, curvy and rounded objects tend to provoke more sensuality while objects containing sharp lines or strong colour contrast tend to appear more aggressive.

Specific styles refer to attributes that are common to a certain period or fashion, for example, high-tech look, style of the 60's, or art deco style. Again, it is possible to identify a number of geometric or physical attributes that are typical for each style. Thus, although there is no one-to-one correspondence between the last and the first two categories of attributes, it is possible to establish a relationship between them in the form of a many-to-one or one-to-many mapping. This would enable the last two categories of attributes to be achieved via geometric and physical properties, which are implementable.

4. REQUIREMENTS FOR AESTHETIC INTENTS REPRESENTATIONS

CAD software tools tend to cater for very precise activities, based on fixed and predetermined assumptions. This is in conflict with the way decisions concerning aesthetics is made. A viable representation for aesthetic intents must be

- able to cope with their *fuzzy* characteristics.
- *well-defined* and *meaningful*: to allow successful matching of the resultant product to the designer's intention
- *intuitive*: to allow the designer to use them with ease and not to hinder his/her creative thoughts
- *unambiguous*: to avoid confusion during interpretation
- *consistent*: to give the same results for the same specification
- *discriminating*: to capture distinct aesthetic factors
- *describing generic information*: to cater for different applications
- *extensible*: to allow further scope for refinement

- *useful for exploration*: to allow the designer to try out and evaluate alternative designs
- *able to be organised in related groupings*: for ease of specification, interpretation and implementation.
- *able to be combined with other types of constraints*: although aesthetic intents are important, products are also governed by other constraints such as functionality and manufacturability.
- *inference rules* may be constructed using logic deduction, induction or abduction. Induction involves with the derivation of logical rules whereas abduction concerns with deriving statements given logical rules and some logic sequences. Hence, both of them are much more powerful than logical deduction, and may enable new aesthetic intents to be evolved and deduced
- *implementable*: to allow integration into a CAD system by automatic matching of the intents to graphical functions (both geometric and non-geometric) for object generation. The following implemetation issues need to be address:

> *(i) reuseable*: to build up a library of aesthetic style (e.g. certain styling characteristic).
>
> *(ii) backtracking*: as the design process involves repeated trial-and-error, it is essential to be able to backstrack a sequence of actions that leads to intermediate or final design.
>
> *(iii) order of priority*: as aesthetic quality is subjective and varies with different applications, it is necessary to allow a designer to specify his or her own order of priority for aesthetic intents.
>
> *(iv) visual representation*: should be used in preference to symbolic descriptors wherever possible for more intuitive interpretation.
>
> *(v) value checking*: needs to be carried out to identify non-existent solutions and suggest other alternatives to designers.
>
> *(vi) inter-relationship of intents*: need to be addressed to resolve dependency and avoid conflicts.
>
> *(vii) agglomeration of intents*: in some cases, it may be more appropriate to treat intents as a class with inter-related properties and inheritance characteristics.
>
> *(viii) efficiency*: must be considered during the construction of inference rules for matching and the design of algorithms for mapping symbolic descriptors to mathematical functions

5. AESTHETIC INTENT REPRESENTATIONS

Single point numerical values are not suitable for depicting aesthetic intents since these intents are inexact and often cannot be measured explicitly by a single continuous scale. Furthermore, the main aim of integrating the aesthetic intents to CAD systems is to assist the designer with exploratory tasks, the least one could offer is an *interval of continuous values* or a *set of discrete values*. An example of the former is a range of orientation angles for the tangent of a curve to allow the designer to decide how flat or curvy a curve section should be during styling stage. An example of the latter is a set of reflective indices which corresponds to the types of paint available for a certain product. These reflective properties governs how the light reflects on a surface and dictates its appearance.

To make the selection more intuitive, *symbolic representations* in a natural language may be used to expressed aesthetic intents. The simplest way is to map each symbolic descriptor to a specific value. For example, a curve section may be described as flat, relatively flat, nearly circular, elliptical, relatively sharp or very sharp.

However, the major drawback is that there can only be a small number of such descriptors. Ideally, we would like to provide the designer first with the choice of a finite set of fuzzy classes for a particular design intent. Once a class is selected, the designer is allowed to explore alternatives in a more exhaustive way. This may be achieved by mapping each class to a specific interval, or a more exhaustive set of discrete values.

In more complex cases, *multi dimensional representations* may be needed. For example, to change the shape of an object represented by a super-ellipsoid, one needs to change the length of the axes, hence the number of parameters to be varied may be one, two or three. In such cases, one needs to explore the interaction of the influence of these parameters on each other. The simplest but most exhaustive way is to vary only one parameter while holding the remaining parameters constant. Another alternative is to allow all parameters to be considered *simultaneously*. This can be achieved by identifying special shape characteristics which correspond to certain ranges of value for the parameters. The shape characteristics could be squarish, rectangular, triangular, oval, asymmetric at one end, etc. The mapping from symbolic descriptors in a natural language to appropriate mathematical expressions must be transparent to users. The ranking and sorting out of the parametric values (hence the type of shape changes) are important.

In general, the aesthetic factors must take precedence over numerical values in relation to any grouping and ranking decision. In other words, classes cannot be sorted by the order of numerical alone. This would have little bearing on the quality or characteristics of a shape in many cases, as the shape often does not vary monotonically according to a linear scale.

Decisions on what to offer designers in terms of descriptors of shape characteristics, parameters to be varied, and the range of values for those parameters, need to be analysed for each type of object representation individually. However, our initial examination shows that the above schemes for representing aesthetic intents are feasible for common types of geometric models (e.g. CSG primitives, B-splines, NURBS, superquadrics, super-ellipsoids), and shading models for display. To introduce the ability to vary shape parameters for geometric models in this manner is equivalent to providing *inexact* geometric models by extending existing *exact* models. A more extensive domain of shape is achieved as a result.

To provide richer shape manipulation or more complex appearance effects which involve more complicated *functions*, each symbolic representation of such intents needs to be mapped into a function or a set of functions, instead of numerical values. Some examples of such transformations are morphing, pulling, twisting and axis transformation. Aesthetic intents may also be realised by combining both numerical values and functions. For example, an irregular spiral twisted shape may be obtained by varying the radial dimension randomly while performing a twisting function.

6. A FUZZY SCHEME FOR REALIZING AESTHETIC INTENTS

To achieve aesthetic intents, we propose a systematic scheme which is composed of 5 main components:

- fuzzy specifications
- fuzzy database
- fuzzy mappings
- precise graphical algorithms
- fuzzy output

Fuzzy specification may be done via symbolic descriptors in a natural language for expressing aesthetic intents. The mapping from these symbols to numerical values, intervals or mathematical functions must be transparent to users. These symbolic descriptors allow designers to convey a fuzzy concept. For example, a shape may be specified as very round, roundish or slightly round; a greyscale intensity may be very dark, medium dark, medium, medium light and very light. The number of descriptors which may be increased to cater for a finer distinction as required, is only restricted by the number of suitable words available in a particular language for communicating different meanings. In addition to this type of fuzzy descriptors, a *fuzzy factor* which depicts the level of fuzziness of the specification, may also be introduced to enrich each specification. For example, a sphere specified with a fuzzy factor of 1% implies that the radius and centre could be anywhere within 1% distance from the specified radius and centre. Multiple outputs resulted from such a specification , would correspond very well with multiple circles often sketched by a designer during the 'doodling' process before the one that appears most optimal is finally selected.

A fuzzy database must be constructed to contain data that can convey fuzzy meanings about aesthetic intents. This database stores fuzzy data types (such as numerical intervals, categories) as well as the hierarchies of these types. Each hierarchy of a fuzzy data type provides a finer description of the variable, hence allows a designer's intents to be better matched. For example, a designer may wish to deal with the sleekness of a curve. In the first level of hierarchy, a number of degrees of sleekness may be specified. In the second level, a number of geometric techniques which can achieve a specified degree of sleekness, may be selected. These techniques may require the specification of a range of the tangential angle at each end point of curves, or the specification of a range of the curvature of the curve. The decision on which parameter to be made fuzzy and with what degree of fuzziness depends very much on the graphical representations and techniques used. These representations and techniques may be either geometrical (ie. concerning with shape and topology) or non-geometrical (e.g. shading, colour, texture).

The fuzzy mapping refers to how each fuzzy aesthetic intent may be realised. It may be achieved by applying a particular graphical technique while varying appropriate parameters, or by applying a number of different graphical techniques. However, for the scheme to be feasible, it is essential that each fuzzy mapping eventually leads to a precise graphical function with appropriate parameters, or to a number of precise graphical functions with appropriate parameters.

In other words, by using fuzzy specifications, fuzzy data and fuzzy mappings, we allow the designer greater freedom during the cognitive process and permit final decisions on design objects to be delayed. This whole process produces fuzzy output in terms of design alternatives from which the designer can select the optimal ones after some comparative analysis.

We have described a fuzzy scheme for achieving designers aesthetic intents in general. We now demonstrate how this scheme can be applied to enrich the power of a 2D artistic activity.

7. BRUSH PAINTING

One area of aesthetic design is in the modelling of brush strokes in calligraphy and in Chinese or Japanese brush painting. In these activities, every brush stroke has a definite trajectory along which a brush is moved. The amount of ink absorbed by the paper on either side of the trajectory varies mainly with the pressure and the amount of ink put on the brush by the artist. There are two main approaches: to model painting as a physical process or to simulate the effects of the painting process.

The modelling in most cases has been carried out for both geometric shape and the shade (or colour) of a brush stroke. Ghosh and Mudur [9] used fourth-order parametric equations for describing the outlines of brushstrokes for some specific brushes and trajectories. Although their techniques are efficient, they are not intuitive and not suitable for interactive use as it is not easy to specify required parametric values for desired effects. The Chinese painting system developed by Pang et al. [16] is very restricted because procedures for drawing a number of primitives such as leaves, branches and waves were built into the system. Strassman [23] simulated the physical process of brush painting and used cubic splines to represent the stroke, pressure and ink dip.

This technique manages to capture the expressiveness of Chinese painting brush strokes, but is computationally expensive and not intuitive. In particular, it is not easy to simulate pressure.

Pham's method [18] aimed to rely mainly on human visual perception and intuition to provide a simple and efficient technique for producing interactively expressive brush strokes. The representation consists of four major components: *trajectory, thickness, shade* and *scratchiness*. The trajectory was modelled as a cubic B-spline whose knots may be input by using a mouse.

In general, only a small number of knots were needed for a fairly complicated trajectory. Variable offset B-splines [17,18] were used to represent the brush strokes, where the thickness of each stroke is defined by the offset distances from the trajectory at the knots. Each bristle was represented as an offset B-spline curve whose knots can be automatically calculated from the specified number of bristles and the offset distances at the knots of the trajectory.

Shading and intensity is treated in a manner similar to that for spatial coordinates. B-splines which were used to describe intensity, allowed both smooth and abrupt changes in intensity to be simulated. Control nodes for the B-spline functions also provided users with a tool to specify and manipulate different shading effects.

Both the thickness and the number of bristles may differ on each side of the trajectory, hence the effects of variable pressure and density of the brush were simulated. Further irregular effects such as scratchiness due to drier brush and spreading bristles due to stronger pressure could also be simulated by varying the offset distance of each bristle from the trajectory.

Realistic shading was achieved by modelling each bristle as a 3D offset B-spline curve whose first two coordinates were the spatial coordinates, and whose third coordinate represents the shade value (or intensity) of the bristle.

To facilitate the inputting of shade values at the knots on the trajectory, a bar containing available range of shades was displayed on the screen, from which shades were selected using a mouse. The main advantage of modelling bristles as 3D B-splines was that the shape and shades of brush strokes could be manipulated locally in a flexible and intuitive manner.

This representation has a further advantage that it facilitates the animation of brush strokes. It was shown that basic transformations such as scaling, rotation and translation, which were essential for producing in-between frames, could be performed on the control vertices instead of the curves themselves [18].

This system can produce expressive brushstrokes and provide improvements over previous methods in terms of interactive and intuitive specifications of relevant parameters based on human visual perception. However, it still does not provide users with a suitable environment for exploration of different alternatives before settling with the most optimal choice. Ideally, users would like to be able to give fuzzy (or imprecise) specifications for each parameter via a natural or symbolic language, and the system would map these fuzzy specifications to appropriate graphical functions or algorithms in a transparent manner to users. To achieve this, we devise schemes for imprecise specifications for each of the four parameters, based on the guidelines described in Sections 4-6.

Associated with each parameter, we first define a variable f which denotes the degree of fuzziness in specification and has a value in $[0,1]$. The value of this variable once specified by the user, will allow the system to provide a number of solutions, instead of one exact solution.

For example, if the position of control vertices which define a brush trajectory is specified with $f = 0.01$, these vertices could be anywhere within 1% distance of the location where the mouse is clicked. The system thus returns a number of locations for each vertex. This number which is proportional to f, denotes the number of alternatives required by the user.

Such a specification removes the need for the user to decide on a solution at the early stage. Instead, alternatives are offered, from which the user can compare, manipulate and select the optimal one at a later stage. This fuzzy factor may have the same value or a different value for each different parameter (position of trajectory, brush thickness, shade and scratchiness).

The thickness on either side of the trajectory may be categorised as thin, medium thin, medium, medium thick and thick. Alternatively, a finer scale which corresponds to the real size of available brushes may be used (e.g. brush size 4-14). The offset distance from each control vertices on the trajectory would be computed automatically according to the specifications.

A similar scheme may be used for the shade and the degree of scratchiness. The effects may be enriched by introducing categories for different types of ink (e.g. block or liquid), paper (e.g. thin, medium, thick weight) and brush (e.g. sable, hog hair or synthetic material).

These techniques can be readily extended to produce different shades of colour by using the same types of splines with three variables. HSV colour space should be used because it is more intuitive to perceive a colour in terms of hue, saturation and brightness.

Furthermore, it is very difficult to predict a colour resulted from a specified amount of Red, Green and Blue components because although RGB is a convenient model for computer screen display, it has little bearing on how humans perceive colour.

Although the above-mentioned techniques were developed as independent systems for specific purposes, it would be quite easy to integrate them to a CAD system as they are based on different types of spline curve and offset curve generation already available in common CAD systems. This would expand the capability to allow the creation of more artistic drawings with user-specified range of dynamics and variability.

7. CONCLUSION

We have presented a framework based on fuzzy logic to allow designers to specify and realise their aesthetic intents. By applying fuzzy concepts to appropriate tasks, a much more powerful CAD system could be developed, which is better suited to creative and artistic work. Such a system facilitates the cognitive process in design by allowing fuzzy thoughts to be realised and by offering alternatives instead of a restricted single precise input or output. On-going research is being carried out to find matching geometric functions and models to achieve a commonly-used set of aesthetic intents for some specific styles and for generic cases. One important question needs to be addressed is how to evaluate the success of this task. Do we have to rely solely on subjective tests that record responses from a number of users, or is it possible to design some objective evaluation measures for intent satisfaction?

REFERENCES

[1] Barr A.H., Superquadrics and angle-preserving transformations, *IEEE Comput, Graph. Appl.* 1, 1981, 11-22.

[2] Barr A.H., Global and local deformation of solid primitives, *Comput. Graph.* 18(3), 1984, 21-30.

[3] Blackmore D., Leu M.C., Shih F., Analysis and modelling of deformed swept volumes, *Computer-Aided Design* 26(4), 1994, 315-325.

[4] Coyne R.D., Rosenman M.A., Radford A.D., Gero J.S., Innovation and creativity in knowledge-based CAD, in *Expert Systems in Computer-Aided-Design*, Gero J.S. (Ed. J.S. Gero), 1987, 435-470, Elsevier Science Publishers (North-Holland).

[5] De Martino T., Falcidieno B., Feature-based modelling by integrating design and recognition approaches, *Computer-Aided Design* 26(8), 1994, 646-653.

[6] Dutta D., Martin R.R., Pratt M.JCyclides in surface and solid modelling, IEEE *Comput. Graph. Appl.* 13(1), 1993, 53-59.

[7] Ferguson H. & C., *Mathematics in stone and bronze*, Meridan Creative Group, 1994.

[8] Gero J.S.(Ed.), *Artificial Intelligence in Design*, Kluwer Academic, 1992.

[9] Ghosh P.K., Mudur S.P., The brush-trajectory approach to figure specification: some algebraic solutions, *ACM Trans. Graph.* 3, 1984, 110-134.

[10] Hanson A.J., Hyperquadrics: Smoothly deformable shapes with convex polyhedral bounds, *Comput. Vis. Graph. Img. Proces.* 44, 1988, 191-210.

[11] Kurangano T., Fresdam system for design of aesthetically pleasing free-form objects and generation of collision-free tools path, *Computer-Aided Design* 24(11), 1992, 573-581.

[12] Lazarus F., Axial deformations: an intuitive deformation technique, *Computer-Aided Design* 26(8), 1994, 607-613.

[13] Laako T., Mantyla M., Feature modelling by incremental feature recognition, *Computer-Aided Design* 25(8), 1993, 479-492.

[14] McNeill D., Freioberger P., *Fuzzy Logic*, Simon & Scuster, NY, 1993.

[15] Negotia C.V., Ralescu D., *Simulation Knowledge-based Computing and Fuzzy Statistics*, Van Nostrand Reinhold, NY, 1987.

[16] Pang Y.J., Yang S.X., Chi Y., Combining computer graphics with Chinese traditional painting, *Comput. Graph.* 11, 1987, 63-68.

[17] Pham B., Offset approximation of uniform B-splines, *Computer-Aided Design* 20 (8), 1988, 471-474.

[18] Pham B., Expressive Brush Strokes, CVGIP Graphical Models & Image Processing 53 (1) 1991, 1-6.

[19] Pham B., From CAD to automation of the design process, *Annals of Numerical Mathematics* 3 , No. 1-4 (1996) , *Computer-Aided Geometric Design*, (Eds. C.A. Micchelli, H.B. Said).

[20] Pham B., CAD and Creativity, *invited talk, Fourth SIAM Conference on Geometric Design*, 1995, Nashville, Tennessee.

[21] Sittas E., 3D design reference framework, *Computer-Aided Design* 25(5), 1993,380-384.

[22] Smither 1989] Smither T., AI-based design versus geometry-based design, *Computer-Aided Design* 21(3), 1989, 141-150.

[23] Strassman S., Hairy brushes, *Proc. SIGGRAPH* 22, 1986, 225-232.

[24] Takala T., Woodward C.D., Industrial design based on geometric intentions, in *Theoretical Foundation of Computer Graphics and CAD* (Ed. B.A. Earnshaw), NATO ASI Series. Vol F40, 1988, 953-963.

[25] Terzopoulos D., Metaxas D., Dynamic 3D models with local and global deformations: deformable superquadrics, *IEEE Trans. Patt. Anal. Mach. Intell.* 13(7), 1991, 703-714.

[26] Tong C., Sriram D, *Artificial Intelligence in Engineering Design*, Acad. Press 1992

[27] Tomiyama T., Ten Hagen P.J.W., Organization of design knowledge in an intelligent CAD environment, in *Expert Systems in Computer-Aided-Design*, Gero J.S. (Ed. J.S. Gero), 1987, 119-147, Elsevier Science Publishers (North-Holland).

[28] Tyugu E.H., Merging conceptual and expert knowledge in CAD, in *Expert Systems in Computer-Aided-Design*, Gero J.S. (Ed. J.S. Gero), 1987, 423-434, Elsevier Science Publishers (North-Holland).

[29] Wallace D.R., Jakiela M.J., Automated product concept design: unifying aesthetic and engineering, *IEEE Comput. Graph. Appl.* 13, 1993, 66-75.

[30] Zadeh L., Fuzzy Sets, *Information and Control* 8(3), 1965, 338-353.

[31] Zucker J., Demaid A., Object-oriented representation of qualitative engineering properties, *Computer-Aided Design* 26(10) 1994, 722-734.

INCLUDING SOCIOECONOMIC ASPECTS IN A FUZZY MULTISTAGE DECISION MAKING MODEL OF REGIONAL DEVELOPMENT PLANNING

Janusz Kacprzyk

Systems Research Institute, Polish Academy of Sciences
ul. Newelska 6, 01-447 Warsaw, Poland
E-mail: kacprzyk@ibspan.waw.pl

Summary. We consider the planning of socioeconomic regional development considered from economic, social, infrastructural, etc. points of view. The development is chracterized by 7 life quality indicators concerning: economic, environmental, housing, health service, infrastructure, work opportunity, and leisure time qualities. Both objective and subjective aspects are accounted for, the latter mainly in terms of resulting social satisfactions. We formulate the problem in terms of Bellman and Zadeh's (1970) multistage decision making (control) under fuzzy constraints on investments applied and fuzzy goals on the life quality attained and their resulting social satisfaction, seeking optimal investments over a planning horizon. We also include some stability requirement.

Keywords: multistage decision making under fuzziness, fuzzy constrain, fuzzy goal, regional development planning, social satisfaction

1. Introduction

In this paper we consider the planning of socioeconomic regional development over a planning horizon. The development is considered from economic, social, infrastructural, etc. points of view. The development is chracterized by a life quality index consisting of 7 life quality indicators concerning: economic, environmental, housing, health service, infrastructure, work opportunity, and leisure time qualities. Both objective and subjective aspects are accounted for, the latter mainly in terms of resulting social satisfactions. We formulate the problem in terms of Bellman and Zadeh's (1970) multistage decision making (control) under fuzzy constraints on investments applied and fuzzy goals on the life quality attained and their resulting social satisfaction, seeking optimal investments over a planning horizon. We also include some stability requirement to reflect a human preference to operate under possibly stable conditions. We have presented a fuzzy model of socioeconomic regional development planning. The model takes as its point of departure Bellman and Zadeh's (1970) general model of multistage decision making (control) under fuzziness further extended by Kacprzyk (1997). The model may be viewed as an example of a "soft" approach to the modelling of a complex socioeconomic problem.

First, we outline Bellman and Zadeh's (1970) model which is our point of departure, and then formulate in its terms a fuzzy regional development planning model. We present a simple example.

2. Bellman and Zadeh's general approach to decision making and control under fuzziness

2.1 The basic model of decision making under fuzziness

If $X = \{x\}$ is some set of possible *options* (alternatives, choices, ...), then:

- a *fuzzy goal* is defined as a fuzzy set G in X, characterized by its membership function $\mu_G : X \longrightarrow [0,1]$ such that $\mu_G(x) \in [0,1]$ specifies the grade of membership of a particular option $x \in X$ in the fuzzy goal G;
- a *fuzzy constraint* is similarly defined as a fuzzy set C in the set of options X, characterized by $\mu_C : X \longrightarrow [0,1]$.

The general problem formulation is: "Attain G <u>and</u> satisfy C" which leads to a *fuzzy decision* D defined as a fuzzy set in X such that

$$\mu_D(x) = \mu_G(x) \wedge \mu_C(x), \qquad \text{for each } x \in X \qquad (2.1)$$

where $a \wedge b = \min(a, b)$ may be replaced by another operation, e.g., a t-norm (cf. Kacprzyk, 1997).

The *maximizing decision* is defined as an $x^* \in X$ such that

$$\mu_D(x^*) = \max_{x \in X} \mu_D(x) \qquad (2.2)$$

If we have $n > 1$ fuzzy goals, $G_1, \ldots, G_n$, and $m > 1$ fuzzy constraints, $C_1, \ldots, C_m$, all defined in X, then the *fuzzy decision* is

$$\mu_D(x) = \mu_{G_1}(x) \wedge \ldots \mu_{G_n}(x) \wedge \mu_{C_1}(x) \wedge \ldots \wedge \mu_{C_m}(x), \qquad \text{for each } x \in X \qquad (2.3)$$

If, on the other hand, C is defined in $X = \{x\}$, G is defined in $Y = \{y\}$, and a function $f : X \longrightarrow Y$, $y = f(x)$, is known, the *fuzzy decision* is

$$\mu_D(x) = \mu_G[f(x)] \wedge \mu_C(x), \qquad \text{for each } x \in X \qquad (2.4)$$

and, for $G_1, \ldots, G_n$ defined in Y, $C_1, \ldots, C_m$ defined in X, and $f : X \longrightarrow Y$, $y = f(x)$, we have

$$\mu_D(x) = \mu_{G_1}[f(x)] \wedge \cdots \wedge \mu_{G_n}[f(x)] \wedge \\ \wedge \mu_{C_1}(x) \wedge \cdots \wedge \mu_{C_n}(x), \qquad \text{for each } x \in X \qquad (2.5)$$

In all the above cases the *maximizing decision* is defined as (2.2), i.c. $\mu_D(x^*) = \max_{x \in X} \mu_D(x)$.

2.2 The model of multistage decision making (control) under fuzziness

In the control-related setting, suppose now that the control space is $U = \{u\} = \{c_1, \ldots, c_m\}$, the state space is $X = \{x\} = \{s_1, \ldots, s_n\}$, and both are finite. The system under control is assumed to be deterministic, governed by a *state transition equation*

$$x_{t+1} = f(x_t, u_t), \qquad t = 0, 1, \ldots \qquad (2.6)$$

where $x_t, x_{t+1} \in X = \{s_1, \ldots, s_n\}$ are the states at t and $t+1$, respectively, and $u_t \in U = \{c_1, \ldots, c_m\}$ is the control at t.

At t, $t = 0, 1, \ldots$, $u_t \in U$ is subjected to a *fuzzy constraint* $\mu_{C^t}(u_t)$, and on $x_{t+1} \in X$ a *fuzzy goal* is imposed, $\mu_{G^{t+1}}(x_{t+1})$.

The *initial state* is $x_0 \in X$ and is assumed to be known, and given in advance. The *termination time* (planning, or control, horizon), $N \in \{1, 2, \ldots\}$, is assumed to be finite, and fixed and specified in advance [for other types of the termination time, see Kacprzyk (1997)].

The *performance* of the multistage fuzzy control process is evaluated by the fuzzy decision

$$\mu_D(u_0, \ldots, u_{N-1} \mid x_0) =$$
$$= \mu_{C^0}(u_0) \wedge \mu_{G^1}(x_1) \wedge \ldots \wedge \mu_{C^{N-1}}(u_{N-1}) \wedge \mu_{G^N}(x_N) \qquad (2.7)$$

and the problem is to find an optimal sequence of controls $u_0^*, \ldots, u_{N-1}^*$ such that

$$\mu_D(u_0^*, \ldots, u_{N-1}^* \mid x_0) = \max_{u_0, \ldots, u_{N-1} \in U} \mu_D(u_0, \ldots, u_{N-1} \mid x_0) \qquad (2.8)$$

For extensions to this basic formulation we refer the reader to Kacprzyk's (1997) book.

Problem (2.8) can be solved using the following two basic traditional techniques:

– dynamic programming [Bellman and Zadeh (1970)], and
– branch-and-bound [Kacprzyk (1978, 1979)],

and also using the following two new ones:

– a neural network [Francelin and Gomide (1992, 1993)], and
– a genetic algorithm [Kacprzyk (1995a, b, c)],

though we will not consider in detail any of them since the purpose of the paper is to present how socioeconomic aspects can be reflected in the model considered, and not to discuss technicalities related to any solution technique.

We will only briefly present the traditional dynamic programming technique.

The application of dynamic programming for the solution of (2.8) was proposed in Bellman and Zadeh (1970). We begin by rewriting (2.8) as to find $u_0^*, \ldots, u_{N-1}^*$ such that

$$\mu_D(u_0^*, \ldots, u_{N-1} \mid x_0) =$$
$$= \max_{u_0, \ldots, u_{N-1}} [\mu_{C^0}(u_0) \wedge \mu_{G^1}(f(x_0, u_0)) \wedge \ldots$$
$$\ldots \wedge \mu_{C^{N-1}}(u_{N-1}) \wedge \mu_{G^N}(f(x_{N-1}, u_{N-1}))] \qquad (2.9)$$

and then, since $\mu_{C^{N-1}}(u_{N-1}) \wedge \mu_{G^N}(f(x_{N-1}, u_{N-1}))$ depend only on u_{N-1}, then the maximization over $u_0, \ldots, u_{N-1}$ in (2.9) can be split into: the maximization over $u_0, \ldots, u_{N-2}$, and the maximization over u_{N-1}, which may be continued for u_{N-2}, u_{N-3}, etc.

This backward iteration implies the set of fuzzy dynamic programming recurrence equations:

$$\begin{cases} \mu_{\overline{G}^{N-i}}(x_{N-i}) = \\ \quad = \max_{u_{N-i}} [\mu_{C^{N-i}}(u_{N-i}) \wedge \mu_{G^{N-i}}(x_{N-i}) \wedge \mu_{\overline{G}^{N-i+1}}(x_{N-i+1})] \\ x_{N-i+1} = f(x_{N-i}, u_{N-i}); \qquad i = 0, 1, \ldots, N \end{cases}$$
$$(2.10)$$

where $\mu_{\overline{G}^{N-i}}(x_{N-i})$ is a fuzzy goal at $t = N - i$ induced by the fuzzy goal at $t = N - i + 1$, $i = 0, 1, \ldots, N$; $\mu_{\overline{G}^N}(x_N) = \mu_{G^N}(x_N)$.

The $u_0, \ldots, u_{N-1}$ sought is given by the successive maximizing values of u_{N-i}, $i = 1, \ldots, N$ in (2.10) obtained as functions of x_{N-i} (an *optimal policy*).

3. Socioeconomic regional development planning under fuzziness

Regional development is problem of general importance. It is, however, difficult as it involves various aspects (political, economic, social, environmental, technological, etc.), different parties (inhabitants, authorities of different levels, formal and informal groups, etc.), many tangible and intangible aspects, etc. It is therefore a typical "soft" problem.

As an attempt to overcome these difficulties, the use of fuzzy sets has been proposed by Kacprzyk and Straszak (1982a, b, 1984). They considered a (rural) region plagued by some severe difficulties exemplified by out-migration of the younger population to neighboring urban centers, the resulting aging of the rural population, and – as a final consequence – a socioeconomic decay. The out-migration is mainly caused by a poor *life quality* perceived. To stop this decay, life quality should therefore be (considerably) improved, and some (mostly external) funds (investments) are needed whose amount should be determined.

Notice that virtually all levels of life quality indicators attained are not only objective but also subjective in the sense that their human perception is what counts. This will be of utmost importance in our next discussion, and we will show that this important characteristic feature of regional development, as well as virtually all socioeconomic planning problems, can be reflected in an adequate, effective and efficient way in the fuzzy multistage decision making model employed.

3.1 A fuzzy model of socioeconomic regional development planning

For our purposes, the essence of socioeconomic regional development considered may be meant as in Figure 3.1.

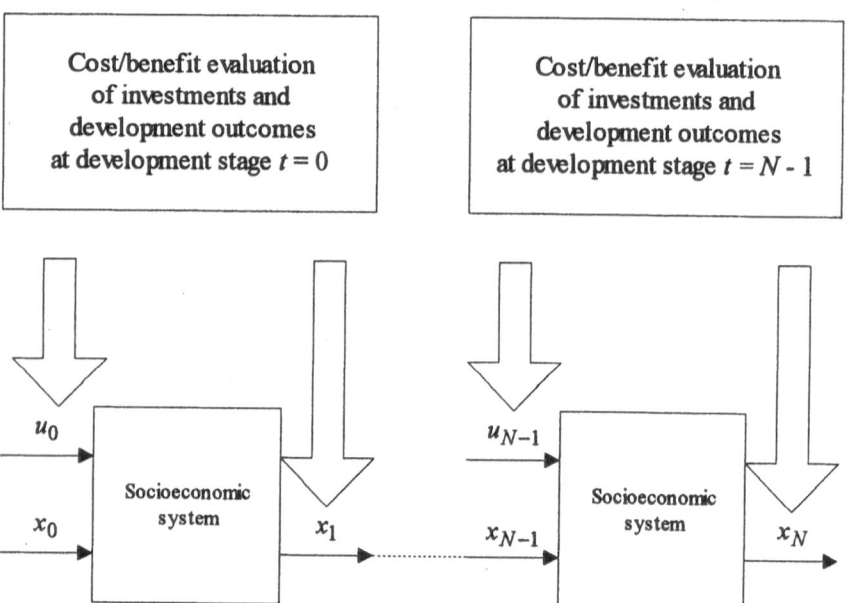

Fig. 3.1. The essence of socioeconomic regional development

The region onsidered is represented by a dynamic system under control whose state (equated with the output, for simplicity) at development (planning) stage $t - 1$, X_{t-1}, is characterized by a set of relevant socioeconomic life quality indicators. Then, the control (decision), equated with investment, at $t - 1$, u_{t-1}, changes X_{t-1} to X_t. This is repeated for $t = 1, \ldots, N$; N is a prespecified planning horizon.

The assessment of a development stage t is performed by accounting for both the "goodness" of the u_{t-1} applied (i.e. costs), and the "goodness" of the X_t attained (i.e. benefits); the former has to do with how well some constraints are satisfied, and the latter with how well some goals are attained.

The main problem is how to properly formulate and evaluate those costs and benefits. Needless to say that both involve human perception and satisfaction.

First, we present the socioeconomic system as in Figure 3.2. Its state (output) X_t is equated with a *life quality index* that consists of the following

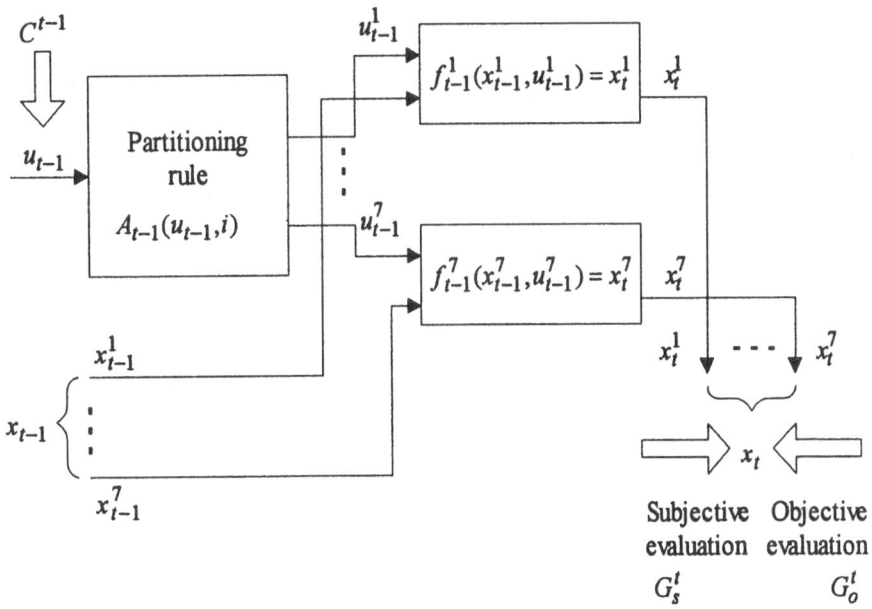

Fig. 3.2. Region represented as a system under control

seven *life quality indicators* (i.e. $X_t = [x_t^1, \ldots, x_t^7]$):

- x_t^1 – economic quality (e.g., wages, salaries, income, ...),
- x_t^2 – environmental quality,
- x_t^3 – housing quality,
- x_t^4 – health service quality,
- x_t^5 – infrastructure quality,
- x_t^6 – work opportunity,
- x_t^7 – leisure time opportunity,

The u_{t-1}, is an investment devoted to the development of the region, and is subjected to a limitation which is in practice not clear-cut and abrupt. In practice this investment is predominatly from outside sources (e.g., provided by central authorities) because its amount needed usually exceeds by far what is available in regions.

We impose on u_{t-1} a fuzzy constraint $\mu_{C^{t-1}}(u_{t-1})$. As in virtually all practical applications of fuzzy logic, exemplified by fuzzy (logic) control, trapezoid and/or triangular fuzzy number are employed for the representation of fuzzy quantities. And also in our case we assume the fuzzy constraint to be in a piecewise linear form as shown in Figure 3.3 to be read as follows. The in-

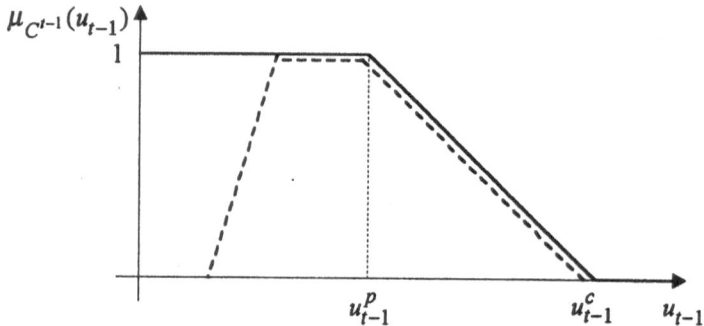

Fig. 3.3. Fuzzy constraint on the investment at stage $t-1$, u_{t-1}

vestment may be fully utilized up to the amount u_{t-1}^p, which is planned in advance, and hence $\mu_{C^{t-1}}(u_{t-1}) = 1$ for $0 < u_{t-1} < u_{t-1}^p$. In practice, however, this is usually insufficient and some additional contingency investment is needed. Evidently, the higher the amount requested, the more difficult to obtain. Therefore, the $\mu_{C^{t-1}}(u_{t-1})$ diminishes as u_{t-1} exceeds u_{t-1}^p; maximally, a prespecified amount u_{t-1}^c may be spent.

The fuzzy constraints are often as shown in the dotted line in Figure 3.3 in that too low a use of available investments should also be avoided. This occurs in virtually all cases and is related to a common bureaucratic practice of lowering in the future the amount of funds to be alotted in the future when it was used lower than alotted in the past.

The u_{t-1}, is now partitioned into parts $u_{t-1}^1, \ldots, u_{t-1}^7$, devoted to the improvement of the respective life quality indicators, $x_{t-1}^1, \ldots, x_{t-1}^7$. This partitioning proceeds due to a (prespecified) partitioning rule, $A_{t-1}(u_{t-1}, i)$. Issues related to the determination of an optimal (or maybe rational) partitioning rule are very specific and not trivial, and will not be discussed in this paper.

The temporal evolution of the particular life quality indicators is governed by the state transition equation

$$x_t^i = f_{t-1}^i(x_{t-1}^i, u_{t-1}^i), \qquad i = 1, \ldots, 7; t = 1, \ldots, N \qquad (3.1)$$

which may be derived from, e.g., data by traditional identification or parameter estimation methods, using experts' opinions, etc.

Since the regiional development is a goal oriented task aimed at the satisfaction of some (human) needs, measures are needed of how well some predetermined goals are fulfilled, i.e. of how *effective* the development is. This should then be related to the investment spent, to find how *efficient* it is. Second, some extra measures may be added of, say, how *stable* (smooth) it is. These two important aspects, referred to as:

– the *effectiveness of development,* and
– the *stability of development,*

will now be briefly discussed.

4. Effectiveness of development

The effectiveness of a particular development stage t has both an *objective* and *subjective* aspect. The objective evaluation (of a particular development stage t) is the determination of how well the fuzzy constraints are fulfilled, and fuzzy goals are attained. The objective fuzzy goals concern desired values of the life quality indicators, i.e. they concern objective entities.

For each life quality indicator at stage t, x_t^i, we define an *objective fuzzy subgoal* $\mu_{G_o^{t,i}}(x_t^i)$ as shown in Figure 4.1 to be read as follows: $G_o^{t,i}$ is fully

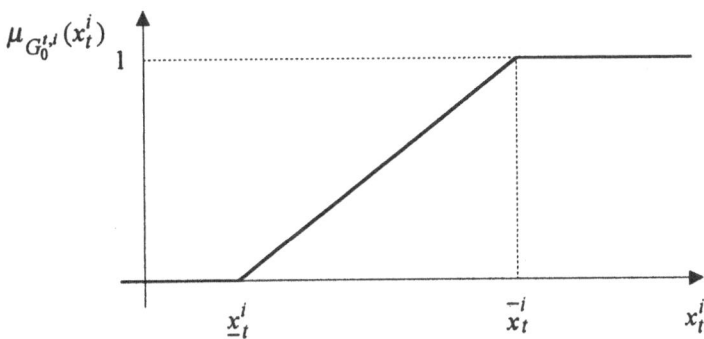

Fig. 4.1. Objective fuzzy subgoal

satisfied for $x_t^i \geq \overline{x}_t^i$, where $\overline{x}_t^i$ is some *aspiration level* for the indicator x_t^i; therefore, $\mu_{G_o^{t,i}}(x_t^i) = 1$, for $x_t^i \geq \overline{x}_t^i$. Less preferable are $\underline{x}_t^i < x_t^i < \overline{x}_t^i$ for which $0 < \mu_{G_o^{t,i}}(x_t^i) < 1$, and $x_t^i \leq \underline{x}_t^i$ are impossible, hence $\mu_{G_o^{t,i}}(x_t^i) = 0$.

The objective evaluation of the life quality index, $X_t = [x_t^1, \ldots, x_t^7]$ is obtained by aggregation of the above, i.e.

$$\mu_{G_o^t}(X_t) = \mu_{G_o^{t,1}}(x_t^1) \wedge \ldots \wedge \mu_{G_o^{t,7}}(x_t^7) \tag{4.1}$$

where "$a \wedge b = \min(a, b)$".

Notice that "$\wedge$" (minimum) reflects a pessimistic, safety-first attitude, and a lack of substitutability in the sense that a low value of one life quality indicator cannot be compensated by a higher value of another one, which is often adequate. Clearly, such a pessimistic-type aggregation is not always adequate, and "$\wedge$" may be replaced here and later on by another suitable operation as, e.g., a t-norm [cf. Kacprzyk (1997)]. Moreover, for a more so-phisticated aggregation Yager's (1988) ordered weighted aggregation (OWA) operators may be employed [cf. Yager and Kacprzyk (1997)]. Note that the objective evaluation concerns more the authorities than the inhabitants.

The inhabitants' assessment concerns in fact the (perception of) *social satisfaction* resulting from the life quality index attained. This is clearly of a *subjective* type. The attained value of x_t^i implies its corresponding partial *social satisfaction* s_t^i derived analogously as in Figure 4.1 for the objective evaluation. However, the lowest possible value, $\underline{z}_t^i$, and the aspiration level, $\overline{z}_t^i$, are now assumed to be some functions of the trajectory (history) of development

$$H_t = [(X_1, s_1, \mu_{G_o^1}(X_1), \mu_{G_s^1}(s_1)), \ldots, (X_t, s_t, \mu_{G_o^t}(s_t), \mu_{G_s^t}(s_t))]$$

where $s_k = [s_k^1, \ldots, s_k^7], k = 1, \ldots, t$, is the social satisfaction resulting from X_k.

A "reduced trajectory"

$$h_t = [(X_{t-1}, s_{t-1}, \mu_{G_o^{t-1}}(X_{t-1}), \mu_{G_s^{t-1}}(s_{t-1})), (X_t, s_t, \mu_{G_o^t}(s_t), \mu_{G_s^t}(s_t))]$$

which takes into account the outcomes of the two recent development stage only, t and $t-1$, is usually used in practice.

The social satisfaction at t is now

$$s_t = s_t^1 \wedge \ldots \wedge s_t^7 \tag{4.2}$$

where "$\wedge$" again reflects a pessimistic, safety-first attitude, and a lack of sub-stitutability. Remarks concerning other types of aggregation are essentially the same as given for (4.1).

The social satisfaction s_t is subjected to a subjective fuzzy goal $\mu_{G_s^t}(s_t)$ meant similarly as its objective counterpart shown in Figure 4.1.

The effectiveness of a development stage t is meant as a relation of what has been attained (the life quality indices and their respective social satis-factions) to what has been "paid for" (the respective investments), i.e. is a benefit–cost relationship.

Formally, the (fuzzy) effectiveness of t is

$$\mu_{E^t}(u_{t-1}, X_t, s_t) = \mu_{C^{t-1}}(u_{t-1}) \wedge \mu_{G_o^t}(X_t) \wedge \mu_{G_s^t}(s_t) \tag{4.3}$$

and the effectiveness of the whole development is

$$\mu_E(H_N) = \mu_{E^1}(u_0, X_1, s_1) \wedge \ldots \wedge \mu_{E^N}(u_{N-1}, X_N, s_n) \tag{4.4}$$

4.1 Stability of development

The *stability* of regional development concerns the variability of (crucial) development indicators, characteristics, conditions, etc. It is known from experience, that a lower variability usually implies a higher acceptance of development, by both the inhabitants and authorities.

The stability in the above sense is meant to involve:

- the *stability of development trajectory* that concerns the variability of development outcomes, i.e. the life quality indicators attained and the resulting social satisfactions, and
- the *stability of development "policy"* that concerns the variability of some development prerequisites, i.e. the imposed fuzzy constraints, fuzzy goals, and investment partitioning rules.

The variability of development trajectory has many aspects. Basically, the variability of x_t^i, only meant for simplicity as is $x_t^i - x_{t-1}^i$, is subjected to a *fuzzy limitation* $\mu_{S^{t,1,i}}(x_t^i - x_{t-1}^i)$ as shown in Figure 4.2 and meant as: we are fully satisfied (to degree 1) with $x_t^i - x_{t-1}^i \geq \overline{y}_t^i$, we are partially

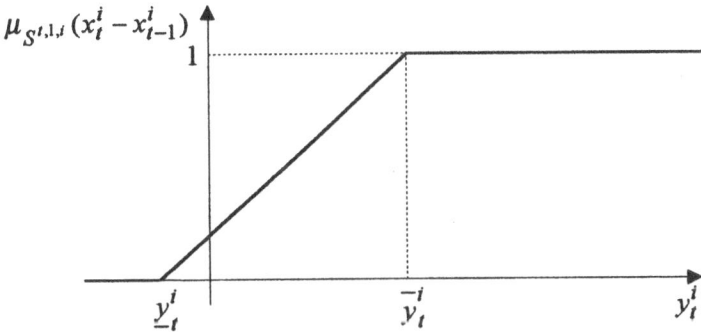

Fig. 4.2. Fuzzy limitation on the variability of the i-th life quality indicator, x_t^i

satisfied (to a degree from $[0,1]$) with $\underline{y}_t^i \leq x_t^i - x_{t-1}^i < \overline{y}_t^i$, and we are fully dissatisfied (satisfied to degree 0) with $x_t^i - x_{t-1}^i < \underline{y}_t^i$, i.e. such values are unacceptable. Note that $\underline{y}_t^i < 0$ may occur, i.e. we allow some interruption in growth (though with a low satisfaction) if it is sufficiently small. So, such a measure of variability may also be viewed as a soft *growth requirement*.

The variability limitation for the life quality index $X_t = (x_t^1, \ldots, x_t^7)$ is

$$\mu_{S^{t,1}}(X_t - X_{t-1}) = \mu_{S^{t,1,1}}(x_t^1 - x_{t-1}^1) \wedge \ldots \wedge \mu_{S^{t,1,7}}(x_t^7 - x_{t-1}^7) \qquad (4.5)$$

where "$\wedge$" reflects mainly a low (lack of) substitutability of life quality indicators; remarks given for (4.1) are also valid.

And analogously for the variability of social satisfaction s_t for which the fuzzy limitation is $\mu_{S^{t,1}}(s_t - s_{t-1})$ defined similarly as above.

The total fuzzy limitation on the stability of t is therefore an aggregation of the above ones, $\mu_{S^t}(X_t, s_t)$, and the fuzzy limitation on the stability of the whole development trajectory is

$$\mu_{S_d}(H_t) = \mu_{S^1}(X_1, s_1) \wedge \ldots \wedge \mu_{S^N}(X_N, s_N) \tag{4.6}$$

Stability of development policy is an important aspect which concerns some development prerequisites, called the *development policy*, meant as

$$B_t = [(C^0, A_0(u_0, i), G_o^1, G_s^1), \ldots, (C^{t-1}, A_{t-1}(u_{t-1}, i), G_o^t, G_s^t)]$$

i.e. how the successive investments are to be limited and partitioned, and how the successive life quality indices and their resulting partial social satisfactions should be.

For simplicity a *reduced development policy* will be used

$$b_t = [(C^{t-2}, A_{t-2}(u_{t-2}, i), G_o^{t-1}, G_s^{t-1}), (C^{t-1}, A_{t-1}(u_{t-1}, i), G_o^t, G_s^t)]$$

which is fully sufficient in practice.

The *stability of policy* concerns the variability of:

- the fuzzy constraints,
- the investment partitioning rules,
- the objective fuzzy goals, and
- the subjective fuzzy goals.

For example, for the stability of a fuzzy constraint (cf. Figure 3.3), high fluctuations of the planned investment, u_{t-1}^p, usually make the intra-regional planning, and then management, more difficult, and are therefore undesirable. The *variability of the planned investment* u_{t-1}^p, i.e. $u_{t-1}^p - u_{t-2}^p$, is hence subjected to a fuzzy limitation $\mu_{S^{t-1,s}}(u_{t-1}^p - u_{t-2}^p)$ shown in Figure 4.3 meant as: some variability, up to $\underline{v}_p$ in both directions (i.e. between $-\underline{v}_p$ and

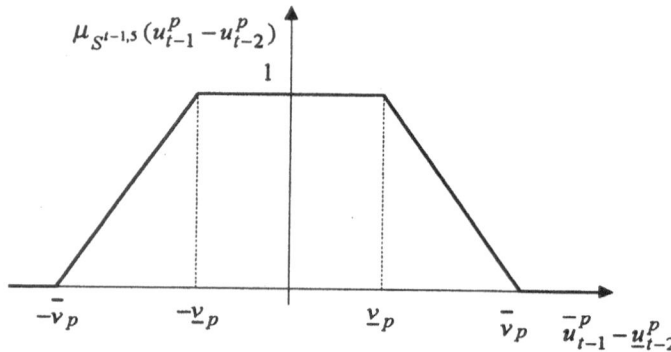

Fig. 4.3. Fuzzy limitation on the variability of the planned investment u_{t-1}^p, i.e. $u_{t-1}^p - u_{t-2}^p$

$\underline{v}_p$), is fully allowable and acceptable. A higher variability, i.e. between $\underline{v}_p$ and $\bar{v}_p$ (and between $\underline{v}_p$ and $\bar{v}_p$) is possible though less desirable, and the variability above $\bar{v}_p$ and below $-\bar{v}_p$ is unacceptable.

An analogous argument can be applied for the objective and subjective fuzzy goals. For other aspects of variability of the development policy, see Kacprzyk (1997).

Hence, the *stability of development policy* at t is $\mu_{St,l}(b_t)$, and the *stability of development policy* over the whole planning horizon $(t = 0, 1, \ldots, N-1)$ is therefore

$$\mu_{S_p}(B_N) = \mu_{S^1,p}(b_1) \wedge \ldots \wedge \mu_{S^N,p}(b_N) \tag{4.7}$$

4.2 A multistage fuzzy decision making (control) model of socioeconomic regional development planning

The fuzzy decision is used for the evaluation of both the effectiveness and stability of the regional development, and hence plays the role of a performance (quality) function. It is as follows

$$\begin{aligned}
\mu_D(u_0, \ldots, u_{N-1} \mid X_0, B_N) &= \\
&= \mu_E(H_N) \wedge \mu_S(H_N, B_N) = \\
&= \mu_{C^0}(u_0) \wedge \mu_{G_o^1}(X_1) \wedge \mu_{G_s^1}(s_1) \wedge \mu_{S^1,p}(b_1) \wedge \ldots \\
&\quad \ldots \wedge \mu_{C^{N-1}}(u_{N-1}) \wedge \mu_{G_o^N}(X_N) \wedge \mu_{G_s^N}(s_N) \wedge \mu_{S^N,p}(b_N) \tag{4.8}
\end{aligned}$$

The fuzzy decision expresses therefore some crucial compromises between, first:

— the fuzzy constraints,
— fuzzy goals, and
— fuzzy stability limitations,

second, between:

— the effectiveness, and
— stability of development,

third, between

— the interests of the authorities, and
— the interests of inhabitants,

etc.

The problem is now to find an optimal sequence of controls (investments) $u_0^*, \ldots, u_{N-1}^*$ such that (under a given policy B_N)

$$\mu_D(u_0^*, \ldots, u_{N-1}^* \mid X_0, B_N) = \max_{u_0, \ldots, u_{N-1}} \mu_D(u_0, \ldots, u_{N-1} \mid X_0, B_N) \tag{4.9}$$

This problem may be solved by any of the techniques mentioned in Section 2.2, i.e. by using dynamic programming, brach-and-bound, a neural network approach or a genetic algorithm. We will not consider these solution techniques in this paper and will refer the reader to Kacprzyk's (1997) book for details.

4.3 Example of application to the development planning of a rural region

To illustrate the use of the model presented in the previous section, we will apply it now [cf. Kacprzyk (1983, 1997)] to the planning of development of a rural, or – better – a predominantly agricultural region. By necessity, this example will be considerably simplified to make the issues, entities, relations, etc. involved understandable to a wider audience. Moreover, since we will assume a 3 year planning horizon and two scenarios (values of "control") only, the corresponding neural network is straighforward and will not be explicitly given.

The region considered, predominantly agricultural, has a population of ca. 120,000 inhabitants, and its arable land is ca. 450,000 acres. For simplicity, the region's development will be considered over the next 3 development stages (years, for simplicity).

The life quality index consists of the four life quality indicators:

– x_t^I – average subsidies in US$ per acre (per year),
– x_t^{II} – sanitation expenditures (water and sewage) in US$ per capita (per year),
– x_t^{III} – health care expenditures in US$ per capita (per year), and
– x_t^{IV} – expenditures for paved roads (new roads and maintenance of the existing ones) in US$ (per year).

Suppose now that the investments are partitioned into parts devoted to the improvement of the above life quality indicators due to the fixed partitioning rule $A_{t-1}(u_{t-1}, i)$:

– 5% for subsidies,
– 25% for sanitation,
– 45% for health care, and
– 25% for infrastructure.

Let the initial, at stage $t = 0$, values of the particular life quality indicators be:

$$x_0^I = 0.5 \qquad x_0^{II} = 15 \qquad x_0^{III} = 27 \qquad x_0^{IV} = 1,700,000$$

For clarity, and what is usually the case in virtually all practical applications, we will not consider here the optimization of the investment policy, but will only take into account the following two *scenarios* (policies):

– Policy 1: $u_0 = \$8,000,000$ $u_1 = \$8,000,000$ $u_2 = \$8,000,000$
– Policy 2: $u_0 = \$7,500,000$ $u_1 = \$8,000,000$ $u_2 = \$8,500,000$

Applying Policy 1 and Policy 2, the values of the life quality indicators attained are:

Policy 1:	Year(t)	u_t	x_t^{I}	x_t^{II}	x_t^{III}	x_t^{IV}
	0	\$8,000,000				
	1	\$8,000,000	0.88	16.7	30	\$2,000,000
	2	\$8,000,000	0.88	16.7	30	\$2,000,000
	3		0.88	16.7	30	\$2,000,000

Policy 2:	Year(t)	u_t	x_t^{I}	x_t^{II}	x_t^{III}	x_t^{IV}
	0	\$7,500,000				
	1	\$8,000,000	0.83	15.6	28.1	\$1,875,000
	2	\$8,500,000	0.88	16.7	30	\$2,000,000
	3		0.94	17.7	31.9	\$2,125,000

Moreover, while evaluating the above two development trajectories, for simplicity and readability we will only take into account the *effectiveness* of development, and the *objective evaluation* only.

The consecutive fuzzy constraints and objective fuzzy subgoals are assumed piecewise linear, i.e. their definition requires two values only (cf. Figure 3.3, and Figure 4.1): the aspiration level (i.e. the fully acceptable value) and the lowest (or highest) possible (still acceptable) value which are:

t

0 C^0: $u_0^p = \$7,500,000$
$\quad\quad u_0^c = \$8,500,000$

1 C^1: $u_1^p = \$7,750,000$
$\quad\quad u_1^c = \$9,000,000$
$\quad\quad\quad\quad\quad\quad G_o^{1,I} : \underline{x}_1^I = 0.6 \quad\quad\quad\quad \overline{x}_1^I = 0.85$
$\quad\quad\quad\quad\quad\quad G_o^{1,II} : \underline{x}_1^{II} = 14 \quad\quad\quad\quad \overline{x}_1^{II} = 16$
$\quad\quad\quad\quad\quad\quad G_o^{1,III} : \underline{x}_1^{III} = 27 \quad\quad\quad\quad \overline{x}_1^{III} = 29$
$\quad\quad\quad\quad\quad\quad G_o^{1,IV} : \underline{x}_1^{IV} = \$1,800,000 \quad \overline{x}_1^{IV} = \$1,900,000$

2 C^2: $u_2^p = \$8,000,000$
$\quad\quad u_1^c = \$10,000,000$
$\quad\quad\quad\quad\quad\quad G_o^{2,I} : \underline{x}_2^I = 0.7 \quad\quad\quad\quad \overline{x}_1^I = 0.9$
$\quad\quad\quad\quad\quad\quad G_o^{2,II} : \underline{x}_2^{II} = 15 \quad\quad\quad\quad \overline{x}_1^{II} = 17$
$\quad\quad\quad\quad\quad\quad G_o^{2,III} : \underline{x}_2^{III} = 28 \quad\quad\quad\quad \overline{x}_1^{III} = 30$
$\quad\quad\quad\quad\quad\quad G_o^{2,IV} : \underline{x}_2^{IV} = \$1,900,000 \quad \overline{x}_2^{IV} = \$2,000,000$

3
$\quad\quad\quad\quad\quad\quad G_o^{3,I} : \underline{x}_3^I = 0.75 \quad\quad\quad\quad \overline{x}_3^I = 1$
$\quad\quad\quad\quad\quad\quad G_o^{3,II} : \underline{x}_3^{II} = 16 \quad\quad\quad\quad \overline{x}_1^{II} = 18.5$
$\quad\quad\quad\quad\quad\quad G_o^{3,III} : \underline{x}_3^{III} = 29 \quad\quad\quad\quad \overline{x}_1^{III} = 31$
$\quad\quad\quad\quad\quad\quad G_o^{3,IV} : \underline{x}_3^{IV} = \$1,950,000 \quad \overline{x}_3^{IV} = \$2,100,000$

Using the minimum operator ("$\wedge$") to reflect a safety-first attitude, which is clearly preferable in the situation considered (a rural region plagued by the aging of the society, out-migration to neighboring urban areas, economic decay, etc.), the evaluation of the two investment policies is:

– Policy 1

$$\mu_D(\$8,000,000; \$8,000,000; \$8,000,000 \mid .) =$$
$$= \mu_{C^0}(\$8,000,000) \wedge (\mu_{G_o^{1,I}}(0.88) \wedge$$
$$\wedge \mu_{G_o^{1,II}}(16.7) \wedge \mu_{G_o^{1,III}}(30) \wedge \mu_{G_o^{1,IV}}(\$2,000,000)) \wedge$$
$$\wedge \mu_{C^1}(\$8,000,000) \wedge (\mu_{G_o^{2,I}}(0.88) \wedge$$
$$\wedge \mu_{G_o^{2,II}}(16.7) \wedge \mu_{G_o^{2,III}}(30) \wedge \mu_{G_o^{2,IV}}(\$2,000,000)) \wedge$$
$$\wedge \mu_{C^2}(\$8,000,000) \wedge (\mu_{G_o^{3,I}}(0.88) \wedge$$
$$\wedge \mu_{G_o^{3,II}}(16.7) \wedge \mu_{G_o^{3,III}}(30) \wedge \mu_{G_o^{3,IV}}(\$2,000,000)) =$$
$$= 0.5 \wedge (1 \wedge 1 \wedge 1 \wedge 1) \wedge 0.8 \wedge$$
$$\wedge (0.9 \wedge 0.85 \wedge 1 \wedge 1) \wedge 1 \wedge (0.52 \wedge 0.28 \wedge 0.5 \wedge 0.33) =$$
$$= 0.5 \wedge 0.8 \wedge 0.28 = 0.28$$

– Policy 2

$$\mu_D(\$7,500,000; \$8,000,000; \$8,500,000 \mid .) =$$
$$= \mu_{C^0}(\$7,500,000) \wedge (\mu_{G_o^{1,I}}(0.83) \wedge$$
$$\wedge \mu_{G_o^{1,II}}(15.6) \wedge \mu_{G_o^{1,III}}(28.1) \wedge \mu_{G_o^{1,IV}}(\$1,875,000)) \wedge$$
$$\wedge \mu_{C^1}(\$8,000,000) \wedge (\mu_{G_o^{2,I}}(0.88) \wedge$$
$$\wedge \mu_{G_o^{2,II}}(16.7) \wedge \mu_{G_o^{2,III}}(30) \wedge \mu_{G_o^{2,IV}}(\$2,000,000)) \wedge$$
$$\wedge \mu_{C^2}(\$8,500,000) \wedge (\mu_{G_o^{3,I}}(0.94) \wedge$$
$$\wedge \mu_{G_o^{3,II}}(17.7) \wedge \mu_{G_o^{3,III}}(31.9) \wedge \mu_{G_o^{3,IV}}(\$2,125,000)) =$$
$$= 1 \wedge (0.92 \wedge 0.8 \wedge 0.55 \wedge 0.75) \wedge 0.8 \wedge$$
$$\wedge (0.9 \wedge 0.85 \wedge 1 \wedge 1) \wedge 0.75 \wedge (0.76 \wedge 0.68 \wedge 1 \wedge 1) =$$
$$= 0.55 \wedge 0.8 \wedge 0.68 = 0.55$$

Policy 2 is therefore better.

5. Concluding remarks

We have presented a fuzzy model of socioeconomic regional development planning. The model takes as its point of departure Bellman and Zadeh's (1970) general model of multistage decision making (control) under fuzziness further extended by Kacprzyk (1997). The development is considered from economic, social, infrastructural, etc. points of view. Both objective and subjective aspects are accounted for, the latter mainly in terms of resulting social satisfactions. The model may be viewed as an example of a "soft" approach to the modelling of a complex socioeconomic problem.

Bibliography

Baldwin J.F. and B.W. Pilsworth (1982) Dynamic programming for fuzzy systems with fuzzy environment. *J. of Math. Anal. and Appls* 85: 1–23.

Bellman R.E. and L.A. Zadeh (1970) Decision making in a fuzzy environment. *Management Sci.* 17: 141–164.

Esogbue A.O. (1983) Dynamic programming, fuzzy sets and the modelling of R&D management control systems. *IEEE Trans. on Systems, Man, and Cybernetics* SMC-13: 18–30.

Esogbue A.O. (1989) *Dynamic Programming for Optimal Water Resources Systems Analysis.* Prentice-Hall, Englewood Cliffs, NJ.

Francelin R.A. and F.A.C. Gomide (1993) A neural network for fuzzy decision making problems. *Proc. of 2nd IEEE Int. Conf. on Fuzzy Systems*–FUZZ-IEEE'93 (San Francisco), Vol. 1, pp. 655–660.

Francelin R.A., F.A.C. Gomide and J. Kacprzyk (1995) A class of neural networks for dynamic programming. *Proc. of 6th IFSA World Congress* (Saõ Paolo), Vol. II, pp. 221–224.

Kacprzyk J. (1978) A branch-and-bound algorithm for the multistage control of a nonfuzzy system in a fuzzy environment. *Control and Cybernetics* 7: 51–64.

Kacprzyk J. (1979) A branch-and-bound algorithm for the multistage control of a fuzzy system in a fuzzy environment. *Kybernetes* 8: 139–147.

Kacprzyk J. (1983) *Multistage Decision Making under Fuzziness*, Verlag TÜV Rheinland, Cologne.

Kacprzyk J. (1995a) Multistage control of a fuzzy system using a genetic algorithm. *Proc. of Int. Joint Conf. of 4th IEEE Int. Conf. on Fuzzy Systems and 2nd Int. Fuzzy Engineering Symp.* (FUZZ-IEEE/IFES'95) (Yokohama), Vol. III, pp. 1083–1088.

Kacprzyk J. (1995b) Multistage fuzzy control using a genetic algorithm. *Proc. of 6th IFSA World Congress* (Saõ Paolo, Brazil), Vol. II, pp. 225–228.

Kacprzyk J. (1995c) A modified genetic algorithm for multistage control of a fuzzy system. *Proc. of 3rd Europ. Congress on Intelligent Techniques and Soft Computing*-EUFIT'95 (Aachen), vol. 1, pp. 463–466.

Kacprzyk J. (1997) *Multistage Fuzzy Control.* Wiley, Chichester.

Kacprzyk J. and A.O. Esogbue (1996) Fuzzy dynamic programming: main developments and applications. *Fuzzy Sets and Systems* 81: 31–46.

Kacprzyk J. and A. Straszak (1982a) A fuzzy approach to the stability of integrated regional development. In G.E. Lasker (Ed.): *Applied Systems and Cybernetics*, Vol. 6, Pergamon Press, New York, pp. 2997–3004.

Kacprzyk J. and A. Straszak (1982b) Determination of 'stable' regional development trajectories via a fuzzy decision-making model. In R.R. Yager (Ed.): *Recent Developments in Fuzzy Sets and Possibility Theory*, Pergamon Press, New York, pp. 531–541

Kacprzyk J. and A. Straszak (1984) Determination of stable trajectories for integrated regional development using fuzzy decision models. *IEEE Trans. on Systems, Man and Cybernetics* SMC-14: 310–313.

Nijkamp P. (1986) *Handbook of Regional and Urban Economics*. North-Hoilland, Amsterdam.

Yager R.R. (1988) On ordered weighted averaging aggregation operators in multi-criteria decision making. *IEEE Transactions on Systems, Man and Cybernetics*, SMC-18: 183–190.

Yager R.R. and J. Kacprzyk (1997) *The Ordered Weighted Averaging Operators: Theory and Applications*. Kluwer, Boston.

Part 2

Fuzzy System Design
for Social Applications

FUZZY LOGIC AS AN EVOCATIVE FRAMEWORK FOR STUDYING SOCIAL SYSTEMS

Robert Woog[1], Vladimir Dimitrov[2] and Lesley Kuhn-White[1]

[1]School of Social Ecology
[2]Centre for Research in Healthy Futures
University of Western Sydney
Hawkesbury, Bourke St.
Richmond NSW2753
Australia
E-mail: {r.woog,v.dimitrov,l.kuhnwhite}@uws.edu.au

Abstract. This study is about the fuzzy and unpredictable nature of social life. Two themes are developed. The first describes and critiques a methodological approach to social systems intervention for improvement. The second theme comprises a propositional questioning about the nature of social systems from a fuzzy logic framework. The crucial role of fuzzy logic is demonstrated in such fields of social inquiry as conversation mapping, heuristic pattern formation, emergence of meaning, multy-layered interpretation, study of temporality and non-foundational thinking.

Keywords: fuzzy logic, social complexity, emergence of meaning, value dark zone, divergence syndrome, temporality, phronesis

1. Introduction

As social/community researchers, we have plied our practice and have learnt from our experiences in many countries, with different social groups and a diversity of cultures. We have worked with vastly dissimilar corporate organisational groups. Our research has been in fields as diverse as: participative community based rural adjustment in the Australian wheat and dairy industries [1, 2]; management of the learning of senior executives undergoing and managing change in multi national corporations [3]; and, facilitation of the resolution of conflict between nomadic tribes and government bureaucracy [4].

We now look back and make sense of past events, proposing generalisations about what we have found to be characteristic in our interventions. Projections are offered about the future with a sense of optimism and excitement, and strategies for meaningful intervention and development are proposed.

From 'Crisp' Technology to Fuzzy Social Systems

Fuzzy set theory, or *fuzzy logic*, first proposed by Zadeh in 1965 [5], represents an attempt to construct a conceptual framework for the systemic treatment of vagueness and uncertainty both qualitatively and quantitatively. Fuzzy logic has developed many branches, such as, fuzzy arithmetic, fuzzy typology, fuzzy graph theory, fuzzy mathematical programming and fuzzy data analysis. Any 'crisp' theory, according to Zadeh, may be 'fuzzified' by generalising the concept of a set within that theory, to the concept of a fuzzy set. Zadeh goes on to describe the benefits of this 'fuzzification':

The impetus for the transition from a crisp theory to a fuzzy one derives from the fact that both the generality of a theory and its applicability to real-world problems are substantially enhanced by replacing the concept of a set with that of a fuzzy set [6].

In the social sciences, fuzzy logic was first applied to the problem of social choice and self organisation in the early 1970's [7, 8, 9, 10, 11]. The application of fuzzy logic to social systems creates opportunities to examine:

- contradictions and inconsistencies pre-emtive to the emergence of critical social situations
- issues that have been repressed under critical social dynamics
- that which is concealed and beyond observed social phenomena.

Fuzzy logic is suited to studying such subtleties in social systems because of its ability to:

- deal with vague, ambiguos and uncertain qualitative ideas and judgements
- concentrate on paradoxical and enigmatic aspects of decision situations
- focus on the margins of any decision making 'space'
- appreciate the uniqueness in any decision making act.

Fuzzy logic based social analysis parallels the deconstructivist approach characteristic of post-modern intertextuality [12]. Fuzzy logic provides an alternative way of understanding uncertainty. From this new way of understanding can be derived innovative approaches and strategies for working with the uncertainty that so often characterises social systems.

2. Fuzzy Logic and Three Forms of Social Inquiry

In this section we describe our initial thinking in terms of the application of fuzzy logic to three fields of social inquiry commonly encountered: community inquiry, social and technological conflict, and ideological conflict.

2.1 Community Inquiry

Community inquiry, held within a variety of communal settings, is often framed in terms of a feeling that there is nothing specifically wrong, but that there is scope for improvement [13]. If only the issues were better understood by the stakeholders and the direction of improvement better identified and agreed upon, the situation could be beneficially developed. The context is always multi-variate and uncertain, involving tangled socio-political and socio-technical interactions. There is often a mismatch between the presenting issues and the prevailing theories; a mismatch not often overtly recognised, let alone explored. Utilisation of the paradigm of fuzzy logic enables such situations to be investigated. Fuzzy logic, in allowing the simultaneous expression of contradictory statements provides a useful communicative framework, albeit a framework full of opposing forces.

2.2 Social and Technological Conflict

In this situation there is a clash of approaches to a problematic situation. The planned implementation of what are thought to be technically feasible and desirable strategies conflicts with the socially constructed norms and priorities of the broader community. The context for the development of these strategies is usually characterised by a fragmented rationalistic approach to problem solving which sees problems as being best understood by reducing them to their smallest rational parts. While the ideas are powerful and the arguments compelling these strategies are fundamentally limited in scope and non-systemic. The framework of fuzzy logic enables problems to be approached in their wholeness, without imposing needless premature reduction nor risking becoming lost in their complexity [14]. We consider it is useful to approach such situations without seeking exactness (as is presumed to result from a reductionist approach) and to remind ourselves that often only approximations are possible.

2.3 Ideological Conflict

Often individuals or groups take strong adversarial positions based on vastly different world views, which are then forcefully put and vigorously defended. The crux of the problem in this situation often lies in the mistaken belief that the knowledge and information available to a particular person or group about the issue unequivocally validates the position they take. The 'other side' is construed as either poorly or misinformed [15]. The main theorem of fuzzy logic denies absolutism, instead arguing that what we know is true and false in parallel - true up to some degree and also false up to some degree. Such an approach provides a means of transcending ideological conflicts. We take the position that our knowledge of social systems is always partial and at best it represents a useful approximation. The framework of fuzzy logic in handling the coexistence of opposing forces allows individual views to be retained in spite of the ambivalence that they bring to the collective view. An enabling change in emphasis then becomes possible in working with such ideological conflicts. The interpretive transition in this regard, is away from correctness, from the presumed clarity of 'right' and 'wrong', towards accommodating the vital ambiguity of reality.

3. A Crucial Role for Fuzzy Logic in Social Inquiry

We propose that notions of causality, predictability and equilibrium that so often form major pillars of social research methodology are proving to be inadequate as a means of researching social life. Under 'normal' circumstances, social systems do not exist as an equilibrium. A fuzzy logic approach to social analysis by way of contrast, does not begin with this assumption, but with an awareness and acceptance of the chaotic unpredictability of social life.

One of the major contributing strengths of a fuzzy logic approach to social inquiry is its ability to cope with multiple constructions of reality, as well as other forms of circumstantial and constructed complexity (such as changes in political or economic circumstances) by describing social systems as normally in this dynamic state.

Fuzzy logic can accommodate a variety of methodologies be they directed towards analysis, or intervention for improvement. This compatibility with a variety of methodologies may be explained by fuzzy logic's inherent tolerance for imprecision and temporality. With fuzzy logic, paradox and contradiction, so often considered as unsettling and unmanageable in social inquiry, are legitimised as the normal rather than aberrant conditions of the system. These characteristics are viewed as inviting inquiry and action, and may even be accepted as legitimate outcomes of the intervention. Fuzzy logic allows social realities to be what they are: pluralistic, dynamic and ambiguous. Social realities can be viewed as vivid representations of how people live in and make sense of their individual and collective worlds, simultaneously participating in many co-generative change processes. Fuzzy logic warns us that in social analysis, for every well defined approach for dealing with a problem, there will be an opposite and equally well defined approach possible. In the application to research/inquiry methodology, fuzzy logic helps us to:

- avoid absolute statements and deny the legitimacy of all dichotomies
- embrace ambiguity and ambivalence, and accept processes without finality and completion
- seek freedom from the myopia of hyper-determined research projects and formulations, and be aware of the tentativeness of causal explanations.

Social Complexity: Ethics, Aesthetics and Phronesis

Generally speaking we consider that in social research attempts have been made to recreate the scenario that occurs in researching with non-human objects of research. Parameters are drawn to restrict and control as much as possible complicating variables. The variability of free will and the relative social independence of human decision making and creativity however remain to be dealt with.

In our research, we purposefully aim to preserve concern with ethics and aesthetics which we view as undergirding all human behaviour. We believe that in social research there should be a balance between individual needs, systemic trends and associated needs as well as deep intrinsically spiritual concerns, and concerns of an applied nature.

It is not surprising then that we include concern with *phronesis* - 'practical wisdom' [16] in our researching. What is surprising however, is that so often ethics, aesthetics and phronesis are sought to be deliberately left out of the research process. The reason for this we believe, is that these concerns inevitably introduce an element of irritating unmanagability to social analysis. Phronesis is hard to cope with methodologically and cognitively, given our deterministic, reductionist, and 'always directed towards the best solution' approach to social systems and their improvement. Fuzzy logic in contrast, adopts as 'normal' the variability, uncertainty and complexity that is inherent in the expression of social phronesis.

4. Research Approach

Our research approach utilises critical conversation as the means of generating interpretive, contextually rich data. Our style of inquiry falls within the broad descriptive parameters of constructivist participative inquiry. We find that when people dialogue together, there is a revelation of the paradoxical, complex and fuzzy nature of humankind. This nature is expressed by means of our 'languaging' activities, and indeed is inherent in language itself.

We postulate that *words are useful only in so far as their meaning is fuzzy*. If the meaning of words was to be absolutely clear and distinct, it would be difficult, if not impossible for communication to occur between people with differing experiences. So paradoxical, so rich, so uncertain and multifaceted is this reality that 'precise' descriptions, even if thought possible, turn out to be senseless. For this reason, our major inquiry approach is the mapping of the fuzzy narratives that people use to make sense of and describe their experience of reality.

4.1 A Starting Point

We argue that a methodology for inquiry into chaotic systems needs to be developed, rather than proceeding by attempting to represent such systems as if they were non chaotic and more stable and predictable than they are.

The metaphoric term *value dark zone* is used to describe the complex, puzzling, ill understood and mysterious character which forms the often unacknowledged starting point of much social inquiry. As researchers, although we are motivated by a desire to understand, we do so with a sense of humility and awe when faced by the presenting complexity and fuzziness of life. It is agreed and accepted from the beginning that instead of seeking to reduce and control complexity we will work with this and retain the sense of mystery. In this regard we become contributers to, as well as inquirers into, the multi-faceted construction that is reality.

4.2 Mapping the Conversation

We begin by mapping in a variety ways conversations between the participants in the inquiry. It is at this stage that the heuristic nature of the methodology comes to the fore. Thematic 'fractal' patterns of dialogue emerge and are recorded. The fractal lines reveal the unpredictable, somewhat chaotic directions that dialogue takes when contextual boundaries are loosely set. The unfolding dynamics of the conversation provides a glimpse of the way in which conversations are saturated with a plethora of perspectives. Our experience has shown that along the fractal lines, nodes of conflict and consensus may be identified.

We do not aim to seek clear explanations and precise answers either from the thematic fractal lines, nor even from the overall map. Rather, we view this mapping as a means of promoting a climate of inquiry which engenders fresh insights and interpretations. We find that individual and collective understanding comes from the interpretive reflection of all participants involved following their examination of the mapped dynamics of the conversation. Reflection ranges from the level of cognition (knowing about; engagement and understanding of the presenting issues at a somewhat superficial level) to meta-cognition (knowing about the factors affecting our sense making of the situation) and from meta-cognition to epistemic cognition (knowing about our means of knowing and perceiving). With this approach fuzziness endures and serves to explain.

We have learnt to look for specific indicators ('value rich zones') wherein there is a plethora of issues of portent and substance. Such areas of controversy, which we term 'nodes of consensus and conflict' are relatively easy to identify. These nodes are useful for indicating where to pursue further exploration and interpretation. Encouraging dialogue in these areas often generates higher order interpretations and understanding. Analytical exploration of the nodes coupled with heightened understanding often leads to specific recommendations for actions and improvement. It is from this process that the 'outputs' of the inquiry are generated.

4.3 The Formation of Heuristic Patterns

Usually, the fractal lines of dialogue conclude with a paradoxical statement or identification of a paradoxical position. These end points form what we term as 'peripheral paradoxes' which when joined together allow us to define the fuzzy boundary and set the conceptual and contextual field of the inquiry. The peripheral paradoxes accommodate and contain within them the seeds of conflict and consensus already identified as nodes along the thematic fractal lines.

Peripheral paradoxes lead to higher order understanding more in the realm of ontology and epistemology, rather than acting as the basis of applied practical explanations.

A cumulative systemic exploration of the fuzzy patterns of the peripheral paradoxes may and should lead to a profound re-thinking of the issues and understandings, the world views and positions taken by the participants. These perspective rich explorations provide the inquiry 'outcomes'. This process, be it tentative and humble, brings some illumination to 'value dark zones'. This process begins to unravel the mystery of what drives the participants to think and do in different ways and to behave in ways others might label as inconsistent. Such unravelling allows us to work with and be less fearful of mystery and complexity.

4.4 The Emergence of Meaning

As we have implemented our approach to social and community based inquiry projects, we have come across applied and conceptual problems regarding meaning construction.

The first problem concerns the influence of the epistemological perspective of the researchers on the process and outcomes of the research. We recognise that we have to be aware of our epistemological position because how we think and what we believe in influences how we research and ultimately, in constructivist terms, what we find.

Despite declaring that our approach acknowledges the inherently unpredictable complex and at times chaotic nature of social systems, we recognise that we seek to develop more or less tangible thematic outcomes. We may be predisposed to notice such thematic outcomes because of the way in which the research objectives have been conceptualised and expressed. Particular thematic outcomes may also result from common but not necessarily predominant mind sets of what is conceived of as normal, desirable or good.

If the goal of the research is known or assumed through commitment to a desired state, then the process of interpretation can end-up in a *creation of meaning* which is manifestly goal-oriented. The surprising and the emergent may be overlooked via this process.

Both an emphasis on goal oriented outcomes and a predisposition for macro level systemic interpretation, results in much of the individual richness evident in the fractal patterns at a micro level being ignored. With a systems approach in particular, we end up seeing the whole or parts of the whole system in terms of the goals as stated in the research submission or framed in ways that are generally comfortable to us and our near partners.

What appears to emerge from this form of systems inquiry is an accurate representation of community issues, needs and priorities. However, a more critical interpretation would suggest that what is identified is usually a construction based either upon scientific, or other expert opinion. The resultant creation of meaning in this case could be described as an expert systems oriented creation. A corollary problem may then become, that during the process of research the interpretive meanings created, develop a presumed credibility, certainty and predictability. Meaning creation ceases to be seen as a product of the research or as the production of useful transient metaphors for sense-making. Rather the meaning created is seen to be, and begins to be talked about and acted upon, as if it was *the* representation of the world of reality.

The process of creating meaning in social/community research is, we believe, more tenuous, fragile and troublesome than generally acknowledged. Specifically, we have observed three conceptual problems common to the creation of meaning in community inquiry. The problems are that:

1. Temporality is ignored. It is forgotten that self-organising systems are dynamic and move towards a critical state.
2. Useful but propositional explanations begin to be used more and more as authoritative theories which eventually gain a foundational status [17].
3. Individuality is neglected. Preoccupation with repeatable, predictable explanations and foundational certainty contributes to this neglect, as does emphasising the whole rather than the parts as in a systemic approach.

The association between these three problems and meaning making as a research process is explored below.

4.4.1 A Reminder of Temporality: the Divergence Syndrome

Our perspective is that the social complexity of life is changing at an accelerating rate and that the intensity and direction of change will continue to be beyond the ability of researchers and society itself, to predict and control.

Chaos theory applied to social systems builds a general picture of society as being constituted by interactive fuzzy dynamics giving rise to a range of emerging types of social behaviour. We observe that a multitude of transient (temporary) equilibrium emerge in social dynamics, that there is a constant evocation of increasingly complex forms of order in evolving systems in the life of the society, and that spontaneity is inherent in the temporality of this life. All of these effects are a manifestation of what we term 'the divergence syndrome'[18] where behaviours may change as result of extreme sensitivity to initial conditions (a very tiny change in the initial stimulus may rapidly give rise to an enormous change in the outcome). The divergence syndrome describes the 'self-feeding' chaotic acceleration of energy flow that takes place in human systems.

4.4.2 Foundational Thinking

Foundational thinking refers to the way in which humans commonly attempt to base their ideas and theories on what are thought of as indubitable foundations of certainty; foundations seemingly located outside of human meaning construction. It appears to be a predisposition of humans to search for certainty when transience is the prevailing condition in dynamic self-organising systems. Foundational thinking usually manifests as a search for understanding, stability and certainty in inherently complex unstable systems. Elaborate theories are built upon beliefs to support an inner human need for certainty. This predisposition for foundational thinking, we think, better reveals humans as creatures of faith, who although desiring certainty, may only ever take as a matter of faith their theories.

4.4.3 Reclaiming Individuality

Individuality has been subordinated to the greater system which has collectively reduced community to mere society: the social system.

What if contemporary history tells us that the crowd, mass movements, the influence of collective responsibility or the collective response to the social environment is at best inadequate or at worst responsible for all manner of evil action? What if deep history tells us that persistence is concealed but carried in radical changes in individuals? What if we were to shift from fascination with the whole to consideration of the parts? What if we were to make the speculation, which would appear monstrous to systems thinkers, that the integrity of the parts seen in dynamics is something greater than the whole built on these parts?

In developing such arguments, attention is directed away from the systemic whole towards its constituents (the parts). The development of the whole is entirely dependent on the ability of each individual part to develop and grow; it is the part as primordial bearer of creative potential energy that can support the existence and growth of the whole - in this sense, we say that the sum of the individual parts can be greater than the whole. Following the growth of its constituents, the whole also changes and grows. In each moment of time the whole can be more than the whole considered a moment ago. In the framework of fuzzy logic these seemingly monstrous statements have a simple and transparent explanation: the whole exists based on the creative support of its parts in so far as the parts exist because of their interconnectedness manifested as the whole.

4.5 Multi-layered Interpretations

The divergence syndrome and non foundationalism warn of the risk of misinterpretation of complex social systems. Meaning can be different depending upon whom is doing the meaning making, but also can differ according to the scale which is used to set the limits of interpretation.

A social system considered within a short time interval with all of its details of individuality, may appear cluttered and murky. In the process of moving to focus at a more understandable generalisable level, the so called 'individual' or insignificant details may be brushed away. However, the abstract generalisable interpretation may remain incomplete and the potential for improvement, development or other form of positive intervention is abrogated if we are unable to deal simultaneously with the details.

Our dilemma concerns the oscillation between models of reality and assumed certainty, and the reality of temporality and obdurate individuality. Neither perspective is adequate in itself. Rather, what is needed is ongoing oscillation of interpretation between two mirror images of knowing: *the knowing of certainty and the uncertainty of knowing.*

The interpretation of social systems should take place simultaneously at multiple levels of granulation. Sense making may be studied at different levels of focus. With a detailed, micro focus, such as that of structural analysis, there is revealed in relation to the sense making process, the individual's predictive interpretive processes.

At a macro focus, collective generalisations and interpretations may be taken to reveal and represent community and societal values. At the macro level we should allow for the identification of macro patterns recognising that no further detail can be discovered because further details are smaller than is allowed by the particular measure of resolution.

For micro patterns the converse applies; we should recognise we are deliberately focussing on the minute, and that at this level of resolution macro patterns will not be apparent.

5. Conclusion

What we have proposed is a form of social inquiry that metaphorically resembles a vortex. The key elements inherent in this vortex approach to social inquiry are attention to:

1. Changes in granulation: alternations in focus between generality of the whole and singularity of the part.
2. Dynamics: capturing the idea that changes in focus and scale are constantly occuring.
3. Emergence: manifestation of human potential for transformation and transcendency.

Finally, a number of observations regarding the application of fuzzy logic to social research may be highlighted. While not arguing for these on the basis only of strong rational evidence, we consider that on the basis of our experiences, the following interesting and innovative proposals are worthy of consideration. We advocate that the researchers should:

- Allow the process of inquiry into social situations to be permissive, liberal and multi disciplinary; promiscuous.
- Work with a sense of surprise; be prepared for the emergent, and be tolerant and alert towards the unexpected.
- Be deeply reflective, and prepared to learn at applied, meta, and epistemic levels.
- Seek to understand and be prepared to intervene at each of these levels of learning.
- Hold their ideas, including those proposed in this paper lightly - with less dogma and more in a spirit of exploration. Retaining humility they should be prepared to pass on.

Fuzzy logic would suggest and demand as much.

References

1. Woog, R. A., Kelleher F. M., Turner A. S. and Blaydon D. L. (1993) Improving Research Adoption By Wheat Producers, Research Report, University of Western Sydney-Hawkesbury.
2. Kuhn-White L. J., Gamble D. and Blunden S. (1993) An Examination of the Epistemological Foundations of an Action-Research Project: The Transfer of the Family Farm Business. Paper presented to „Constructivism: The Intersection of Disciplines", Deakin University, Victoria, Research report, University of Western Sydney-Hawkesbury.
3. Woog R. A. and Bawden R. (1994) All You Ever Needed to Learn About Management in Your MBA But Probably Didn't, Proceedings of the Australian Management Congress for Management Development Practitioners.

4. Khatoonabadi A. and Woog R. A. (1992) Drama as a Researching Tool With Nomadic People, Proceedings of the International Conference, Nomadism and Development: Survival Strategies and Development Policies. Isfahan, Iran.

5. Zadeh L. (1965) Fuzzy Sets, Information and Control, Vol 8, pp. 338-359.

6. Zadeh L. (1994) Neural Networks and Soft Computing, Communications of the ACM, Vol 37, No.3, pp. 77-84.

7. Dimitrov V. (1970) Heuristic Generator in Human-Operator Systems, Technical Cybernetics, Vol 2, 1970 (published in Russian by the Institute of Cybernetics of the Ukranian Academy of Sciences, Kiev).

8. Barnev P., Dimitrov V. and Stanchev P. (1974) Fuzzy System Approach to Decision-Making Based on Public Opinion: Investigation Through Questionnaires, Proceedings of the Stochastic Control Symposium, Budapest, Hungary, pp. 637-642.

9. Dimitrov V. and Wechler W. (1975) Optimal Fuzzy Control of Humanistic Systems, Proceedings of the Sixth World Congress of IFAC, Boston.

10. Dimitrov V. (1976) Social Choice and Self Organisation Under Fuzzy Management, Kybernetes, Thales Publ.Ltd., vol. 6, pp.153-156.

11. Dimitrov V. (1983) Group Choice under Fuzzy Information, Fuzzy Sets and Systems, North Holl.Publ.Co., vol.9. pp. 25-39.

12. Culler J (1982) On Deconstruction: Theory and Criticism after Structuralism, Cornell University Press.

13. Woog R. A. (1994) Understanding the Dairy Industry: Knowledge, Beliefs and Values, Research Report, University of Western Sydney-Hawkesbury.

14. Woog R. A. and Dimitrov J. (1994) Social Ecology: A Post Modernist Neo-Positivist Methodology, Humanecologi, vol 13, pp. 7-11.

15. Bawden R. and Woog R. A. (1994) Changing Minds For Changing Ways: Hawkesbury's Quest For Better Agriculture. Proceedings of Congress De L'Orde Des Agronomes Du Quebec.

16. Guthrie W. K. C. (1981) A History of Greek Philosophy Vol. VI, Cambridge University Press.17. Kuhn-White L., Woog R. and Dimitrov V. (1996) An Exploration of Human Knowing: From Beyond the Foundational Position on Towars Humility Through Irritating Ideas, presented at The 1996 Scholarly Conference for Post-Graduate Students and Academics, Sydney, July 17-20.

17. Dimitrov V., Woog R. and Kuhn-White L. (1996) The Divergence Syndrome in Social Systems, Complex Systems 96, IOS Press Amsterdam, Netherlands, pp. 142-146.

FUZZY LOGIC AND THE MANAGEMENT OF SOCIAL COMPLEXITY

Vladimir Dimitrov[1] and Kalevi Kopra[2]

[1]Centre for Research in Healthy Futures
[2]School of Social Ecology
University of Western Sydney
Hawkesbury, Bourke St.
Richmond 2753
Australia
E-mail: V.Dimitrov@uws.edu.au

Abstract. To manage social complexity means to manage: (1) the paradoxes inherent in social systems and their effects, thus avoiding the danger of double bind paralysis, and (2) the chaotic dynamics of social life as well as the effects of these dynamics, thus avoiding the danger of destruction and collapse. Fuzzy Logic helps in describing, analysing, understanding and eventually dealing with the paradoxical and chaotic nature of social systems. Research results related to a study of a fuzzy logic based consensus seeking enterprise are discussed in detail. It is shown how this enterprise can help people to deal with six dangerous maladies of today's society. One of the maladies relates to the 'butterfly effect' manifested in politician narratives: a subtle variation in the use of fuzzy hedges creates the potential for a wide variety of interpretations reflecting divergent, even opposite, socio-economic consequences. Fuzzy logic also offers a useful framework for understanding the oscillations between disaggregation and communion in society: what matters in a 'healthy community' is the mutual acceptance of people such as they are. This is the starting point for any consensus seeking process.

Keywords: social complexity, consensus seeking enterprise, fuzzy management, social maladies, 'butterfly effect' in narratives

1. Introduction: Paradoxical and Chaotic Nature of Real-Life Complexity

Human concepts, opinions, judgements and expectations change and evolve in the dynamics of real-life complexity. This complexity appears paradoxical and chaotic. Paradoxical, because it is the source of many contradictory and opposing forces acting together, and at the same time, the product of the forces. Chaotic, because its dynamics are both unpredictable and sensitive to changes in the magnitude and location of these forces.

Social complexity is strongly influenced by the paradox of independency: it is only when social and ecological dependencies are established that the interdependence emerges, and it is this collective interdependence (between people and between people and their environment) that provides the notion of individual independence with meaning.

Complexity of social systems is characterised by the difference-similarity paradox: while differences in interests and values are important for the survival of any human system, they have meaning only because of the similarities that also exist and provide a basis for any collective endeavour.

Another basic paradox inherent to social complexity is the paradox of self-knowing: we cannot understand ourselves without understanding others; at the same time, we cannot accumulate comprehensive knowledge about others without constantly trying to understand ourselves. This paradox reflexes human nature - we are both creators and products of social reality at the same time.

Paradoxical (contradictory and self-reflexive) forces - emotions, thoughts and actions coexist inside every human being, every group of people, every society. As a rule, any attempts to unravel these forces create a circular process that can paralyse further individual or social actions. "The more the members of a group seek to pull the contradictions apart, to separate them so that they will not be experienced as contradictory, the more enmeshed they become in the self-referential binds of paradox" [1]:

To manage social complexity means to manage its inherent paradoxes and their effects avoiding the danger of double bind paralysis.

The dynamics of complexity are chaotic in the sense that the aggregate fluctuations of any complex social process (or behaviour) represent an endogenous phenomenon that persists even in the absence of 'stochastic shocks'. The emergence of complex irregular behaviour depends on both the initial conditions under which the process dynamics evolve and the critical values of the parameters characterising this evolvement.

Every time we deal with mathematical representations of social reality as a whole, we deal with chaos. Closed social systems are dissipative; their dynamics are described by strange attractors. Open social systems are symplectic [2]; their dynamics are chaotic without the occurrence of strange attractors.

Emergent phenomena are typical of the dynamics of complexity. In social systems emergent phenomena manifest themselves any time when collective behaviour transcends the behaviour of its 'components' behaviour; because it transcends individual behaviour, it cannot be 'in' it before the emergence takes place.

To manage social complexity means to manage its chaotic dynamics as well as its effects, thus avoiding the danger of destruction and collapse.

2. Role of Fuzzy Logic in Managing Social Complexity

With Fuzzy Logic we can describe, analyse, understand and eventually manage the paradoxical and chaotic nature of complexity.

<u>How does Fuzzy Logic help deal with social paradoxes?</u>

Simply by tolerating opposites, by balancing them to such degrees that they cease to cancel each other out, and become complementary. With Fuzzy Logic we can create an imprecise and easy to re-shape and modify framework in which the "either/or" approach to the contradictory concepts expressed in the paradox is replaced by a "both/and" relationship of their parallel acceptance. For example, a fuzzy framework created in the management practice can meaningfully transform expressions like "collaboration OR competitiveness" into "collaboration AND competitiveness", "re-organisation or stability" into "re-organisation AND stability".

The 'independency paradox' becomes easily manageable any time we apply the following fuzzy rule:

IF *there is interdependence between A and B* AND *their relationship is one of a* <u>*high enough*</u> *degree of trust, mutual understanding and tolerance* THEN *both A and B are able to act* <u>*quite*</u> *independently.*

The 'difference-similarities' paradox is handled without any problem, if both similarities and differences are balanced by fuzzy logic in a way that excludes too much emphasis on them. When a suitable fuzzy framework is created, differences between people and the viewpoints they represent become complementary and necessary to one another, instead of being contradictory and opposite. In this way, the difference-similarity paradox simply is dissolved. One can find many examples of this transformation in the practice of negotiation.

Considered in various fuzzy frameworks, the paradox of self-knowing opens up an array of opportunities for grasping better the richness of human experience: by trying to understand as much as possible how and why people behave in a certain way in situations that have never happened to us, we increase substantially our preparedness to cope well with such situations when they happen. Partly, this is because the improved understanding of each others' points of view enables us to 'turn' dependence into interdependence, and through that into a wider scope of actions.

<u>How does Fuzzy Logic help deal with chaotic patterns of behaviour in social systems?</u>

By generating fuzzy rules (heuristics) which focus on the local, the decentred and the marginal in social behaviour. Often what is excluded, what is not seen in the patterns, 'what is not there' becomes important in constructing a list of fuzzy rules to describe and grasp 'what is there', what is in the core of the system, what makes the system chaotic.

A chaotic social reality paradoxically seems to unite people in a common desire to be different, unique and creative. Plural descriptions, open for change, based on people's personal experience and shared through dialogue, or derived by collaborative inquiry, are meaningful in such reality - they help to disclose the tensions in it, not to resolve them, but to examine the contradictions and inconsistencies, and by the same token to reveal the conditions which could trigger the emergence of some kind of order (transitory though it usually is) in social systems.

The principle of non-exclusion and non-isolation is fundamental for the use of fuzzy logic in any turbulent social (or socio-ecological) environment.

Non-exclusion means that no options or alternative, however improbable it seems to be for inclusion in the future scenarios, should be excluded from consideration; it might turn out to be of crucial significance for the survival of society and its environment.

Non-isolation means that chaotic behaviour does not privilege any economic optimality: to isolate only one option, alternative or strategy by describing it as the best, the optimal, or the most efficient, from whatever point of view, is senseless; turbulent dynamics do not tolerate any pre-imposed isolation, however 'optimal' it may appear to the decision-maker.

The paradoxical and chaotic nature of social reality causes a great deal of uncertainty and vagueness in human decision-making. Under conditions of uncertainty and vagueness, when no ultimate answers or best solutions exist, the *search* for understanding and consensus between people becomes crucial for the management of social complexity.

2.1 Fuzzy Logic based consensus seeking enterprise

In the turbidity of human interactions consensus ceases to be a peaceful long-term commonality of stakeholders' interests. Such commonality grows on determinacy and stability. Unfortunately, neither determinacy nor stability are features of social reality. The more we reach for commonality in human interactions, the farther away it seems to be.

'Consensus is a horizon that is never reached' [3].

An irreducible indeterminacy constantly emerges when we explore more deeply both variety and uncertainty of group decision-making. Paradoxically, instead of consensus being the power house of common social action, it is 'dissensus' which operates in consensus seeking enterprise, permanently implanting chaotic vibrations in the process of communication.

However the chaos does not cause the communication network to dissipate. Rather, it eventually gives birth to an emerging order in the form of a new type of dynamic consensus between stakeholders: consensus for seeking a consensus.

2.2 Second order consensus

This can be defined as *'second order consensus'* [4] - people 'agree to disagree' or 'disagree to agree'. They seek consensus by exploring different ways that might lead to mutual understanding and preparedness to move together, to make the next step into the fuzziness of common expectations.

It does not matter that consensus in our society is 'condemned' to be momentary and transient - what can endure in time is human anticipation and aspiration, the impulse to act together, the natural desire to interact and communicate, to share with and care for others. In other words, not only a search for common actualisation of meaning but strong
emotional factors (sharing and caring) catalyse the emergence of second-order consensus out of the chaos of dissent and disagreement, contradictions and conflict.

Consensus seeking differs from *consensus building*. When seeking consensus stakeholders do not necessarily look for a 'common ground' in the form of full agreement. On the contrary, they underline and study the differences between them, trying to understand social mechanisms which make stakeholders differ in their interests, values, goals, etc. No constraints on stakeholders' views and opinions, no changes of their values and beliefs are required as preliminary conditions for seeking a consensus.

The process is entirely open for emergence of new features and unpredictable situations - *spontaneity* is an important characteristic of this process. No preliminary assigned goals exist - pre-imposed goals, constraints or requirements can narrow the scope of the stakeholders' search.

The search for consensus is motivated by the stakeholders' drive to be *mutually complementary* in their efforts to more fully understand the complexity of the issues and of their concern, to find out how to act together in order to benefit from the differences in their knowledge. While conducting their inquiry, the stakeholders are aware of the irreducible fuzziness and uncertainty of this knowledge, yet they agree to explore it together, and *construct* it anew. Thus, a second order, dynamic type consensus emerges. This kind of consensus is not simply an overlap of stakeholders' interests, values, goals, positions, views, etc.

Second order consensus means that there is a shared acknowledgment that there are diverse, changing and only partially shared views about complexity among the stakeholders, that it is full of zones of uncertainty (*'value-dark zones'*) in which neither the causes nor the effects of what occurs is clear or even can be known. Also, there is an agreement to explore the complexity together in order to arrive at a better understanding of it by using not only your own but each other's experience, expertise and ideas, and, through this better understanding, arrived at an improved *preparedness to act together*, i.e. to engage in joint, collaborative action to manage the complexity.

What occurs in the zones of uncertainty is at least in part influenced by the joint action of the participants (stakeholders), and so the complexity is being partly made by them. On the other hand, as the complexity evolves in time, it also exerts an influence on the stakeholders, and trigger them to reconstruct their views. What the new complex situation will be, contains both intentional and unintentional elements as the participants in the joint action coordinate their activities and respond to each other's constructions of the reality of the situation and to each others' actions in it [5,6].

2.3 Preparedness to Act Together

Stakeholder's preparedness to act together (i.e. 'consensus for seeking a consensus') can be expressed as a fuzzy composition of three major components:

* willingness to engage in dialogue (*communicativeness*)
* deservingness of trust or confidence (*trustworthiness*)
* ability to create options (*creativity*).

The willingness to engage in dialogue (both to listen and to converse) implies willingness to acknowledge the validity of different statements or positions (on an issue of stakeholders' concern) - not closure in a pre-defined rigid conceptual framework but search for a fuzzy logic based context, where stakeholders' intentions and anticipations have fuzzy, easily changeable formulations - able to be re-shaped, to move into opposites, to be set aside or to grow within a tree-like prolific structure.

What matters in the streamline of the dialogue is stakeholders' willingness to keep moving together - to explore options for consensus, to share knowledge and experience, to learn together how to create and implement group decisions when tolerating, appreciating, and even 'celebrating' the differences in people's thoughts and actions.

The second, the trustworthiness is renewed in acting together; it ceases to be a derivative of the past only, and appears as a property of stakeholders' involvement in collaborative activity, based on their shared responsibility and accountability.

The third, the creative urge (creativity) is a psychological process that involves innovative use of fresh ideas and new formulations (including use of metaphors, metaphorical imagery, metonymy, synecdoche, paradoxes, humour, jokes, story telling, etc.) helping stakeholders not only to enrich and extend variety of their potential options for consent, but also to discern the sense of being in a group, the sense of 'moving' and acting together in a world where relativity, complexity and uncertainty are inevitable companions.

In the free flow of unanchored, constantly changing and shifting individual values, beliefs, and expectations, the higher the degree of willingness for dialogue, trustworthiness, and creativity, as expressed by a group of stakeholders, the higher their preparedness to act together.

Fuzzy Logic helps to transforms the group profiles of the above consensus seeking parameters into characteristics of stakeholders' preparedness to take actions together (PAT).

The fuzzy logic rules are of the following form:

IF D = "low" AND T = "low" AND C = "low", THEN PAT = "low"

IF D = "low" AND T = "low" AND C = "moderate", THEN PAT = "low"

IF D = "low" AND T = "low" AND C = "high", THEN PAT = "low"

IF D = "low" AND T = "moderate" AND C = "low", THEN PAT = "low"

IF D = "low" AND T = "moderate" AND C = "moderate", THEN PAT = "low"

IF D = "low" AND T = "moderate" AND C = "high", THEN PAT = "moderate"

IF D = "low" AND T = "high" AND C = "low", THEN PAT = "low"

IF D = "low" AND T = "high" AND C = "moderate", THEN PAT = "moderate"

IF D = "low" AND T = "high" AND C = "high", THEN PAT = "high"

IF D = "moderate" AND T = "low" AND C = "low", THEN PAT = "low"

IF D = "moderate" AND T = "low" AND C = "moderate", THEN PAT = "low"

IF D = "moderate" AND T = "low" AND C = "high", THEN PAT = "moderate"

IF D = "moderate" AND T = "moderate" AND C = "low", THEN PAT = "low"

IF D = "moderate" AND T = "moderate" AND C = "moderate", THEN PAT = "moderate"

IF D = "moderate" AND T = "moderate" AND C = "high", THEN PAT = "high"

IF D = "moderate" AND T = "high" AND C = "low", THEN PAT = "moderate"

IF D = "moderate" AND T = "high" AND C = "moderate", THEN PAT = "high"

IF D = "moderate" AND T = "high" AND C = "high", THEN PAT = "high"

IF D = "high" AND T = "low" AND C = "low", THEN PAT = "low"

IF D = "high" AND T = "low" AND C = "moderate", THEN PAT = "moderate"

IF D = "high" AND T = "low" AND C = "high", THEN PAT = "high"

IF D = "high" AND T = "moderate" AND C = "low", THEN PAT = "moderate"

IF D = "high" AND T = "moderate" AND C = "moderate", THEN PAT = "high"

IF D = "high" AND T = "moderate" AND C = "high", THEN PAT = "high"

IF D = "high" AND T = "high" AND C = "low", THEN PAT = "high"

IF D = "high" AND T = "high" AND C = "moderate", THEN PAT = "high"

IF D = "high" AND T = "high" AND C = "high", THEN PAT = "high"

where: D, T, W and PAT denote the following fuzzy classes: willingness to engage in dialogue, trustworthiness, creativity, and preparedness to act together, respect. The values of the corresponding membership functions are assigned by the facilitator who participates (observes, facilitates, helps) in negotiation between stakeholders [6].

The above fuzzy rules have been used in a software product called FLOCK (Fuzzy Logic based Consensus Knowledge) [7] to help in practical realisation of the idea of second order consensus - when the facilitator's role is not to foster unanimous decisions based on commonality and conformity of stakeholders' thinking, but instead to foster individual differences, creativity and innovation, cooperative attitudes and willingness to be engaged in dialogue, responsibility and trustworthiness, and thus to keep alive stakeholder's drive to seek consensus, to explore new ways that might lead to consensus, to get prepared to move and act together.

3. Role of Fuzzy Logic in Understanding the Socio-Political Maladies of Democratic Societies

Malady one: the false face of representative politics

'Fuzzy thinking' is at the very basis of politics, it lies in its 'genes'. The fuzziness of political statements and narratives is constructed to be endlessly reproduced. The expressions politicians use are chosen with a view to persuasion and pleasing. They project an image of the politician as someone 'good' - likeable and close to the people, dedicated to the great principles of democracy. The fuzzy umbrella of words is used to cover the inevitable struggle for power and compulsory deceptions inherent in it. The pursuit of popularity and a high approval rating is an inherent part of representative democratic politics from its beginning, but it has been given new, dramatic dimensions by the power of electronic mass-media.

Great skills in performance and media manipulation are required of politicians to make society, or at least the 'mainstream' of the electorate, accept policies it would not necessarily support, were these revealed in detail and with all their consequences. In all of this, a deliberate linguistic fuzzification is used as a tool for deceptive 'selling' of policies and decisions based on hidden agendas of ideology and power. In their efforts to hold on to power, politicians and political parties rely also on crucial financial support from wealthy individuals and organizations. While most of the contributions are legitimate, the danger of political and economic corruption is ever present in these exchanges, often influencing them in some subtle way but sometimes taking outrageous forms (e.g. Mafia's support of the Andreotti governments in Italy).

The process of healing this malady involves a continued *public disclosure* of the ambiguity and deception in policy narratives and decisions as well as in the charisma and behaviour of politicians and political parties; it requires a permanent exposure and denouncement of political actions in conflict with the public interest. Under conditions of freedom of expression, the 'logical component' of fuzzy logic is strong enough to be useful in social inquiry for revealing deceitful political charisma and behaviour.

3.1 'Butterfy effect' in politicians' narratives

It is the 'butterfly effect' manifested in fuzzy statements and narratives of the politicians, that can help the community to reveal how serious is the social malady described above.

Slight changes in the ideological platform of a political party or a politician, expressed by means of fuzzy hedges (i.e. words used in statements or narratives to intensify or dilute the fuzzy set's membership functions or to change the degree of fuzziness in a fuzzy set, such as: 'more or less', 'very', 'quite', 'somewhat', 'slightly', 'extremely', 'positively', 'generally', 'around', 'about', 'near', etc.) can bring forth enormous changes in their interpretation as a basis for action.

For example, during an election campaign politicians can promise 'to keep the defence potential of the state at a *more or less* stable level', but once in power, they can use this statement as justification of a large program for testing new nuclear weapons with enormous negative consequences for people and environment.

The butterfly effect in polititical narratives serves as a signal for community actions aimed at revealing the false face of representative politics.

Malady two: misconception of liberty and equality as mutually exclusive

The Enlightenment quest for a rational society based on knowledge and reason instead of tradition, dogma and faith led in the 18th century to the genesis of representative democracy and the nation state. The sovereignty of the state began to be seen as derived from the sovereignty of its citizens, and not any more from the ruling sovereign monarch. This shift established, in principle, the equality and liberty of the citizens. Although there was general agreement on this, citizens came to be divided over the political interpretations given to these terms, fuzzy as they are.

At one theoretical extreme is the interpretation of *absolute democracy*, meaning power of the collective over the individual. At the other extreme is the interpretation of *absolute liberalism*, i.e. limitless power and freedom of the individual vis a vis the collective. In practice, the political systems in most industrialised countries today lie somewhere between these extremes, and can be called *liberal democracies*. The ideas of the political parties of the Left favour policies which give the government and the public sector a major role in society, by providing public services to citizens and their organizations as well as by regulating the private sector. The Right favours free markets, with economic decision-making by individuals, to function in most sectors of society, without interference by government or collective decision-making. At the present time of economic globalization, the ever moving pendulum of government policy emphasis has swung in most countries towards the Right, affirming uninhibited competetive individual endeavour and achievement, and disregarding the interdependencies and cooperative relationships between individuals, groups and societies.

We hope this trend is not definitive. It is our contention that liberty and equality are not as mutually exclusive as they are seen to be in the dichotomy underlying political decisions in today's society. Instead, we believe liberty and equality to be *complementary* prerequisites for a genuine establishment of each other. There is no such thing as a self-contained individual, a separate self-possessed being, abstracted from relationships with others and the collective. Correspondingly, there is no collective, no community which does not consist of its individual members and their relationships with each other and with the community. And to be authentic, these relationships have to be established in conditions of freedom and equality, not domination and exploitation, however subtle.

The healing process from the described malady of the misconception of liberty and equality as mutually exclusive comprises a permanent search for complementarity between the individual and the collective, between individual well-being and collective welfare. By providing vital conditions for the functioning of any consensus-seeking enterprise, fuzzy logic plays an essential role in this search for complementarity.

Malady three: destructive competition based on short-sighted economic rationalism

The political processes of the modern nation state favour competition, unilateral use of power, domination, monologue, debate and argument at the expense of dialogue, cooperation and learning. Arguments are conducted in a way which inevitably leads to an exaggeration of differences. The global economy and financial markets exert a relentless pressure on the management of national economies by governments.

The language of politics and government has increasingly become that of economics and economic-managerial rationalism, frameworks which largely excludes social, human and ecological points of view. The restrictedness of the framework for political discourse and decision-making, and its inherent predominance by aspects of destructive competition, is pulling apart the world web of social and ecological relationships.

In line with what we have said about liberty and equity, our contention is that competition and cooperation should be considered not as mutually exclusive characteristics but as complementary. In the process of individual, social, economic and political progression, the role of competition can be compared to that of dialogue in the development of ideas, learning and innovation. But these processes have to take place in a framework in which the participants see each other not only as competitors but also, in the larger scheme of things, as partners in cooperation.

Progress occurs where different and competing assumptions, facts and ideas are given free play in a creative collaboration between colleagues who, although competitors in their field are united and interdependent in their aim of making a meaningful and significant contribution to it. Something similar happens in economic life: firms competing in an industry form complex networks of relationships both on the level of their ideas, know-how and technologies and on the level of their personal and institutional relationships. Even in industry these relationships are characterised by simultaneous competition and cooperation (whether explicit and admitted, or not).

The healing process from the malady of destructive competition, derived from a short-sighted economic rationalism, requires moderation of the forces and processes of competition, and support to the forces and processes of cooperation. The 'black' or 'white' approach of losing or winning needs to be replaced by dialogue-oriented and multi-alternative approaches of fuzzy logic, aimed at creating integrative solutions and conducted through processes which combine features of both competition and cooperation in complementarity.

Malady four: disconnection of economy from society.

Reduced levels of economic growth, globalisation of markets and increased competition have caused most, if not all, advanced industrial countries to reconsider their allocation of resources between the private and the public sectors, and to do so in favour of the former. More than ever, government policy in all areas is dominated by instrumental economic rationality and free market-thinking, while social and ecological issues are seen as secondary. The aim is to create favourable conditions for economic growth in general, and in particular to meet the request of transnational big business. Consequently, the anticipated responses to government policy by world financial markets largely dictate the directions and limits of national policies and programmes. The democratic ideal of 'government for the people by the people' has become increasingly empty of meaning.

This social malady leads to two critical consequences: enormous inequality in distribution of wealth and the break-down of the political system. The political system as a whole starts to work for people who are the privileged in society.

The first step towards weakening the symptoms of this social malady is to keep the community's eyes open for the dangers to society of a too close-knit alliance between politics and business. The second is re-thinking the meaning of community.

3.2 What does it mean to *be-in-common*?

Neither politics nor business are formations isolated from the rest of society - the very possibility of their being in an alliance depends on their *ontological* potential to share existence within society [9]. In Heidegger's definition, 'being-there' ("Dasein") is none other than a 'being-with' ("Mitsein").

Again, fuzzy logic offers a framework for understanding the oscillating between "being-there" and "being-with", i.e. between disaggregation and communion. For to *be-in-common*, it does not matter much what our social status is (politician, businessperson, factory worker, farmer, academic) - what matters is the mutual acceptance of one another *such as we are*. According to Agamben [10], the 'others' matter to us not because we are attracted by their specific characteristics (succesful, rich, clever, influencial) nor because we identify them with some favoured social formation but because we are appreciative of them with all of their traits, such as they are, warts and all. We cannot seek consensus without accepting each other as we are.

In the context of fuzzy logic, such an actualization of the meaning of 'being in community' makes sense and helps when dealing with the forces of social disintegration. Any act of acceptance of the other is preceded by some kind of fuzzification of any separating boundaries; this allows people to act together. Neither government nor business can function without accepting the rest of society as it is. Economic, social and political systems should be seen as evolving interrelated networks *within* society, and not as separated systems. National and global survival of humanity crucially depends on social integrity

Malady five: no-alternatives strategy.

The political elites tend to develop policies and political programs within the circle of their own party or coalition of parties, without wider input from others. While based on a set of world views, values and assumptions, these are only implicit in the stated policies. After their formulation, policies are announced and presented to the public as the only viable ones in the circumstances of the situation. Objections and differing or complementary views from representatives of other parties or stakeholders in the wider community are dismissed without much public discussion by characterising them as unrealistic because of non-availability of resources or irresponsible because of their undesirable consequences.

There are many *examples* of this line of argument in the history of various countries. When Finland joined Germany in attacking Soviet Union in 1941, the war was presented to the population, and accepted by its majority, as 'inevitable and unavoidable'. Finnish governments in power after the war argued, again successfully, that coupled with peacetime neutrality the official policy of close cooperation with the Soviet Union was the 'only realistic foreign policy option' available to Finland.

In the Australia of the present era of global markets, both the former Labor and the current Liberal/National Party government argue that a far-reaching deregulation of markets, coupled with various measures to improve the competitiveness of the Australian economy, is 'the only viable way' to maintain and develop prosperity in Australia.

A social awareness and attitude towards treatment of this malady can be sought through fuzzy logic: it is incompatible with any approaches praising only one 'best' alternative.

Malady six: a false concept of personal identity.

The idea of the person as a bounded, self-contained and unique self separated from other such selves, in possession of physical and mental capacities which facilitate independence of thought and action, is a construct peculiar to western culture.

This western abstraction of the individuals from their social relationships overlooks the fact that a person achieving anything like this level of 'ideal independence' is often a member of a privileged, dominant *in*-group. Individualism is therefore founded on an implicit distinction between the *in*-group or supposedly self-contained individuals and the *out*-group whose members do not possess the above qualities to a sufficient degree. The standards of the *in*-group come to determine the nature of proper personhood. The *in-/out*-group distinction underlying individualism is concealed, and the erroneous assumption arises that individuality is the property of persons rather than of in-group dominant and out-group submissive relationships. With this conception of the individual, differences and otherness are suppressed.

The politics of domination of one group over others emerges in the name of finding a unifying, single perspective for evaluation of the socio-political situation and the human experience at large. All forms of otherness, be they cultural, ethnic, sexual, intellectual or political are denied, repressed or transformed to be merged into the qualities of the dominant self-contained individual.

The German historian J. Rüsen argues that in order to avoid repeating the terrible mistakes of nationalism and ethnocentrism, it is necessary for Germans to develop a new, more open, multidimensional identity. Nationalism and ethnocentrism are based on the black and white distinctions of us versus others, good versus bad, and so on. As a concrete step towards an open multidimensional German identity, J. Rüsen has suggested 9 November be made the German National Day of Unification (9 November marks the dismantling of the Berlin wall in 1989, the Kristall-Nacht of the Nazis in 1938, and the abdication of Kaiser Wilhelm II in 1918).

This is an example how such a fuzzy term as 'unification' can be defined in a way that helps to avoid associations with false dominative patterns of national identity. Thus, fuzzy logic helps in the search for a remedy against the social malady based on false concepts of personal identity.

4. Conclusion

When Fuzzy Logic is applied to the management of social complexity, "management" denotes an activity which all the participants perform, regardless of their formal organisation or social position. As shown above, to achieve a high degree of preparedness to act together, at least moderate but preferably high degree of communicativeness, trustworthiness and creativity are required. As social systems are highly interdependent, it is difficult to imagine that these kinds of qualities could be achieved without the presence of a fair degree of mutuality, reciprocity and equality of power and influence on both the content and the process of the inquiry. This in turn requires a positive appreciation of differences, a "celebration of the other".

Unfortunately, western culture in to-day's world is traditionally more inclined to exercise an 'enlightened suppression of the other'. The maladies of our days' society clearly demonstrate the consequences of such 'entlightened suppression'. Seen against this background, the introduction and diffusion of Fuzzy Logic as a tool for the management of social complexity is a task of enormous challenge and profound constructive potential.

References

1. Smith K. and Berg D., Paradoxes of Group Life, Jossey-Bass Publ., 1987
2. Dimitrov V., Fuzzy Symplectic Systems: A New Framework for Multistakeholder Decision-Making, Proceedings of NAFIPS/IFIS/NASA Conference, December 18-21, 1994, San Antonio, Texas
3. Lyotard J.-F., The Postmodern Conditions: A Report of Knowledge, Manchester University Press, 1984
4. Dimitrov V. and Russell D., The Fuzziness of Communication, Seized by Agreement, Swamped by Understanding, Eds. L.Fell, D.Russell, A.Stewart,

University of Western Sydney, Hawkesbury Printing, 1994

5. Maturana H. and Varela F., The Tree of Knowledge, Shambala, 1992
6. Turner B.S., Citizenship and Social Theory, SAGE Publ., 1993
7. Social Ecology of Waste Management, Report on Project 9.1 of the Cooperative Research Centre for Waste Management and Pollution Control Limited, Sydney, 1994
8. Sampson E.E., Celebrating the Other: A Dialogic Account of Human Nature, Westrien Press, 1993
9. Chang H., Postmodern Communities: The Politics of Oscillation, Postmodern Culture, vol.4, No1, September 1993
10. Agamben G., The Coming Community, University of Minnesota Press, Minneapolis, 1993
11. Tuomas Savonen, The Heavy Burden of History: an interview with the German historian J. Rüsen, Yliopisto/Acta Universitat Helsingiensis, 11, 1996 (in Finnish).

USING FUZZY SETS TO EXTEND HOLLAND'S THEORY OF OCCUPATIONAL INTERESTS

Michael Smithson[1] and Beryl Hesketh[2]

[1]Division of Psychology
Australia National University
Canberra A.C.T. 0200
Australia
E-mail: Michael.Smithson @anu.edu.au

[2]School of Behavioural Sciences
Macquarie University
Sydney, NSW 2109
Australia
E-mail: bhesketh@bunyip.bhs.mq.edu.au

Abstract. In psychology there are several examples of ways in which new methodological or analytic approaches have stimulated theory development. This chapter illustrates and applies ideas arising from fuzzy sets to highlight the importance of measuring both approach and avoidant reactions to activities, and to clarify and extend theoretical constructs in Holland's (1982) typological theory of careers. The framework developed here combines concepts associated with 1) avoidance in decision-making, 2) Holland's theory, and 3) fuzzy set theory. Data from two samples (N= 363 and N= 256) are used to illustrate the conceptual advantages of using fuzzy set theory. Three hypotheses arising from this framework are supported by findings from both samples. Finally, the chapter outlines directions for future research and some counseling applications of a fuzzy set approach to interests and interest types.

Keywords: Occupations, occupational interests, personnel selection, career counselling.

1. Background and Theory

1.1. Holland's theory of careers

The use of interest inventories, particularly within the context of Holland's (1982) typological theory of careers, has dominated approaches to careers counseling in many areas of the western world (Lokan et al., 1986, Hansen & Campbell, 1985; Hesketh & Rounds, 1995). Holland (1985a) has presented a theory of personality applied to careers that outlines six personality types and six work environments (Realistic, Investigative, Artistic, Social, Enterprising and Conventional; RIASEC). Holland suggests that the RIASEC types are arranged in a hexagon, with adjacent types being most similar, while those types on opposite ends of the hexagon are most dissimilar.

However, Holland's notion of type is more complex, as he argues that individuals develop a hierarchy of orientations toward each of the six types (Holland 1959). Hence a person may be primarily Investigative, but also have orientations toward Artistic and Realistic types providing a 'summary code' of IAR. Typically interest inventories are used to obtain measures of the Holland types. However in most cases the emphasis is on the measurement of positive orientations towards the types of activities rather than on avoidant reactions, with most attention given to the first three orientations highest on the hierarchy. In this chapter, we use concepts and analytic approaches from fuzzy set theory to highlight the advantage of explicitly making use of the negative or avoidant aspect of interests. These concepts are also used to clarify and extend theoretical constructs within Holland's (1985) typological theory of careers.

The concept of types

The conceptualization and use of types within Holland's (1985) theory has received very little critical appraisal. Despite the use of sub-types, most recent discussions have tended to assume that individuals can be classified primarily as belonging to one of the six types. In practice, the initial formulations of Holland's theory (1959), which made reference to the hierarchy of orientations toward each of the six types, is ignored. In part, this may be due to the use of conventional analysis which is unable to deal with graded membership in types and instead assumes an 'all-or-nothing' grade of belonging to a type. In order to obtain a complete picture of an individual's work personality it is necessary to retain and use information about orientation to all six types.

The various measures associated with Holland's theory (VPI, and SDS Holland, 1985b and 1985c; and the Strong Interest Inventory , Hansen & Campbell, 1985) provide a basis for obtaining scores on all six types. The Position Classification Inventory (PCI; Gottfredson & Holland, 1991) provides a job analytic approach to describing jobs in terms of the six Holland types.

Applying fuzzy types to Holland's categories will help retain information about the 'grade of membership' that an individual has of each of the six types. This is an approach implicit in the use of fuzzy ratings to obtain measures of Holland's interests (Hesketh, Elmslie & Kaldor, 1990; Hesketh, McLachlan & Gardner, 1992; Hesketh & Gardner, 1992).

Congruency, consistency and differentiation

The hexagonal relation among types is used to define two important constructs in Holland's theory. The first of these is 'congruency', which is the degree of match between people and their work environments. Congruency is highest if a person is in an occupation of the same type, namely a Realistic person in a Realistic job or a Social person in a Social job. Congruence is lowest if a person is in an occupation that is on the opposite end of the hexagon, namely a Realistic person in a Social job or vice versa. Realistic people in Investigative jobs have intermediate levels of congruency, although congruency is higher than if a Realistic person is in an Artistic job. The example provided here offers the simplest way of operationalizing the concept of congruency. Other methods of assessing congruency have been developed (e.g. Iachan, 1984), but few pay attention to the avoidant component of congruency.

Traditional approaches to defining congruency also fail to take into account the influence of the second major construct in Holland's theory, namely consistency. Consistency is defined by the hexagonal relation among types. An individual with a summary code of IA has higher consistency than one with a code IS, while a code of IE would be highly inconsistent. Typically only the first two letters of the summary code are used to define 'consistency' although in principle the idea can be extended to incorporate more information. An individual or an environment may have a high level of inconsistency which, by definition permits a greater degree of flexibility with respect to achieving congruency. A person who has high Social and Realistic interests is much more likely to find an outlet with some degree of congruency than is a highly consistent person who may have strong Realistic interests with no Social orientation. This point has practical implications, yet it has received little attention as the relation between consistency and congruency has not been acknowledged in previous literature. The fuzzy set approach discussed below provides a possible way of dealing with this, by distinguishing person-environment incongruency from inconsistency within the person and/or within the environment.

The construct of differentiation, while not dependent on the hexagon for its definition, is also worth noting. As in most tests providing a profile of scores, differentiation, which can be obtained from Holland's measures, refers to the distance between the most preferred orientation and the least preferred orientation. An individual with a highly differentiated profile has clear peaks and troughs, while an undifferentiated profile is flat.

The constructs described above apply equally to environments, which Holland (1985) argues derive their 'climate' from individuals, and can be measured using an environmental assessment technique. This simply means obtaining the average orientations of the people in an environment or an occupation, and using the averages to derive a 'summary code' for the environment or occupation. Permitting environments to belong partly to more than one Holland type is more compatible with averaging and opens up new conceptual and theoretical possibilities, as will be argued later. It also has important counseling implications as highlighted in the discussion.

1.2. Seeking and avoiding areas of work

Interest measurement provides a way of summarizing orientations towards or away from different types of activities, people, occupations or roles. Although interest inventories often include like and dislike responses, attention is seldom given to the avoidant component. There are a few exceptions in the careers literature. Tyler (1964) and Gottfredson (1981) have drawn attention to strong avoidant reactions to highly sex-typed activities or occupations, and Harmon (1971) also includes notions of avoidance. Hesketh, Hesketh, Hansen and Goranson (1995) used fuzzy variables to develop scales for the Strong Interest Inventory based on both the distinctive like and the distinctive dislike responses on the Occupational Scales (Hanson & Campbell, 1985). On a more general level, several writers have been interested in the issue of 'avoiding the negative' in career decision-making (Tversky, 1972; Zytowski, 1965).

In the light of these asymmetries, there are many situations where it may be valuable to retain specific information about both approach and avoidance orientations in interest measurement and decision making. The lack of attention to dislikes or the avoidant aspects of interests may be due to the atheoretical nature of interest measurement generally, and the limitations of current definitions of interests (Hesketh & Rounds, 1995). Pryor's (1991) definition, and many others, do little more than identify interests with preferences. Nevertheless, as suggested by Hesketh and Rounds (1995), identifying interests with preferences permits interests to be viewed within an operant perspective such as the Premack principle (Premack, 1965).

Using a fuzzy graphic rating scale, Hesketh, Gardner and Lissner, (1992) found that avoidant reactions were the only aspect of interests differentiating engineer managers from technical specialist engineers. Engineer managers did not dislike Enterprising activities as much as did their technical specialist counterparts, who placed Enterprising activities lowest on their hierarchy. The summary code, represented typically in terms of just the top three orientations was identical for both groups, namely Investigative first, followed by Realistic activities and Artistic activities (IRA). The traditional emphases on positive orientations and the first three letters of the code would have obscured this important difference.

2. A fuzzy set extension of Holland's framework

Fuzzy set theory and fuzzy logic both have seen fairly widespread application in psychology, particularly in recent years (for surveys and overviews see Smithson 1987; Smithson & Oden, 1996; Zetenyi, 1988). Applying the concept of fuzzy sets to Holland's theory has several advantages. First, it permits one to focus on the relation of the individual to each of the six types, and not only those highest on the hierarchy. Second, the unique information contributed by measuring avoidance reactions also is better handled by fuzzy sets than many conventional approaches, since it permits the exploration of the extent to which people both prefer and avoid a job-type and the extent to which they do neither. Finally, and most important, viewing Holland's types as fuzzy sets permits the use of different types of analysis that both sharpen and extend the theoretical meaning of consistency and congruency (and inconsistency and incongruency) while also clarifying the impact of consistency on congruency. What follows is an outline of ways in which fuzzy sets may extend Holland's ideas.

To begin with, although this paper does not focus on differentiation, fuzzy sets provide a reasonable and sensitive measure of it. Maximal differentiation requires that the person have full membership in one set and absolute nonmembership in another. It therefore requires that the person's degrees of membership of type include at least one that is 1 and one that is 0 (i.e., 'crisp' rather than fuzzy). Measures of fuzziness (Klir 1987, Smithson 1987) could therefore be used as indicators of differentiation.

Both consistency and congruency may be conceptualized in terms of joint membership in sets, which is a well-defined concept in fuzzy set theory. In the case of consistency this would relate to the extent to which a person has joint membership in sets that are adjacent to one another on the hexagon. Inconsistency, on the other hand, arises when the person has joint membership in nonadjacent sets. In the case of congruency it would be the extent to which the individual is a member of a type and the environment is a member of the same type. The value of fuzzy set theory becomes apparent when dealing with the obverse of concepts such as congruency and consistency. Both incongruency and inconsistency may be characterized as joint membership in 'incompatible' sets. This joint membership, in turn, may be given greater weight for pairs of sets that are further apart on the hexagon, and the weighted memberships may be summed to form a measure of incongruency or inconsistency. We will set aside the issue of the actual assignment of weights. Qualitative versions of these weights for inconsistency are shown in Table 1.

It is noteworthy that if a person's total membership in all types exceeds 1 then that is sufficient to guarantee a nonzero degree of inconsistency. Consider the case example in Table 1. The total memberships sum to 14/6, so some inconsistency is inevitable. This person's joint memberships are distributed across the weights as follows: 4/6 in "small", 6/6 in "moderate", and 2/6 in "large".

There is, as previously indicated, an interdependency between inconsistency and incongruency which arises when both the person and environment are inconsistent (i.e., when the person has more than one type of interest and when the environment makes more than one type of demand in a job). Suppose, say, that the person has interests of type R and S and the occupation also has characteristics of types R and S. Then both the person and environment are inconsistent, but they are not incongruent with each other. Fortunately, this interdependency may be dealt with by reconceptualizing incongruency as the residual membership shared by the person and environment in R and S when the person's joint membership and the environment's joint membership in R and S have been removed.

Table 1: Qualitative weights for inconsistency

	R	I	A	S	E	C	Person
R	zero						
I	small	zero					
A	mod.	small	zero				
S	large	mod.	small	zero			
E	mod.	large	mod.	small	zero		
C	small	mod.	large	mod.	small	zero	
Person							

Case example

	R	I	A	S	E	C	Person
R	2/6						2/6
I	0	0					0
A	0	0	0				0
S	2/6	0	0	6/6			6/6
E	1/6	0	0	1/6	1/6		1/6
C	2/6	0	0	5/6	1/6	5/6	5/6
Person	2/6	0	0	6/6	1/6	5/6	14/6

Membership of a person, P, in Holland type R, say, will be denoted by $m_P(R)$. Membership of an environment, E, in R will be denoted by $m_E(R)$. We need to consider four kinds of inconsistency: person, $P_{RS} = \min(m_P(R), m_P(S))$; environment, $E_{RS} = \min(m_E(R), m_E(S))$; and two involving person-environment, $PE_{RS} = \min(m_E(R), m_P(S))$ and $PE_{SR} = \min(m_E(S), m_P(R))$. To the extent that PE_{RS} or PE_{SR} exceeds both E_{RS} and P_{RS}, we may say that there is an additional amount of inconsistency in the combination of the person and environment not accounted for by either inconsistency within the person or within the environment. The positive residuals $T_{RS} = PE_{RS} - \max(E_{RS}, P_{RS})$ and $T_{SR} = PE_{SR} - \max(E_{RS}, P_{RS})$, then, are indicators of person-environment incongruency. A simple algebraic argument shows that if one of these is positive then the other cannot be.

Unfortunately when we are comparing person and environment membership in the same type (R, say), neither of these measures is positive. For this special case we may use the absolute difference between the person and environment membership in R, $|m_P(R) - m_E(R)|$. A formal definition of incongruency for any types X and Y is therefore

$$G_{XY} = G_{YX} = \max(0, T_{XY}, T_{YX}) \text{ if X and Y are different types}$$
$$G_{XX} = , |m_P(X) - m_E(X)|, , \text{otherwise} \tag{1}$$

This measure also has a fuzzy set-theoretic justification (details are available from the first author). Returning to our case example, suppose the job selected by this person has a membership of 1 in Social. Her membership in Social is also 1, so $I_{SS} = 0$. Her membership in Conventional is 5/6, so $T_{SC} = PE_{Sc} - \max(E_{SC}, P_{SC}) = 0 - \max(0,5/6) = -5/6$, and $T_{CS} = PE_{CS} - \max(E_{SC}, P_{SC}) = 5/6 - \max(0,5/6) = 0$. Therefore, $G_{SC} = 0$. The same results follow for all other pairs, which is reasonable given that there is no inconsistency within the environment so the sole source of mismatch between this person and the environment is her own inconsistency.

Now we turn to avoidance, a topic related to inconsistency and incongruency. If a person does not have an interest in R, then this lack of interest is not sufficient to predict whether they would find R intolerable or wish to avoid it. Such a person might be merely indifferent to R. The fact that avoidance is not entirely predictable from mere lack of interest underscores the need to investigate avoidance in its own right. The most obvious approach-avoidance inconsistency is seeking and avoiding the same Holland category, say, R. However, this restriction is asymmetric: If a person does not seek R then they may either avoid R or not. It seems reasonable to give this kind of inconsistency a different name, so we will call the tendency to both seek and avoid R 'ambivalence' towards R. Conversely, the tendency to neither seek nor avoid R will be called 'indifference'. Here, we develop a measure of ambivalence, which is the tendency to both seek and avoid the same type.

Such a measure may be based on simultaneous membership in the seeking and avoiding sets for each type. The min operator for intersection, however, counts even the slightest tendency to both avoid and seek R, say, as evidence of ambivalence towards R. Given our hypothesis that people obey a somewhat looser consistency principle, i.e., their memberships in seeking and avoiding sum to 1 or less, a more appropriate measure would be one that 'penalizes' only those whose memberships sum to more than 1. An alternative version of fuzzy set theory, based on Łukasiewiczian logic (cf. Smithson, 1987), does exactly that. Denoting membership in the avoiding set for R by $m_P(!R)$, Łukasiewiczian intersection gives the following measure for ambivalence:

$$M_{R|R} = \max(0, m_P(R) + m_P(!R) - 1), \tag{2}$$

which is also known in the fuzzy set literature as a "bold" intersection. Its most important property is that it gives a value of 0 unless $m_P(R) + m_P(!R)$ exceeds 1 (i.e., unless a case lands in the upper right-hand triangular region in Figure 1). The case example recorded $m_P(R) = 2/6$ and $m_P(!R) = 3/6$ whose sum does not exceed 1, so $M_{R!R} = 0$. We will present analyses of two data-sets that address the issue of ambivalence.

Avoidance also yields an important extension of incongruency. Clearly a person who wishes to avoid R is poorly matched with a job strongly characterized by R. It would seem reasonable to assume that the degree of 'incongruency' here (let's call it 'antipathy') should be considered greater than if the person were merely indifferent to R. For diagnostic and counseling purposes, it might be desirable to distinguish incongruency based on indifference from that based on antipathy.

Antipathy is a form of incongruency, so it is based on the joint membership of the environment in R, say, and the person in the avoidance-set for R: $\min(m_E(R), m_P(!R))$. However, this measure ignores the extent to which the person may also seek R. A reasonable approach is to use the difference $m_P(!R) - m_P(R)$, bounding it below by 0, so that we have $\max\{0, \min[m_E(R), m_P(!R) - m_P(R)]\}$. This definition may be extended to pairs of nonidentical types such as S and R, by using $\max\{0, \min[m_E(R), m_P(!R) - \min(m_P(!R), m_P(S))]\}$ which reduces to:

$$N_{S!R} = \max\{0, \min[m_E(S), m_P(!R) - m_P(S)]\}. \quad (3)$$

The same kind of weighting scheme may be used for pairs that are nonadjacent on the hexagon as was suggested for inconsistency and incongruency. For the case example, since $m_P(S) = 1$, $N_{S!R} = 0$. The same is true for any other pair involving S. The other pairs also yield antipathy values of 0 because the environment has 0 membership in everything except S, so her total antipathy is 0. As might be expected, job-choices for both data-sets entail very little antipathy because they are made under rather unrestricted conditions. Antipathy is undoubtedly best researched in environments where occupations are not freely chosen or where occupations have undergone rapid and radical change.

Thus far, we have extended Holland's concepts of incongruency and inconsistency to a typology with five kinds of incompatibility:

(1) Inconsistency, joint membership in nonadjacent types on the hexagon;
(2) Ambivalence, the tendency to both seek and avoid a particular type;
(3) Indifference, the tendency to neither seek nor avoid a particular type;
(4) Incongruency, the membership of a person in one type located in an environment of an incompatible type;
(5) Antipathy, the tendency of a person to avoid the type that characterizes their environment.

Inconsistency, ambivalence, and indifference apply either to the person or the environment, while incongruency and antipathy apply to person-environment fit.

3. Empirical studies of seeking and avoiding

3.1. Aims and hypotheses

We report here two studies in which ratings were collected on the extent to which individuals seek each Holland type and the extent to which they avoid each type. The monopolarity of the 7-point scale used implies that the data may be treated as degrees of membership in fuzzy sets. The data-sets also contain Holland-typing of respondents' post-tertiary education job and course choices. However, those data are not fuzzy since the courses were assigned only a single Holland code. In light of this, the empirical investigations focus on inconsistency, ambivalence, and indifference within the person and not the environment.

The overall aim here is to assess the utility and relevance of the fuzzy extension of Holland's framework. Accordingly, three aims and hypotheses are listed below:

(1) The relationship between avoidance and seeking. We hypothesize that very few respondents will display ambivalence (both seeking and avoiding) but many will be indifferent towards a given type. This statement has a fuzzy-logical interpretation that if people seek a type, they will not avoid it (but not vice versa).

(2) Holland's hexagon, seeking, and avoidance. We hypothesize that joint membership in types (sets) declines for types that are further apart on the hexagon. This should be the case for both seeking and avoiding. Although this structure has been replicated for seeking (Tracey & Rounds 1992, Rounds & Tracey 1993) it is not a foregone conclusion that avoidance responses will also support the hexagon.

(3) The joint impact of seeking and avoidance on choice. We hypothesize that avoidance as well as preference will exert an influence on choice. We predict an interaction effect of seeking and avoiding, whereby ambivalence (both seeking and avoiding) suppresses the likelihood of choosing a type.

3.2. Data and Methods

In the state of NSW in Australia, students submit preferences for university courses to a central University Admissions Centre at the end of the year in which they sit their final examination. Two cohorts of Year 12 students (1993 and 1994) were surveyed three times to assess the decision processes involved in making choices about tertiary courses, and to ascertain their final destination. Entrance into university courses is determined by a state wide Tertiary Education

Rank based on a state examination whose results are rescaled to optimize the prediction of academic potential. The data reported in this paper relate to the first questionnaire sent to both cohorts shortly after sitting their final examination, but before they obtained their actual TER. The initial preferences submitted to the University Admissions Centre, which formed the basis of sampling, relied on students' estimates of their likely TER. Data from the longitudinal component of the research dealing with such issues as accuracy of TER estimates, decision processes and evaluation of help received are reported in Hesketh & Whiteley, 1994, 1995 and 1996).

In 1993 the initial sample consisted of 513 respondents who selected university courses that required a high TER cutoff. In 1994, the initial questionnaire went to 497 HSC students chosen to represent those who selected university courses that had a low TER cutoff and where the area of study was also available at technical colleges. In both 1993 and 1994 the selected courses used as a basis for sampling could be categorized in terms of each of Holland's (1985) six themes while also ensuring that both males and females would be represented in courses sampled for each of the themes.

In 1993 questionnaires were returned by 363 students, yielding a 71% response rate with 148 males (41%) and 215 females (59%). In 1994 among the students applying for courses with TER requirements (lower ability students), the response rate was not as high. Of the 497 students, 256 (52%) returned the questionnaire (108 males (42%) and 148 females (58%)). The large majority of the respondents in both years was 17 or 18 years of age (mean age = 17.63 years),. In light of the multicultural nature of the New South Wales student population, respondents were asked where they were born. Of the 619 students who returned the first questionnaire in both years, 478 (77%) were born in Australia, with the remainder citing Asia, Europe, USA, the Middle East and Africa and South America as their places of birth. In response to a question about nationality, 494 (80%) responded that they were Australian, 45 (7.2%) that they were Asian, 38 (6%) that they were European and 26 (4%) that they were from the Middle East. The remaining 3% of the responses were distributed among several different nationalities.

Seek and Avoid Scales for Holland areas

Single item statements of each of the Holland types used in previous research (Hesketh, Pryor & Gleitzman, 1989; Hesketh et al., 1995) formed the basis of the seek and avoid ratings. Using a 7-point importance scale (Extremely Unimportant to Extremely Important), respondents were asked to indicate how important it was to seek each of the six areas of work, and then how important it was to avoid each of the six areas of work. Previous research has revealed reliability estimates for these Holland statements ranging from .67 for Investigative to .87 for Artistic. Validities (correlations with the appropriate scale on the Vocational Preference Inventory: VPI) ranged from .42 for Social to .51

for Realistic (Hesketh et al., 1989). These correlations compare favorably with those obtained between the UNIACT Holland themes and the VPI (.35 to .65; American College Testing Program).

Job and Course Preference

Participants were asked to nominate their preferred job after they finished their tertiary study. If they were unsure, they were given the option of checking an undecided box. The job preferences of the 227 respondents in 1993 and 174 in 1994 who listed a job were given a Holland code, while the remaining 'undecided' students were coded as 'uncertain'. Tertiary course preferences were also given a Holland code, although these data are reported only for the 1993 sample.

3.3. Results and Analysis

In order to provide an introduction to the approach used to test the hypotheses, a detailed illustrative example is given in relation to the first hypothesis dealing with the construct of ambivalence. The tests of the second and third hypotheses build on the analytic ideas illustrated initially.

Here, we will explain how a model-fitting approach enables a test of the hypothesis that few respondents are ambivalent (both seeking and avoiding) although more may be indifferent (neither seeking nor avoiding an interest areas). Hypothesis 1 implies that for each of the RIASEC types, the data should be confined to the lower left-hand triangular area in a scatterplot of the seeking and avoidance scales for each of the same types, such as R and !R which are shown in Figure 1.

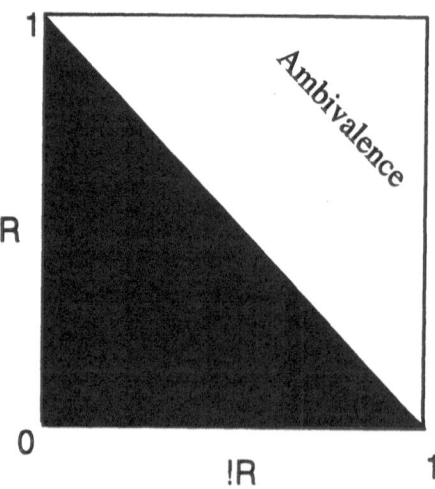

Figure 1: Ambivalence region.

Given that hypothesis 1 predicts that the data should lie in the lower left-hand triangle, a reasonable way to model the data is by cumulative logit models (Agresti, 1990 and Smithson, 1995). We will use an example of one type (R) from the first data-set to illustrate our explanation of the models.

Table 2 shows the joint frequency distribution for ratings on seeking of R and avoidance of R. Inspection should suffice to persuade the reader that most of the data lie in the lower left-hand triangular region of Table 2. The data in Table 2 have been used to create Table 3, which presents the information in a manner that both permits a test of the hypothesis that data do lie in the lower left-hand triangular region while also allowing a test that the odds ratios at each point on the diagonal are the same and close to 1. Accumulating frequencies from the lower right-hand corner of Table 2, and moving backwards up the diagonal (see result in Table 3), it can be seen that the numbers on the diagonal are almost as large as the column totals. This can happen only when the data mostly lie in the lower left-hand triangular region.

Table 2: Seeking vs avoidance of Realistic type

	S 0	e 1/6	e 2/6	k 3/6	i 4/6	n 5/6	g 6/6	totals
6/6	8	3	2	1	1	1		16
A 5/6	8	4	3	2				17
v 4/6	17	11	7	10	1			46
o 3/6	13	11	22	30	6		1	83
i 2/6	5	7	12	23	3	2		52
d 1/6	1	5	10	25	19	9	1	70
0	3	3	2	16	13	13	26	76
totals	55	44	58	107	43	25	28	360

NOTE: The cells in Tables 2 and 3 are indexed in the discussion in the text according to the conventional arrangement for contingency tables, with row 1 being the top row and column 1 being the left-most column. The reader should bear in mind, however, that the Avoid scale values have been arranged from the bottom (not important to avoid) to the top (important to avoid) as in the graph in Figure 1.

We wish to test a model that specifies that the odds-ratios for diagonal and column cumulants are constant across scale values and close to 1, so that we may summarize the entire table with one odds-ratio. In order to do so, we need to calculate odds-ratios for each point on the diagonal. For example, in cell(3,3) in Table 3 the cumulant is 251, so the corresponding odds are 251/(360-251) = 2.32. The 3rd column cumulant is 261, so the odds are 261/(360-261) = 2.63. The odds ratio therefore is 2.32/2.63 − 0.882. The same procedure applied to cell(2,2) gives an odds-ratio of 0.850, which is fairly close to 0.882. The other cells give similar odds-ratios.

Table 3: Joint and marginal cumulative distributions for seek and avoid responses

	S 0	e 1/6	e 2/6	k 3/6	i 4/6	n 5/6	g 6/6	totals
6/6								360
A 5/6		297						344
v 4/6			251					327
o 3/6				187				281
i 2/6					86			198
d 1/6						49		146
0							26	76
totals	360	305	261	203	96	53	28	

In order to test whether the odds-ratios are all equal, we must use a logit model, which converts the odds to log-odds. If we denote the log-odds of the diagonal cumulant in cell(k,k) by L_{kk} and the corresponding kth column cumulant by L_{+k}, our model is

$$L_{kk} = L_{+k} + b, \quad (4)$$

where b must be estimated. We then use that estimate to compute predicted L_{kk} values. Since the diagonal cumulants partition the whole table, we may take the differences between adjacent ones to obtain expected frequencies for each of the L-shaped regions in the table above, and compare those with the observed frequencies using a Chi-square statistic. A good fit is indicated by a small Chi Square and as the statistic is sensitive to degrees of freedom and sample size, dividing Chi-Square by the degrees of freedom provides one indication of fit.

In our example, the observed frequencies are {26, 23, 37, 101, 64, 46, 63} (the 63 results from taking 360-297). The maximum likelihood estimate for b = -.1325, so the odds ratio is .8759, which is fairly close to 1. The expected frequencies from the model are {25.7, 22.7, 39.4, 102.3, 59.2, 47.2, 63.5} and comparing observed and expected frequencies results in Chi-square = 0.691 (5 df), indicative of a good fit. Hence we are justified in claiming that the odds of finding a respondent at or below the diagonal from the kth level onward is about .8759 times the odds of finding a respondent anywhere from the kth column onward. We will now show that this kind of model fits each of the Holland types separately and that a model handling all six types simultaneously also fits the data adequately.

Hypothesis 1: Rarity of ambivalence

Table 4 provides the beta estimates, odds ration and the Chi Square for each of the six types as well as for all six types when comparing the seek and avoid distributions. Both the 1993 and the 1994 data-sets support the hypothesis that ambivalence (both seeking and avoiding) occurs only rarely whereas indifference (neither seeking nor avoiding) is relatively common. Moreover, all six Holland types display the same tendency to obey a constant cumulative odds-ratio law (range from .87 for Realistic to .73 for Artistic in the 1993 data set and from .88 for Realistic to .70 for Conventional in the 1994 data set.

The Chi-Square values all are quite low, given their df and sample sizes. The two models for all six types fit well enough for both data-sets that the single odds-ratio of .7921 for one and .7677 for the other adequately summarizes the relationship between all six seek-avoid pairs. To see this, it suffices to observe that the Chi-Squares divided by degrees of freedom for all six types in the 1993 and 1994 data sets are 14.156/35 = .405 and 13.787/35 = .394 respectively, while the average Chi-Square/df for the six RIASEC models in these data-sets are .370 and .359 respectively, so there is little difference.

Table 4: Summary analysis relating seek and avoid responses to Holland types for the 1993 and 1994 data sets

Type	1993 Data (N=363)			1994 Data (N=256)			
	b estim.	Odds ratio	χ^2	b estim.	Odds ratio	χ^2	df
R	-.1325	.8759	0.691	-.1182	.8885	1.807	5
I	-.2110	.8098	2.007	-.2512	.7779	3.465	5
A	-.1812	.8343	0.966	-.2113	.8095	1.329	5
S	-.3097	.7337	1.209	-.3140	.7305	1.499	5
E	-.2978	.7424	1.735	-.3427	.7099	0.746	5
C	-.2663	.7662	4.506	-.3490	.7054	1.937	5
All six:	-.2331	.7921	14.156	-.2664	.7677	13.787	35

The fuzzy set measures of ambivalence defined earlier also support the claim that ambiguity is rare. Average set-intersections may range from 0 (no intersection to 1 (complete overlap), and these intersections were .023 for R, .031 for I, .032 for A, .049 for S, .042 for E, and .045 for C. The intersections of seek and avoid on the same types are substantially lower than the average intersections for avoid and seek sets of different Holland types, providing a form of construct validity for the measures.

Hypothesis 2: Hexagonal relations and Inconsistency

The cumulative logit models were used to test the hypothesis that joint membership occurs more often for adjacent types on the hexagon, than for those one removed or opposite. Briefly, the model assesses the cumulative logits of respondents' maximal inconsistency for adjacent types, for types separated by one type, and for opposite types. These three sets of cumulative logits are then modelled in the following way:

$$L_{ki} = a_k + b_i, \quad (5)$$

where k denotes the level of the scale as before, and i denotes the adjacent, separated-by-one, and opposite conditions. The a_k are not of interest here (and therefore are not reported), but the b_i provide a test of our hypothesis. We expect the b_i to be largest for adjacent pairs and smallest for the opposite pairs. Estimates for a_k and b_i were obtained using a weighted-least-squares approach. As before, the model is evaluated by obtaining expected frequencies and comparing them with the observed ones via a Chi-Square statistic.

The resultant summaries from the analysis are given in Table 5 which contains the logits for each of the three groups and the Chi-Squares for the models. From these data it can be seen that both the seeking and avoidance inconsistency data follow a Holland structure, at least insofar as their odds-ratios are in the expected order. They also are well-described by a cumulative logit model as indicated by their low Chi-Square values, with the possible exception of the seeking data for the 1994 (lower ability) data-set (although the chi-square value therein is inflated by a poor fit in only three of eighteen cells).

Table 5: Summary of analysis relevant to Hypothesis 2 regarding the respective size of the fuzzy set intersection for types varying in inconsistency.

Group	Seeking 1993 b_i	1994 b_i	Avoiding 1993 b_i	1994 b_i
Adjacent	0.752	1.709	0.559	0.430
Once-sep.	0.120	0.936	0.048	-0.252
Oppos.	-0.874	-0.260	-0.606	-1.085
	$\chi^2 = 9.082$	$\chi^2 = 23.879$	$\chi^2 = 7.747$	$\chi^2 = 7.616$

Note: For all Chi-Squares, df = 11.

Hypothesis 3: Avoidance, Seeking, and Job and Course Choice

Ideally, we should like to examine the joint impact of seeking and avoidance separately for each Holland type. Job-choice data were available for both data-sets, but course-choices had been coded only for the 1993 data. Unfortunately both job and course choices, when cross-classified with pairs of seeking and avoidance scales on any of the Holland types, yield contingency tables whose sparseness pose difficulties for analysis.

However, a simple method for assessing the joint impact of seeking and avoidance on all types of choice overcomes the sparseness problem, albeit at the price of not being able to make this assessment separately for each of the Holland types. The tactic is to form a table of all those who made a choice, locating them according to their seeking and avoidance ratings on the Holland type they chose. For instance, someone who chose an Artistic job is located in the table according to their ratings of the extent to which they wish to seek an Artistic areas and the extent to which they wish to avoid Artistic areas. To obtain reasonable expected counts in the cells, the rating-scales are collapsed into three categories: 0-2/6, 3/6, and 4/6-6/6. The resulting crosstabulations are shown in Table 6.

Table 6: Interaction effect of avoidance and seeking on choice frequency

1993 Data: Course Choice

		Avoid		
		0-2/6	3/6	4/6-6/6	totals
	0-2/6	17	10	11	38
Seek	3/6	43	13	2	58
	4/6-6/6	233	13	15	261
	totals	293	36	28	357

1993 Data: Job Choice

		Avoid		
		0-2/6	3/6	4/6-6/6	totals
	0-2/6	10	2	4	16
Seek	3/6	21	8	3	32
	4/6-6/6	121	8	10	179
	totals	192	18	17	227

1994 Data: Job Choice

		Avoid		
		0-2/6	3/6	4/6-6/6	totals
	0-2/6	1	3	6	10
Seek	3/6	12	6	1	19
	4/6-6/6	132	4	9	145
	totals	145	19	16	174

Chi-square (df = 4) for the 1993 job-choice crosstabulation is 25.343 and for the 1994 table is 64.094, indicating that the effects of seeking and avoidance on choice do not operate independently. The 1993 course-choice data yield a similar picture, with a Chi-Square value of 58.429. Inspection of the subtables reveals that the number of people making a choice increases little as we move down the right-most or central columns, but increases dramatically as we move down the left-most column. In other words, ambivalence suppresses the likelihood of choosing a course type even though the individual indicates a strong tendency to seek it.

4. Discussion and conclusions

In this article, we have had two aims: to elaborate constructs in Holland's theory using tools and concepts from fuzzy set theory, and to provide a partial empirical evaluation of the elaborations. We claim this exercise is theoretically productive in two respects. First fuzzy set theory provides a coherent framework for incorporating avoidant responses along with seeking responses in Holland's system, via the definitions and operationalizations of the concepts 'ambivalence' (both seeking and avoiding), 'indifference' (neither seeking nor avoiding) and 'antipathy' (avoiding an environment in which one is located). Asymmetries in the seek and avoid orientations can be operationalized and dealt with statistically using fuzzy set inclusion and logical implication. Second, by treating Holland types as fuzzy sets, it is possible to sharpen the conceptualization of consistency and congruency as well as inconsistency and incongruency. An individual with high levels of inconsistency (joint membership of non adjacent types) is in many respects more flexible in terms of the probability of locating a matched environmental outlet. Likewise, environments that are inconsistent in some ways offer inconsistent individuals a greater chance of obtaining congruent outlets for their interests. One way of dealing with these relations is to redefine incongruency as that which is left over after taking into account overlapping membership sets per type for the individual and the environment.

The fuzzy set approach seems fruitful. All three hypotheses were supported *by both data-sets*. First, findings from both studies indicate that avoidance is not simply a bipolar opposite of seeking. Instead, many respondents exhibit indifference by neither seeking nor avoiding a particular kind of job. However, the restriction that most of the respondents obey is not to be ambivalent, i.e., not to both seek and avoid the same type. Moreover, the cumulative logit models show that this restriction is quantitatively similar across Holland types.

Second, the fuzzy-set version of inconsistency obeys a Holland-like structure for avoidance ratings as well as for seeking ratings, thereby suggesting that avoidance also has 'hexagon-like' properties. Again, cumulative logit models indicate that this regularity holds independently of Holland type. Given the asymmetric relationship between seeking and avoidance, this finding is not trivial since there is no a priori guarantee that avoidance ratings should obey any kind of consistency principle akin to that found in seeking.

Third, both studies' findings suggest that seeking on its own is not sufficient to account for course or job preferences. In addition to seeking an area, people must also not be ambivalent (i.e., they must not wish to avoid the same area). This outcome provides additional justification for considering avoidant responses along with seeking ones.

These hypotheses and findings underscore the utility of fuzzy set theory. All of them require the use of fuzzy set operations, which in turn are embedded in a coherent conceptual framework. It is worth emphasizing that asymmetric associations such as the seeking-avoidance relationship cannot be evaluated via conventional measures of bivariate association (e.g., correlation) because such measures detect symmetric, one-to-one associations only. Fuzzy set theory, on the other hand, is able to handle asymmetric, one-to-many associations. Moreover, the cumulative logit models developed here are not simply off-the-shelf, but instead are underpinned by fuzzy set tools. We do not wish to argue against the possibility of using more traditional analyses to test some of the hypotheses (e.g. a test of the hexagon on the avoid scales using either the Tracey & Rounds, 1992 method or the Myors, 1996, method). However, the advantages of thinking about Holland types in terms of fuzzy sets, and in applying fuzzy set analyses have been strongly demonstrated in optimizing the use of information in the avoidance scales. Although the examples given in the paper have been comparatively simple, the mathematical discussion in the Appendix provides a basis for tests of these sorts using relations among and between all six types simultaneously.

The current data sets did not permit a full illustration of the ways in which the interaction between congruency and consistency (and between incongruency and inconsistency) can be overcome using a fuzzy set approach. A complete data set is needed that contains information about seek and avoid responses of individuals to each of the six types, as well as ratings of relevant environments in terms of the extent to which each requires/demands or restricts/rejects characteristics associated with the types.

Research needs to be undertaken examining more directly the impacts of inconsistency, incongruency, and antipathy on occupational choice, tenure and job satisfaction. Worthwhile examples of testable hypotheses include the following:

(1) Inconsistency in itself (i.e., without consequent incongruency) is not related to dissatisfaction or turnover.
(2) Moderate levels of inconsistency imbue people with greater adaptability and thereby enable them to handle career or job changes better than people with narrower interests.
(3) Antipathy (active avoidance of a job in which one is located) is more strongly related to dissatisfaction and turnover than incongruency.

The counseling implications arising from the application of the fuzzy set approach to Holland's theory, and from measuring both seek and avoid responses to areas of work, are important. One of the major challenges facing many job seekers in the current climate of changing work structures, higher levels of unemployment and rapid change is to obtain suitable outlets for interests. Rather than harboring unrealistic expectations about finding jobs that are ideally suited, many job seekers must be prepared to be flexible, and accommodate changes in their jobs over time. In this context, for many individuals it may suffice to find outlets in areas of work about which they are indifferent. However, having clear information about actively disliked areas of work, or areas to be avoided is very important. Future counselors may have to help clients clarify what they wish to avoid, and also help them recognize the potential for partial congruency even in areas of work that they are not actively seeking. The fuzzy set concepts discussed in this paper highlight these issues. Although it was traditionally predicted that an individual with a highly inconsistent profile would experience difficulty choosing a career, and may have been more likely to vacillate, now such individuals may be thought of as flexible.

Conceptualizing Holland types as fuzzy sets facilitates a more fine-grained way of thinking about interests and encourages the flexibility that job seekers need. For example, Hesketh, Feiler and Kanavaros (1984) found that many engineering positions in manufacturing settings now required high levels of social skill, communication and people management. The social component of work had not been evident in similar jobs in research undertaken three years previously. Massive restructuring of manufacturing and the introduction of semi-autonomous work teams during the intervening period accounted for the changed profile of the jobs. Post restructuring, the positions had highly inconsistent codes in that they were both Realistic and Social. Individuals with highly inconsistent profiles located in these jobs showed strong congruency. The example illustrates the importance of careers help that strengthens latent flexibilities in individual's preferences and interests rather than precipitating crisp orientations toward only one area. A fuzzy set approach provides one way of capturing that flexibility through gradedness.

Acknowledgments

This research was funded by Macquarie University Research Grant Fund. The research ideas, however, arose from earlier Australian Research Council Funding.

References

Agresti, A.(1990). Categorical data analysis. New York: John Wiley.
American College Testing Program. (1981). Technical report for the Unisex edition of the ACT Interest Inventory. Iowa City, IA.

Gottfredson, G. D., & Holland, J. L. (1991). Position Classification Inventory (PCI): Professional manual. Odessa, FL: Psychological Assessment Resources, Inc.

Gottfredson, L. S. (1981). Circumscription and compromise: A developmental theory of occupational aspirations. Journal of Counseling Psychology [Monograph], 28, 545 - 579.

Hansen, J. I., & Campbell, D. P. (1985). Manual for the Strong Interest Inventory (4th ed). Palo Alto, CA: Consulting Psychologists Press.

Harmon, L. W. (1971). The childhood and adolescent career plans of college women. Journal of Vocational Behavior, 1, 45-56.

Hesketh, B., Elmslie, S., & Kaldor, W. (1990). Career compromise: an alternative account to Gottfredson's (1981) theory. Journal of Counseling Psychology, 37, 49-56.

Hesketh, B., & Gardner, D. (1992). Person-environment fit models: A reconceptualization and empirical test. Journal of Vocational Behavior, 42, 315-332.

Hesketh, B., Gardner, D., & Lissner, D. (1992). Technical and managerial career paths: An unresolved dilemma. The International Journal of Career Management, 4 (3), 3- 8.

Hesketh, B., Hesketh, T., Hansen, J. C., & Goranson, D. (1995). The use of fuzzy variables in developing new scales from the Strong Interest Inventory. Journal of Counseling Psychology.

Hesketh, B., McLachlan, K., & Gardner, D. (1992). Work adjustment theory: An empirical test using a fuzzy rating scale. Journal of Vocational Behavior, 40, 318-337.

Hesketh, B., Pryor, R., & Gleitzman, M. (1989). Fuzzy logic: A measure of Gottfredson's (1981) occupational social space. Journal of Counseling Psychology, 36, 103-109.

Hesketh, B. & Rounds, J. (1995). Cross cultural perspectives on career decision-making. in B. Walsh & S.H. Ospiow (Eds) Handbook of vocational psychology (2nd edition). Manwah, N.J.: Lawrence Erlbaum, 367-390.

Hesketh, B. & Whiteley, S. (1994). University admissions decisions research. HSC Prospects, 2, 10-12.

Hesketh, B. & Whiteley, S. (1995). Students' perceptions of the help received during the tertiary admissions process. Australian Journal of Career Development, 4(3), 62-67.

Hesketh, B. & Whiteley, S. (1996). Accuracy of self-estimates of TER.

Holland, J. L. (1959). A theory of career choice. Journal of Counseling Psychology, 6, 35-45.

Holland, J. L. (1985a). Making vocational choices: A theory of vocational personalities and work environments (2nd ed). Englewood Cliffs, NJ: Prentice Hall.

Holland, J. L. (1985b). Manual for the Vocational Preference Inventory. Odessa, FL: Psychological Assessment Resources.

Holland, J. L. (1985c). The Self-Directed Search: Professional Manual. Odessa, FL: Psychological Assessment Resources.

Iachan, R. (1984). A measure of agreement for use with the Holland classification system. *Journal of Vocational Behavior, 24,* 133-141.

Klir, G.J. (1987). Where do we stand on measures of uncertainty, ambiguity, fuzzines, and the like? *Fuzzy Sets and Systems, 24,* 141-160.

Lokan, J. & Taylor, K. (Eds) (1986). Holland in Australia: A vocational choice theory in research and practice. Melbourne: ACER.

Myors, B. (1996). A simple, exact test for the Holland hexagon. Journal of Vocational Behavior (in press).

Premack, D (1965). Reinforcement theory. In D. Levine (Ed.) Nebraska symposium on motivation. Lincoln: University of Nebraska Press.

Pryor, R. (1991). Assessing people's interests, values and other preferences. in B. Hesketh & A. Adams (Eds.). Psychological perspectives on occupational health and rehabilitation. Marrickville, Sydney: Harcourt Brace.

Rounds, J., & Tracey, T. J. (1993). Prediger's dimensional representation of Holland's RIASEC circumplex. Journal of Applied Psychology, 78, 875-890.

Smithson, M. (1987). Fuzzy set analysis for behavioral and social sciences. N.Y.: Springer-Verlag.

Smithson, M. (1995). Testing the stability of fuzzy subsethood across membership levels. Proceedings of the Sixth International Fuzzy Systems Association Congress, Sao Paulo, Brazil, 323-327.

Smithson, M., & Oden, G. (in press). Applications of fuzzy set theory and fuzzy logic in psychology. In D. Dubois, H. Prade, & H.-J. Zimmermann (Eds.) International handbook of fuzzy systems, vol. 5: Applications. Dordrecht: Kluwer.

Tracey, T. J., & Rounds, J. (1992). Evaluating the RIASEC circumplex using high-point codes. Journal of Vocational Behavior, 41, 295-311.

Tversky, A. (1972). Elimination by aspects: a theory of choice. Psychological Review, 79, 281-299.

Tyler, L. E. (1964). The antecedents of two varieties of vocational interests. Genetic Psychology Monographs, 70, 177-227.

Zetenyi, T. (Ed.) (1988). Applications of fuzzy sets in psychology. Amsterdam: North-Holland.

Zytowski, D. G. (1965). Characteristics of male university students with weak occupational similarity on the Strong Vocational Interest Blank. Journal of Counseling Psychology, 15, 182-185.

A MODEL FOR FUZZY PERSONAL CONSTRUCT PSYCHOLOGY

Alastair Anderson

Deakin University
Faculty of Management
School of Management Information Systems
221 Burwood Highway
Burwood, VIC 3125
Australia

E-mail: alpal@deakin.edu.au

Abstract: This paper examines the natural linkage between Personal Construct Psychology (Kelly, 1955) and Fuzzy Set Theory (Zadeh, 1965). Kelly conceptualised constructs as dichotomous abstractions upon which human beings construe others. In contemporary work constructs are accepted as dichotomous in form. However, they are not only seen as dichotomous in application. It is this feature of construing which engenders the linkage between Personal Construct Psychology and Fuzzy Set Theory. The paper presents a mathematical model that combines the essence of Personal Construct Psychology with Fuzzy Set Theory. The model can be used to generate structural measures for a construct system in terms of the complexity of that system. The model has been incorporated in a software program named FUZZYGRID. An overview of this program is presented as is a case study which demonstrates the application of the model. The way in which model results can be used in a multidimensional scaling analysis is also presented.

Keywords: Cognitive Complexity, Cognitive Schema, Consensus, Fuzzy Sets, Multidimensional Scaling, Personal Construct Psychology, Psychology, Repertory Grid, Work, Non-Work.

1. Introduction

George Kelly's major work, The Psychology of Personal Constructs was anomalous. It presented a countervailing view of human behaviour, as directed by the personal constructs of individuals not by ego emotion, motivation, reinforcement, drive, the unconscious or need [1]. The work stood in stark contrast to the behaviourist paradigm which was the basis for much of the research and practice in psychology. It appeared to have no linkage with the early or contemporary literature. This is evidenced by the scant reference to the work of others when considerable discussion was warranted. Bartlett's work on cognitive schema and the developmental work of Piaget rate no mention [2]. One is also surprised that Osgood achieved only a minor mention by Kelly since the Semantic Differential is a 'kissing cousin' of the Personal Construct [3]. In conjunction with the theory of Personal Construct Psychology (PCP), Kelly also formulated a methodological approach known as the *Repertory Grid* technique. Elaborations of this technique, and its applications can be found in [3] [4] [5] [6] [7] [8].

Notwithstanding Kelly's lack of attribution, conceptually speaking his work was ahead of its time. It has come to the forefront with the emergence of research concerning knowledge acquisition, and cognitive processes particularly as they relate to specialist human knowledge [9]. PCP and the grid technique provide a framework for knowledge acquisition and the representation and analysis of cognitive schema across a range of areas including expert systems, intrapersonal and group processes and human resources management to name but a few. The application of the theory and its accompanying methodology is limited only by one's imagination.

This paper comprises four sections. Section 2 establishes the natural bridge between PCP and Fuzzy Set Theory (FST). The conjunction of PCP and FST will be referred to as Fuzzy Personal Construct Psychology (FPCP)[i]

In Section 3, the mathematics of a model in which constructs are treated as fuzzy subsets is presented. FUZZYGRID, a program which embodies the model is introduced.

Section 4 presents results from the application of the model to a single case. The way in which the model results can be used in a multidimensional scaling analysis is also demonstrated. This enhances the model as data diagnostics and graphical output are available.

Section 5 draws some conclusions from the work to date and discusses future research directions.

2. Fuzzy Personal Construct Psychology

Zadeh conceptualized *fuzzy sets* and developed the theory which surrounds them [10]. Fuzzy set theory is elaborated in considerable detail in [11] [12]. In a classical set membership is binary. The valuation set is constrained to contain only the values 0 and 1 which denote non-membership and membership respectively. In contrast, a set is fuzzy when the valuation set is not constrained to be just 0 and 1 but can contain values in the interval 0 to 1. This relaxation caters for membership values which are crisp (0 or 1) as well as those which are fuzzy (between 0 and 1). Fuzziness is not the same as probabilistic assessments made by individuals. Probability concerns uncertainty, whereas fuzziness relates classes where there may be grades of membership between full membership and nonmembership [12].

2.1. Personal Constructs

Personal Constructs are a means of discrimination between items (usually called elements) in terms of similarity and contrast. They are bipolar in form and comprise an emergent and a contrast pole. The logic is that one cannot identify similarity without difference or contrast. Constructs are usually elicited by presenting a participant with a triad of elements and asking her/him to identify those two which are similar in some way and different from the third. The participant is then asked to indicate the basis of the similarity. For example, a person may be presented with the names of three people with whom they work and asked to group the two people who are similar in some respect and different from the third. This may result in the elicitation of a construct say *Authority/Subordinate*. This construct indicates how the participant views the triad in term of power relations. One might be tempted to think of a construct as always comprising semantic opposites such as *Good/Bad* or *Intelligent/Stupid*, but this need not be so [13]. What is important is that a person understands a construct as a dimension [3]. This gives expression to the notion of personal constructs as meaningful and useful to the individual.

In his major work Kelly assumed that constructs were dichotomous, in form and in application [1]. The assumption of dichotomy forced one to locate elements on one or other of the construct poles, not in any intermediate positions. There was no room for grayness. Kelly tried to repair the situation, as the following indicates : 'thus the construct refers to the nature of the distinction one attempts to make between events, not to the array in which his events appear to stand when he gets through applying the distinction between each of them and all the others' [3]. What one can take from this is that construct poles serve as a basis for discrimination between elements. Discriminations may be crisp (Boolean) but they need not be so. They can be fuzzy [12].

In summary, constructs serve as a basis for discriminations involving similarity and contrast. Elaborating constructs in terms of fuzzy set theory, allows one to entertain discriminations other than those of dichotomy, and to overcome the limitations inherent in using interval level rating scales as the means for representing and analyzing those discriminations [14].

2.2. Cognitive Complexity

Bieri defined a complex person as one who can construe others in a multidimensional way [15]. A simple cognitive structure is one in which there are large correlations between the constructs. In the extreme case, one might refer to a person as exhibiting a monolithic structure, when the constructs are highly correlated. A factor analysis of such a system might uncover one factor which accounts for 80% or more of the variation. Complex structures are indicated by a relatively large number of independent constructs. That is, one expects to find lower correlations amongst constructs in a complex system.

Conceptually speaking Bieri's idea of complexity was a novel one, and a welcome addition to the literature. However, cognitive complexity should not be viewed in isolation. The context to which constructs relate and the content of constructs should be evaluated in tandem with measures of cognitive complexity. For example, when one interviews a doctor about her expertise one expects constructs connected with the context to be primarily instrumental in their content. One would not be surprised to observe a complex system indicative of the subtle discriminations which are the hallmark of experts. In contrast, if interviewing the same person about her interpersonal relationships, one expects to elicit expressive constructs, which are reflective of her as a person and which exhibit a greater degree of interdependence. This idea is consistent with Kelly's Range corollary which states that a construct (or a collection of constructs) is convenient for only a defined range of events [1]. People need to switch in an out of construct systems according to the context in which they find themselves. The FPCP model draws on the concept of cognitive- complexity.

3. The Fuzzy PCP Model

The foregoing has prepared the way for modelling personal constructs as fuzzy subsets and for deriving structural measures for construct systems. A primary idea behind the model is that of intraindividual consensus. The idea was prompted after an analysis of the work in [16] [17] [18] [19] [20] [21] [22] and [23]. The work in [23] is a summative paper. It employs fuzzy set theory as the basis for a measure of group consensus. The notion of group consensus which is interindivdual can be applied to the task of assessing the structure of a construct system. If measures of group consensus can be constructed, then measures of grid consensus in terms of the constructs can be developed by applying similar logic. That is, by viewing constructs as members of a synthetic group. What is derived in this work then is a measure of intraindividual consensus in terms of the constructs. The measure can be used as an indicant of the cognitive complexity of a construct system.

3.1. The Mathematics of the Model

The basis of the model is a *Fuzzy Repertory Grid.*

Let **E** be a set of n elements:

$$\mathbf{E} = \{e_1, e_2, \ldots, e_n\} \tag{1}$$

where **E** is a classical set. By using triadic elicitation, or by a similar method, a Fuzzy Repertory Grid $\underset{\sim}{\mathbf{G}}$ is generated. In general, $\underset{\sim}{\mathbf{G}}$ is an $m \times n$ matrix but is usually square. (The $\sim$ denotes matrices and vectors which are fuzzy).

The columns of $\underset{\sim}{\mathbf{G}}$ are the elements of **E** and the rows are *Fuzzy Construct Subsets* of **E**.

A fuzzy construct subset $\underset{\sim}{\mathbf{C}_i}$ of **E** is a set of ordered pairs:

$$\underset{\sim}{\mathbf{C}_i} = \left\{ \left(e_j | \mu_{\underset{\sim}{\mathbf{C}_i}}(e_j) \right) \right\}, \forall e_j \in \mathbf{E}, \quad 0 < i \leq m, \quad 0 < j \leq n \qquad (2)$$

where $\mu_{\underset{\sim}{\mathbf{C}_i}}(e_j)$ is a membership characteristic function which takes its values in the totally ordered set $\mathbf{M} = [0, 1]$ and indicates the degree or level of membership. If $\mathbf{M} = \{0, 1\}$ then the fuzzy construct subset is understood to be a non-fuzzy construct subset or ordinary construct subset [11][ii]

The procedures which have been developed, to generate a measure of construct consensus, and similarity measures for construct pairs are summarized below:

Procedure 1

Transpose $\underset{\sim}{\mathbf{G}}$

Procedure 2

Create m *Fuzzy Construct Matrices* $\underset{\sim}{\mathbf{F}_k}$, $0 < k \leq m$, where m is the number of columns in

$$\underset{\sim}{\mathbf{G}^t} = \left\{ g^t(i,k), \quad 0 < i \leq n, \quad 0 < k \leq m \right\} \qquad (3)$$

Each $\underset{\sim}{\mathbf{F}_k}$ is an $n \times n$ matrix with elements

$$f_k(i,j) = \begin{cases} g^t(i,k), & \text{if } i=j \\ \min\left(g^t(i,k), g^t(j,k) \right), & \text{if } i \neq j \end{cases} \qquad (4)$$

Procedure 3

Using the Decomposition Theorem [11] create *Hard Alpha Level Matrices* such that:

$$\mathbf{H}_{k_{\alpha_l}} = \left\{ h_{k_{\alpha_l}}(i,j), \ \ 0<k\le m, \ \ 0<\alpha\le 1, \ \ 0<l\le q, \ \ 0<i,j\le n \right\}$$

where

$$h_{k_{\alpha_l}} \begin{cases} 0 & \text{if } f_k(i,j) < \alpha_l \\ 1 & \text{if } f_k(i,j) \ge \alpha_l \end{cases} \tag{5}$$

Procedure 4

(a) First, compare all of these hard matrices pairwise in order to determine the level of agreement between them. Accordingly and following [23] define the agreement measure $A\left(\mathbf{H}_{k_{\alpha_l}}, \ \mathbf{H}_{k'_{\alpha_l}}\right)$, as:

$$A\left(\mathbf{H}_{k_{\alpha_l}}, \ \mathbf{H}_{k'_{\alpha_l}}\right) = \frac{tr\left(\mathbf{H}_{k_{\alpha_l}} \mathbf{H}_{k'_{\alpha_l}}^t\right)}{tr\left(\mathbf{H}_{k_{\alpha_l}} \mathbf{H}_{k_{\alpha_l}}^t\right) + tr\left(\mathbf{H}_{k'_{\alpha_l}} \mathbf{H}_{k'_{\alpha_l}}^t\right) - tr\left(\mathbf{H}_{k_{\alpha_l}} \mathbf{H}_{k'_{\alpha_l}}^t\right)}, \ \ 0 < k, k' \le m, \ k \neq k'$$

(6)

where $tr(\cdot)$ and $(\cdot)^t$ denote the trace and the transpose operations respectively

(b) Second, display the results in an *Alpha Level Consensus Matrix* [23]

$$\left(\mathbf{C}_{\alpha_l}\right)_{ij} = \begin{cases} A\left(\mathbf{H}_{k_{\alpha_l}}, \ \mathbf{H}_{k'_{\alpha_l}}\right) & i \neq j \\ 0 & i = j \end{cases} \tag{7}$$

Procedure 5

Calculate an estimate of the α_l - *level Construct Consensus*, K_{α_l} using

$$K_{\alpha_l} = \frac{tr\left(\mathbf{C}_{\alpha_l}^2\right)}{m(m-1)}, \ \ 0 < \alpha \le 1, \ 0 < l \le q \tag{8}^{iii}$$

Recall that m is the number of hard matrices under consideration.

Procedure 6

Repeat Procedures 3, 4, 5, until $\alpha_i = 1$.

Procedure 7

(a)   First, calculate a measure of *Overall Construct Consensus, K* using

$$K = \sqrt{\frac{1}{2l}\left(K_{\alpha_0} + 2K_{\alpha_1} + 2K_{\alpha_2} + \dots + K_{\alpha_l}\right)}, \quad K_{\alpha_0} = 1 \qquad (9)^{iv}$$

where K can be interpreted as a global measure which indicates the cognitive complexity of a construct system. The minimum and maximum values for K are 0 and 1 respectively. At its minimum K indicates a complex construct system with maximal differentiation between the constructs. At its maximum K is indicative of a simple construct system with no differentiation between the constructs.

(b) Second, calculate a measure of the *Overall Pairwise Consensus* of Constructs, $A(H_k, H_{k'})$, using

$$A(H_k, H_{k'}) = \frac{1}{2l}\left(\begin{array}{c} A\left(H_{k_{\alpha_0}}, H_{k'_{\alpha_0}}\right) + 2A\left(H_{k_{\alpha_1}}, H_{k'_{\alpha_1}}\right) + 2A\left(H_{k_{\alpha_2}}, H_{k'_{\alpha_2}}\right) + \dots \\ + A\left(H_{k_{\alpha_l}}, H_{k'_{\alpha_l}}\right) \end{array}\right),$$

$$A\left(H_{k_{\alpha_0}}, H_{k'_{\alpha_0}}\right) = 1 \qquad (10)$$

where $A(H_k, H_{k'})$ can be interpreted as indicating the similarity of the constructs as pairs. When arranged as a matrix these values can be interpreted as a similarity matrix. The minimum and maximum values for $A(H_k, H_{k'})$ are 0 and 1 respectively. When $A(H_k, H_{k'})$ is 0 a construct pair is dissimilar. When $A(H_k, H_{k'})$ is 1 a construct pair is similar.

3.2. An Overview of FUZZYGRID

The model procedures have been incorporated in a software program named FUZZYGRID. This was necessary since the number of matrix multiplications required, excluding those which relate to the Consensus Matrices $\left(C_{\alpha_i}\right)^{ij}$, is

$$\frac{m!}{x!(m-x)!} \times q$$, where m is the number of rows in $\underset{\sim}{G}$, $x = 2$ and q is the number of *alpha cuts*. For example, if $\underset{\sim}{G}$ has 8 rows and $q = 10$, then the number of matrix multiplications required is 280. FUZZYGRID allows these calculations to be done quickly and accurately.

As output FUZZYGRID provides K_{α_i} values, a measure of construct consensus K and two matrices, a similarity matrix for the construct pairs $A(H_k, H_{k'})$ and a dissimilarity matrix for the construct pairs $A(H_k, H_{k'})$. The dissimilarity matrix can be ported to SPSS as input for a multidimensional scaling analysis of the constructs.

Though not shown in this paper the program also produces a consensus measure for the elements, and similarity and dissimilarity matrices for the element pairs.

4. An Application of the Fuzzy PCP Model

This section introduces a case study as an illustration of how the Fuzzy PCP model can be employed to derive measures of cognitive structure. The emphasis is on the relationship between the constructs. However, it will be shown that it is also instructive to examine the relationship between the elements. Output from FUZZYGRID will be used as a basis for discussion and a classification exercise using multidimensional scaling techniques.

4.1. The Context for the Case Study

The repertory grid reproduced in this paper forms part of a single case from a pilot study. The study addressed the way in which men construe their work and non-work domains. Each participant completed three repertory grids, one relating to work and non-work activities, one relating to significant others at work and one relating to significant others in the non-work world. The term participant is used quite deliberately here, to distinguish this kind of study from more traditional work where the term subject is usually employed. It reflects the idiographic philosophy which underpins PCP. The interest is in finding out how the participant construes the domain of interest, not in imposing a predetermined frame of reference to which the participant is asked to respond. The role of the interviewer is to encourage the participant to articulate their own frame of reference, to be unobtrusive.

David the participant in the case reported here is 37 years old . He has an honours degree in linguistics. At the time of interview, he had just returned to the workforce after completing further studies to qualify as a computer programmer. His partner is 38 years of age and has a doctoral qualification. She is a university lecturer. David's response to the question Who am I? was : a pilgrim, a searcher for meaning, a movement between two infinities, a need, a love. This background helps to illuminate the material which follows. The grid shown is that which was elicited from David about his work and non-work activities.

4.2. Grid Elicitation and Representation

Grids were elicited by the method of triads also known as the minimum context card form [1] [3]. Following [25] [26] and [27] participants were asked to write down pairs of activities, one work and one non-work as follows: two activities which they liked, two activities which they disliked, two activities which were important to them and two activities which they engaged in frequently. Each activity was written on a separate yellow card shaped in the form of an ellipse. The interviewer also wrote down the elements on a form which had been designed for recording the elements, the constructs and the ratings. The cards were used because they helped participants to focus their thoughts. Their shape made them easy to move around and fun to play with. They assisted in developing rapport between the interviewer and the participant.

Triads of elements were selected from a schedule which had been constructed to ensure that they were random in nature. For each triad participants were asked to indicate which two elements were alike and in what way. The response was written down in the leftmost column of the form. This word or phrase formed the emergent pole of the construct. The contrast pole was elicited by asking the participant to indicate in what way the third element was different from the other two. The response was written down in the rightmost column of the form. This process was repeated until eight constructs had been elicited.

Following this participants were asked to rate each element against each construct on a scale of [0, 10]. They were instructed to use a rating of 0 if they felt that the emergent pole of a construct described an element very well. If they felt that the contrast pole was seen as describing an element very well then a rating of 10 was to be awarded to that element. Intermediate unit ratings were also permitted to accommodate non-categorical evaluations. The participants were asked to indicate any constructs which did not make sense to them as a dimension in respect of one or more of the elements.

Following each elicitation session the data for each grid was rescaled on [0, 1] and reverse coded. The data was rescaled because earlier work revealed that participants had difficulty in using decimals to rate elements. Furthermore, participants appeared confused when asked to rate elements on the emergent pole (left pole) as 10 and the contrast pole (right pole) as 0. Consequently, it was decided to use 0 to anchor the emergent pole, 10 to anchor the contrast pole and to subsequently rescale and reverse code the ratings so as to reflect membership values for the fuzzy construct subsets described by the emergent pole. Whilst reverse coding is a fuzzy negation [12] the constructs themselves are meaningful only when viewed in terms of both poles. Kelly said that " both similarity and the contrast are inherent in the same construct" [1] , there is a "tie of opposition uniting the duality of meanings in a construct" [26]. Therefore, one is always mindful of the contrast pole even though mathematical analysis is conducted with respect to the emergent pole of a construct.

In the finalised grids scores of 1 indicate elements which are full members of the fuzzy construct subsets described by the emergent pole. Scores of 0 indicate elements which are non-members of the fuzzy construct subset described by the emergent pole. Values between 0 and 1 indicate differing degrees of membership.

The grid which describes David's work and non-work activities and the related constructs is shown below.

Table 1 David's Work and Non-Work Activities and Constructs

	E_1	E_2	E_3	E_4	E_5	E_6	E_7	E_8	
C_1 : Pleasant	1.0	1.0	0.1	0.1	1.0	0.9	0.2	0.9	Unpleasant
C_2 : Meaningful	0.9	0.9	0.1	0.1	0.9	0.8	0.2	0.9	Meaningless for me
C_3 : What is new	0.9	1.0	0.2	0.2	0.9	0.8	0.4	0.9	Boring
C_4 : Quality	0.9	0.9	0.1	0.1	0.9	0.8	0.2	0.9	Poor quality
C_5 : Haphazard	0.1	0.9	0.5	0.1	0.9	0.8	0.8	0.2	Structured
C_6 : Weak side	0.2	0.2	0.9	0.9	0.2	0.3	0.5	0.1	Long suit
C_7 : Enjoyable	1.0	0.9	0.1	0.1	0.9	0.8	0.2	0.9	Not enjoyable
C_8 : Introverted	0.9	0.9	0.1	0.1	0.9	0.6	0.2	0.9	Extroverted

Note : The emergent poles of the constructs which are the names for the fuzzy constructs subsets are shown in the leftmost column.

The elements are:

E_1 : (a work activity I like) Problem solving
E_2 : (a non-work activity I like) Discovering things
E_3 : (a work activity I dislike) Meetings
E_4 : (a non-work activity I dislike) Shopping
E_5 : (a work activity which is important to me) Creativity
E_6 : (a non-work activity which is important to me) Good conversations
E_7 : (a work activity which I engage in frequently) Computer Browsing
E_8 : (a non-work activity which I engage in frequently) Reading

In some respects the elements in David's grid are as informative as the constructs. The predominant theme which shows through is his liking for things which are cerebral and his dislike for the mundane and run of the mill. His preferred milieu is the intellectual.

4.3. Extracting Structural Measures

Repertory grids should be analysed in terms of both structure and content. Structural measures refer primarily to the relationship between constructs. Content analysis focuses on the meaning of the grid. A comprehensive method for assessing the content of repertory grids can be found in [28]. By using the Fuzzy PCP model an overall indication of the structural properties of a construct system can be developed. The main results for David's grid are shown below.

Table 2 Construct Consensus Values for David's Grid

Alpha Values α_l	Construct Consensus K_{α_l}
0.00	1.00
0.10	1.00
0.20	0.53
0.30	0.46
0.40	0.46
0.50	0.55
0.60	0.55
0.70	0.45
0.80	0.45
0.90	0.44
1.00	0.00

Table 2 shows that when α_l is between 0.10 and 0.40 inclusive the consensus between the constructs K_{α_l}, decreases monotonically. Between 0.50 and 0.60 there is an inflection and the construct consensus values increase. Between the values of 0.70 and 1.00 the construct consensus values decrease monotonically. This pattern is not unexpected. The measure of agreement between pairs of constructs is based on an extension of the Tanimoto Coefficient for binary valued data vectors to binary valued matrices [23]. That is, one is interested in counting up the number of ones which occupy the same position in each of two *Hard Alpha Level Matrices*. Whilst it is usual for the number of unit entries to be decreasing as α_l increases, there can be more of these unit entries in common between two hard matrices. This will cause the measure of agreement to increase. Therefore, since the agreement coefficients $A\left(\mathbf{H}_{k_{\alpha_l}}, \ \mathbf{H}_{k'_{\alpha_l}}\right)$ form the elements of C_{α_l}, K_{α_l} can in fact increase as α_l increases. However, in general, the expected pattern is for K_{α_l} to decline as α_l increases. The *Overall Construct Consensus* value K, for David's grid is 0.73. This value indicates that in relation to his work and non-work activities David employs a *tight* construct system. There is not much differentiation between the constructs. This is indicated by examining the pattern of agreement between the constructs in the similarity matrix shown in Table 3 below.

Table 3 The Similarity of Construct Pairs in David's Grid

0.00	0.94	0.85	0.94	0.47	0.41	0.94	0.88
0.94	0.00	0.88	0.97	0.47	0.41	0.97	0.91
0.85	0.88	0.00	0.88	0.49	0.46	0.88	0.81
0.94	0.97	0.88	0.00	0.47	0.41	0.97	0.91
0.47	0.47	0.49	0.47	0.00	0.42	0.47	0.46
0.41	0.41	0.46	0.41	0.42	0.00	0.41	0.41
0.94	0.97	0.88	0.97	0.47	0.41	0.00	0.91
0.88	0.91	0.81	0.91	0.46	0.41	0.91	0.00

Note: Zero entries are used on the main diagonal since the relationship of a construct with itself is redundant.

The matrix appears to indicate one cluster of constructs which involves constructs 1, 2 , 3, 4 , 7 and 8. Constructs 5 and 6 appear to stand alone. The similarity matrix can be used to generate a dissimilarity matrix for the construct pairs. This is done by subtracting the coefficients for each of the elements in the similarity matrix from 1.00. The resulting matrix of dissimilarities is shown as Table 4 below.

Table 4 The Dissimilarity of Construct Pairs in David's Grid

0.00	0.06	0.15	0.06	0.53	0.59	0.06	0.12
0.06	0.00	0.12	0.03	0.53	0.59	0.03	0.09
0.15	0.12	0.00	0.12	0.51	0.54	0.12	0.19
0.06	0.03	0.12	0.00	0.53	0.59	0.03	0.09
0.53	0.53	0.51	0.53	0.00	0.58	0.53	0.54
0.59	0.59	0.54	0.59	0.58	0.00	0.59	0.59
0.06	0.03	0.12	0.03	0.53	0.59	0.00	0.09
0.12	0.09	0.19	0.09	0.54	0.59	0.09	0.00

The values in this matrix form the basis of a multidimensional scaling analysis the results of which are outlined below. For a metric analysis in two dimensions the *S-Stress, Stress* and *RSQ* coefficients are 0.03, 0.10 and 0.99 respectively. Stress coefficients are a measure of error or lack of fit, high values are undesirable. The RSQ coefficient is interpreted in much the same way as the Coefficient of Determination. [29] It is the best indicator of how well the data fit the model [30]. Based on these statistics the two dimensional solution is excellent.

A graphical representation derived from the multidimensional scaling analysis is shown as Figure 1 below.

Figure 1. A Plot of David's Work Non-Work Constructs

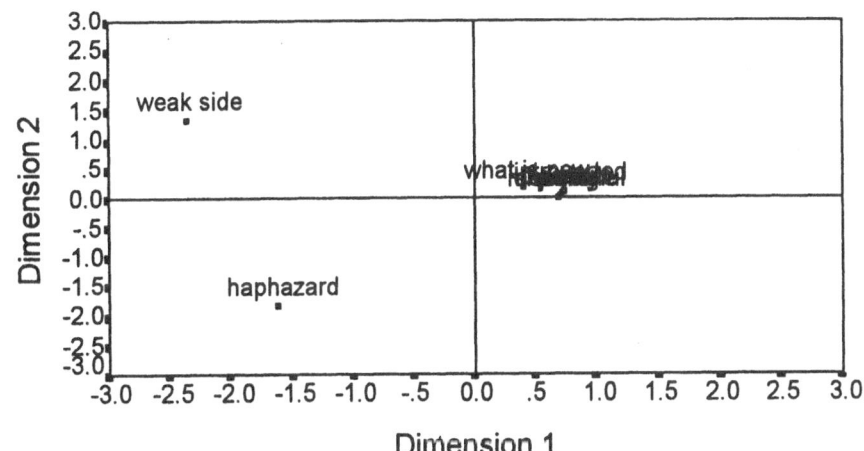

What appears to be a typographical error is a tight construct cluster.
The constructs in the cluster are 1, 2, 3, 4, 7, and 8.

The pattern of constructs in a system can be described in terms of *internal cohesion* and *external isolation* [31]. What is represented here, is a simple construct system in which there is one primary cluster which is cohesive. It stands in isolation from two other constructs which are themselves isolated from one another. The six constructs in the primary cluster 1, 2, 3, 4, 7, and 8 appear to indicate one *superordinate construct* [1]. An umbrella term which describes the emergent pole of this superordinate construct might be *I like*. One notices that in the original grid Meetings and Shopping attract relatively high scores as *Structured* activities (the opposite of *Haphazard* for David) which David aligns with his *Weak side*. They are activities which he dislikes intensely. It would appear that it is the mundane nature of these activities which affects David in a negative way. In contrast activities such as problem solving and reading are construed as bringing out the best in him, they are his *Long suit*. Whilst they are structured they are also intellectual. They appeal to his introverted nature.

4.4. Exploiting the Duality of the Repertory Grid

What has been presented thus far is an analysis of the constructs. Classification studies usually comprise a set of *n objects,* each object being described by several *variables.* Occasionally these 'objects' could possibly be more conveniently described as 'variables' [31]. One can think of a repertory grid as comprising objects (the elements) and variables (the constructs) between which there is a duality. This duality can be exploited, by using the Fuzzy PCP model to generate consensus measures for the elements and the element pairs. This is achieved by omitting Procedure 1 and implementing the other procedures with regard to the elements and not the constructs. As mentioned earlier this process has been incorporated in FUZZYGRID and measures are available for both elements and constructs. The *S-Stress, Stress* and *RSQ* coefficients, from the multidimensional scaling analysis of the elements in David's grid are 0.10, 0.12 and 0.98 respectively. A two dimensional plot of the elements is shown as Figure 2 below.

Figure 2. A Plot of David's Work Non-Work Activities

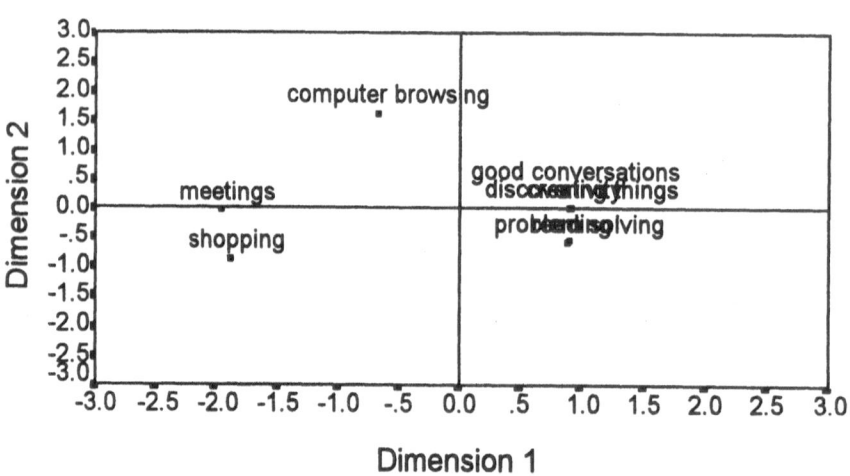

What appears to be a typographical error is a tight element cluster.
The elements in the cluster are 1, 2, 5, 6 and 8.

The plot reinforces what was discovered through the analysis of the constructs. One observes one primary cluster of cohesive elements. They are reflective of the David's intellectual orientation, at work and outside the work environment. *Discovering Things* and *Creativity* appear to be synonymous with one another as do *Problem Solving* and *Reading*. *Good Conversations* are intimately connected with these activities. *Meetings* and *Shopping* stand together as those activities which are disliked. A somewhat related dislike although to a lesser extent is *Browsing on the Computer*.

In summary, the analysis of David's construct system by way of the Fuzzy PCP model articulated with multidimensional scaling methods appears to be very successful. It renders a picture of David which is consistent with the interviewers perceptions. It is supported by many other indicators which were provided by way of a questionnaire and the other grids.

5. Conclusions and Future Research Directions

The primary purposes of this paper were to establish a linkage between Personal Construct Psychology and Fuzzy Set Theory; and to demonstrate that by using this conjunction a model was developed which allows one to assess the structure of a construct system. The model is an attractive one because it incorporates fuzziness. This is more realistic in terms of capturing the way in which human beings construe objects, activities, people, relationships and situations.

The measure of cognitive complexity is also attractive since it is bounded on [0 1] as is the measure of the pairwise similarity of constructs. The duality in a repertory grid has been exploited. A consensus measure for the elements and similarity measures for element pairs can be derived.

It would appear that the model can be applied to the analysis of data emanating from Focus Groups and Group decision making environments. In these contexts the construct dimensions might be product attributes or decision criteria. The elements would represent individual members of a group. It should be possible to identify coalitions in groups as well as isolates. In marketing research applications one might be able to identify consumer types and to target key product attributes.

Further research is being undertaken to allow Weighted Multidimensional Scaling Methods to be employed in conjunction with model results. This would permit one to plot coalitions and isolates in a weight space and to examine temporal changes in perceptions.

George Kelly has been criticised for his lack of academic etiquette [32]. Nevertheless, when viewed in concert with the subsequent development of Fuzzy Set Theory and contemporary work in knowledge engineering, his work should be judged as remarkable and creative.

References

[1] Kelly, G. A. (1955) The Psychology of Personal Constructs, Vols 1 & 2, Norton, New York.

[2] Mancuso, J. C. and Adams-Weber, J. R. (1982) The Construing Person, Praeger, New York.

[3] Bannister, D. and Mair, J. M. (1968) The Evaluation of Personal Constructs, Academic Press, London.

[4] Fransella, F. and Bannister D. (1977) A Manual For Repertory Grid Technique, Academic Press, London.

[5] Easterby-Smith, M. (1981) "The Design, Analysis, and Interpretation of Repertory Grids", in Shaw, M. L. G. (ed), Recent Advances in Personal Construct Technology, Academic Press, London.

[6] Slater, P. (ed) (1977) The Measurement of Intrapersonal Space by Grid Technique, Vols 1 & 2, Wiley, London.

[7] Stewart, V. and Stewart, A. (1981) Business Applications of the Repertory Grid, McGraw-Hill, London.

[8] Huff, A. F. (ed) (1990) Mapping Strategic Thought, Wiley, Chichester.

[9] Shaw, M. L. G. and Gaines, B. R. (1992) "Kelly's Geometry of Psychological Space and its Significance for Cognitive Modeling", The New Psychologist, October, 1-14.

[10] Zadeh, L. A. (1965) "Fuzzy Sets", Information and Control, 8, 338-353.

[11] Kaufmann, A. (1975) Introduction to the Theory of Fuzzy Subsets, Vol 1, Academic Press, New York.

[12] Smithson, M. (1987) Fuzzy Set Analysis for the Behavioral Sciences, Springer Verlag, New York.

[13] Bahm, A. J. (1970) Polarity, Dialectic and Organicity, Thomas, Illinois.

[14] Hesketh, B., Pryor, R., Gleitzman, M. and Hesketh, T. (1988) "Practical Applications and Psychometric Evaluation Of a Computerized Fuzzy Graphic Rating" in Zétényi. T. (ed), Fuzzy Sets in Psychology, North Holland Amsterdam.

[15] Bieri, J. (1955) "Complexity-Simplicity and Predictive Behavior", Journal of Abnormal Social Psychology, 51, 263-268.

[16] Blin, J. M. and Whinston, A. B. (1974) "Fuzzy Sets and Social Choice", Journal of Cybernetics, 3, 28-36.

[17] Blin, J. M. (1974) "Fuzzy Relations in Group Decision Theory", Journal of Cybernetics, 4, 17-22.

[18] Bezdek, J. C., Spillman, B. and Spillman, R. (1977) "Fuzzy Measures of Preference and Consensus in Group Decision Making" IEEE, Proceedings of Conference on Decision and Control, 1303-1308.

[19] Bezdek, J. C., Spillman, B. and Spillman, R. (1978) "A Fuzzy Relations Space For Group Decision Theory", Fuzzy Sets and Systems, 1, 255-268.

[20] Spillman, B., Spillman, R. and Bezdek J. C. (1979) "Fuzzy Relations Spaces For Group Decision Theory: An Application", Fuzzy Sets and Systems, 2, 5-14.

[21] Spillman, B., Spillman, R. and Bezdek, J. C. (1979) "Coalition Analysis With Fuzzy Sets", Kybernetes, 8, 203-211.

[22] Spillman, B., Bezdek, J. C. and Spillman, R. (1979) "Development of an Instrument for the Dynamic Measure of Consensus", Communication Monographs, 46, 1-12.

[23] Spillman, B., Spillman, R., and Bezdek, J. C. (1980) "A Fuzzy Analysis of Consensus in Small Groups", in Wang, P. P. and Chang, S. K. (eds), Fuzzy Sets Theory and Applications in Policy Analysis and Information Systems, Plenum Press, New York.

[24] Zadeh, L. A. (1971) "Quantitative Fuzzy Semantics", Information Sciences, 3, 159-176.

[25] Brook, J. A. and Brook, R. J. (1989) "Exploring the Meaning of Work and Nonwork", Journal of Organisational Behaviour, Vol, 10, 169-178.

[26] Knowles, M. C. and Taylor, D. (1990) "Conceptualizations of Work, Family and Leisure By Managers of Information Technology", International Journal of Psychology, 25, 735-750.

[27] Brook, J. A. (1993) "Leisure meanings and comparisons with work", Leisure Studies, 12, 149-162.

[28] Landfield, A. W. and Epting, R. E. (1987) Personal Construct Psychology-Clinical and Personality Assessment, Human Sciences Press, Inc, New York.

[29] Norusis, M. J. (1994) SPSS Professional Statistics 6.1, SPSS Inc, Chicago.

[30] Schiffman, S. S., Reynolds, M. L. and Young, F. W. (1981) Introduction To Multidimensional Scaling, Academic Press, New York.

[31] Gordon, A. D. (1981) Classification, Chapman Hall, London.

[32] Neimeyer, R. A. (1985) The Development of Personal Construct Psychology, University of Nebraska Press, Lincoln.

---

[i] The term Fuzzy-PCP has been used by Vladimir A. Geroimenko in a paper entitled *Personal Construct Theory and Fuzzy-Set Theory*. The paper was presented at the 2nd European Conference of the European Personal Construct Association, St Andreasberg, Germany, April 19-22, 1994. The author would like to express his appreciation to professor Geroimenko for the helpful correspondence on the potential for a Fuzzy-PCP.

[ii] This is an extension of the work in [11]. In [11] reference is made only to *fuzzy subsets* and *ordinary subsets*. The terms *fuzzy construct subset, non-fuzzy construct subset* and *ordinary construct subset* are new. The extension is supported by Zadeh in [24] as the following illustrates:

Let U be the universe of objects which we can see. Let T be the set of terms *white, gray, green, blue, yellow, red, black*. Then each of these terms, e.g., *red*, may be regarded as a name for a fuzzy subset of elements of U which are red in color. Thus, the meaning of, *red, M(red)*, is a specified fuzzy subset of U.

The ideas articulated above can be applied directly to modelling constructs as fuzzy subsets. It is for this reason that the term *fuzzy construct subset* has been introduced. In terms of PCP the elements of a grid are analogous to the *universe of objects U*. Constructs are analogous to the *set of terms* which can be used to describe the elements of U. For example, suppose a person is asked to name 8 people with whom they work. Constructs are then elicited about those people. One such construct might be *Manager Type/Non-Managerial*. The emergent pole of the construct *Manager Type*, identifies a *fuzzy construct subset* of which the elements of U (the work colleagues) may be full members, partial members or non-members. Thus following Zadeh [24] the meaning of *Manager Type, M(Manager Type)* is a specified fuzzy *construct* subset of U.

[iii] In [23] the expression used to derive an alpha level measure of group consensus K_α has 2 in the numerator. Using the data in [23] a manual recalculation of the K_α produced results which could not be reconciled with those reported in [23]. In all cases the values derived for K_α were double those reported in the paper. In some cases the K_α exceeded 1.0. Eliminating 2 in the numerator rectified this problem. Equation (8) reflects this amendment.

[iv] In [23] the *root mean square* does not appear in the expression for deriving an estimate of group consensus K. The K_{α_i} are found by multiplying the Consensus Matrices $\left(C_{\alpha_i}\right)_{ij}$ by their transpose and averaging the trace. Multiplying the Consensus Matrices by their transpose is equivalent to squaring them as shown in (8). Therefore, it seems sensible to apply the *root mean square* to the K_{α_i} in (9).

Acknowledgment. The author would like to thank Dr Roger Wallace for valuable comments on earlier drafts of this paper and Mr George Nowara for providing programming assistance during the development of FUZZYGRID.

ON THE TOPOLOGY OF UNCERTAINTY

Donald H. McNeil* and Vladimir Dimitrov*

*2721 Estella Avenue, Montoursville, PA 17754 USA
** Centre for Research in Healthy Futures, University of Western Sydney
Bourke St., Richmond, NSW 2753, Australia
E-mail: V.Dimitrov@uws.edu.au

Abstract

The only thing of which we can be certain is that uncertainty is the rule and certainty is the exception. In our technocratic culture, correctness of formulations, exactitude of measurements, accuracy of predictions, and guarantees of control are highly esteemed, but in our reality these hold true only in rather special cases which are rare and very limited in applicability. Certain schemes have been developed to try to reduce our uncertainties, e.g., fuzzy logic, statistics, chaos theory, and indeed the sciences and mathematics generally. This paper offers a perspective of our fabricated certainties which — rather than trying to minimize uncertainty — enlarges upon it to explore how the topology of our construed systems co-determines our uncertainties. Three permutations of certitude are identified: rational, compelling, and systemic. The authors are certain that this paper will raise more questions than it answers, and we invite further elaborations upon uncertainty from readers.

Keywords: certitude, uncertainty, topology, toroid, heterarchy, complementarity, compellors, systems.

1. Introduction

It ain't what we don't know that makes trouble; it's what we know that ain't so.
[Will Rogers]

Although it is convenient and often very useful for us to posit exact formulations and to construct precise geometrical arrangements, it is not in the nature of things that the world can be so neatly known. Certainty is the exception; uncertainty is the rule. Even the techniques which we employ to try to accommodate inexactitudes, errors, tolerances, mutuality, stochasticity, and "fuzziness" have their limits, especially insofar as they fail to place our paradigms into a perspective which appreciates uncertainties as inherent, inevitable, necessary, even desirable.

We are necessarily uncertain about uncertainty. It is uncertain how many different kinds of uncertainty there may be, and indeed there may be uncertainties of which we are ignorant. For example, neither the quantum uncertainties of the microcosm nor the "chaotic" behaviors of the macrocosm were appreciated until rather recently in the history of human sciences. It is possible not only that we may be ca tegorically unaware of an indefinitely large universe of uncertainties but also that there may be categories of uncertainty which human nature can never identify.

There are many possible contexts for uncertainty: underdimensionality of analysis and redundancy of details; changeability and ephemerality; lack of perspective and obscured vantage points; hidden inner mechanisms and cropped outer en vironments; inaccurate measurement and overprecise reporting; lack of standards and incommensurability of measures; random irregularity and regular randomness; irreducible organization and inappropriate partitioning; undecidability of algorithms and unconstructability of models; incalculability and uncomputability; actual variations and observational variances; slow rates of sampling and low bandwidth of communications; poor resolution of instruments and wrong calibration of instrumentation; invariance and undistinguishability; many-to-one mappings and one-to-many mappings; invisible interactions with observers and blurred observation of interactions; ignorance of optimization and amorphous criteria for optimization; lapses of attention and unfocused perception; uncontrollable factors and conflicting controls; unfitting formulations and incorrect calculations; aberrant ends and inappropriate means; servomechanical deviations and judgmental errors; inconsistent orthodoxy and confusing paradoxicality; absence of standards and incompetence of paradigms; contingent events and unknown factors; missing facts and excessive data; contradictory information and outright disinformation; indescribable phenomena and unarticulatable concepts; equivocal expression and incoherent explanation; disorientation and distraction; limited perspective and lack of comprehension; dullness of perception and befuddlement of cognition; drifting indecision and outright ignorance; inconstant purpose and willful deception. One common theme among various uncertainties is that something is missing — be it dimensionality, information, coherency, constancy, etc. Thus uncertainty itself pertains to an absence, a void, something outside, something which is not there. Even where uncertainty arises in a flood of stimuli or an overabundance of conflicting information, there is something missing, namely the dimensionality of representation or the bandwidth of comprehension to organize it all.

We organize our certainties into systems and externalize our uncertainties into the environment. It therefore behooves us to recognize the features of systemicity which matter to appraising our certainties and to take perspective of how certainties are associated, related, and inter-connected as systems in an environment of uncertainty. To do so we can turn to topology as a scheme for representing the systemicity of certainty and uncertainty.

2. Reflections upon the Hole

Topology is the study of position in general and, in particular, of the connectedness among positions. It is primarily concerned with the <u>continuality of orderings</u>, re gardless of geometrical measures. In the realm of topologies which we can readily visualize, i.e., two dimensional surfaces extended in three dimensional space, to pology tells us that there can be only four distinct and fundamental topological schemes: the sphere (S), the torus (T), the Klein bottle (K), and the projective plane (P) as sketched in Figure 1:

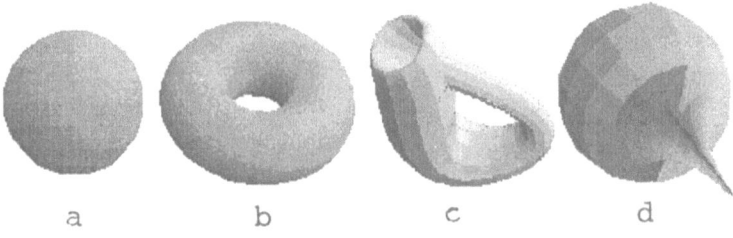

a   b   c   d

Figure 1

The classification theorem of topology tells us that any compact two dimensional surface whatsoever can be constituted of the connected sum of N tori, supplemented by one Klein bottle or by one projective plane for surfaces which are not oriented [4]. In particular, any compact <u>oriented</u> surface is comprised of N tori (where, in the degenerate case that N = 0, the surface is that of a sphere). Thus there is a topological similarity between the cyclonic tempest and the tranquil teacup and indeed between these and a length of water conduit or a wedding ring or a fresh bagel. Moreover, besides brachiation around multiple holes, toroids may be deformed, flattened, extruded, twisted, folded, even knotted while still remaining true to their topological type.

A world of meaning can be attributed to the Klein bottle [9] and to the projective plane [12], but for the purposes of this paper we shall focus upon the oriented sur faces of the sphere and the torus. The sphere has been used and abused as an ar chetype for metaphors in the arts and the sciences for at least as long as anyone has kept notes. Clearly, however, when we consider any generalized oriented surface, a toroid — possibly with more than one hole — is the definitive element. It is not surprising, therefore, that tori are ubiquitous in nature. We see toroidality in the funnel of the tornado and in the body of the cyclone; in the rotational momentum of a gyro and in the field of a magnet; in the whirlpool in the stream and in the dust devil along the road; in the dynamics of a living plant (around its stem) and in the anatomy of a living animal (around its alimentary canal); in every pipe, tube, chan nel, and sieve; and in any manifest circuit, network, web, and lattice. A toroidal topology arises wherever there is an externally constrained flux, e.g., in the con vection of a centrally heated and peripherally cooled fluid in a closed container. An internally self-limited flux is toroidal also, e.g., a vortical flux in where a current and a cross-current interact to form a vortex. Thus the tore is the natural three dimensional topology of asymmetrical, dynamical processes, in contrast to the sphere which is the natural three dimensional topology for symmetrical quiescence and stasis within a minimal surface area.

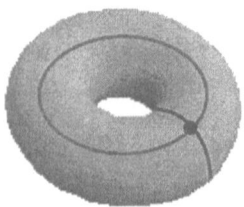

Figure 2

Whatever other significance attaches to it, the torus represents a cyclical ordering of a cyclical order, e.g., the rotation of a closed figure around an axis or the rolling of a plane into a cylinder and then connecting the open ends. It has been asserted that the ordering of order constitutes the very meaning of generalized systemicity [11, 7]. What makes the tore such a powerful image for representing systemic or ganization per se is that the ordering order need not be commensurable with the order which it orders. An image of this is offered by Figure 2 where the different textures of the two intersecting cyclical lines suggest how the annular order around a toroidal hole may be incommensurable with the meridial order through a toroidal hole. Such a topological accommodation of complementary incommensurables makes it possible to rationalize about the irrational and be irrational about the ra tional both at once and without paradox in a systemically toroidal context. It also allows us to transcend the bounds of linear imagery, since every annular and me ridial trace upon a toroidal topology forms an enclosing loop.

The varieties of systemic consequences which flow in, from, and through a toroidal topology are too rich to elaborate upon in this paper. It is, however, of the utmost importance to observe that the tore is not a simply-connected surface such as a plane or a sphere is. Thus, upon a the surface of a torus, two closed cyclical traces may intersect one another at only one point, as suggested by Figure 2. This means that any system which has a toroidal topology is organized heterarchically, i.e., has two or more complementary influences at any site upon its surface. As Warren

McCulloch pointed out so very long ago, this means that no simple hierarchy can be used to explicate even the most elementary of dynamical phenomena, least of all a system of values such as a person might hold [6] ... or any system of putative certainties. The irreducible heterarchical organization of the tore helps us to visualize, for example, how we can talk about the monetary value of an emotion, yet cannot reduce emotion to money in any fully-dimensioned system of values.

A degree of uncertainty is inevitable — indeed necessary — if there is to be room for variegation, creativity, willful decisions, and corrective feedback, but there is a difference between uncertainty and bewilderment. While there can be no such thing as an omniscient "superobserver" [11], there are better and worse perspectives from which to appreciate any system, including systems of certainty and uncertainty. For example, it is easier by far to understand the nature of a whole cyclonic storm from the vantage point of a satellite than from within the gale of the storm itself. Our taking a topological perspective can reduce the ambiguities associated with underdimensioned characterizations of systemic organization and thus helps us to clear away superfluous uncertainties so as to make room to consider the substantial uncertainties which matter.

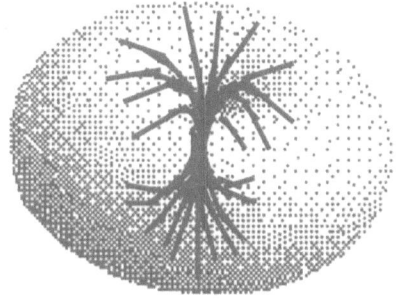

Figure 3

3. Ascertaining the Topologies of Certain Certitudes

On a topology of certitude, where might a *rational certitude* lie? From Figure 3 we can imagine an inward (inductive) branching and an outward (deductive) branching through the core of a central logical premise. Such brachiations might constitute the more or less consistent structure of a particular rationality and can be as elaborate as ratiocination can devise. In the whole picture, however, the rational structure is necessarily suffused in a bath of diffuse meta-rational (abductive) cognitions which are more or less valid in their own right, even though they cannot be reduced to the rational scheme of certitude. Moreover, insofar as the rationalized branchings are only piecewise continuous, as suggested by the dashed lines in the figure, threads of rational certainty are complemented throughout by meta-rational certainties and by uncertainties as well. Thus the rational and the meta-rational are essential to one another, even if they are mutually incommensurable, and they can be represented together in one image such as Figure 3 without paradox. This provides us one way to assign a visual meaning to Gödel's theorem about the deep complementarity of consistency and completeness. It also offers a visualization of how our certitudes of <u>thought</u> mutually complement the certitudes which we <u>feel</u> and how our quantitative certitudes mutually complement our qualitative certitudes. Such a representation is not unique to the overall topology of certitude, however, for it reappears in the topology of the mechanisms by which we generate certitude insofar as our brachiated nervous system is everywhere mutually complemented by our hormonal system which conditions the neuronal surroundings, e.g., through the synaptic gaps [10].

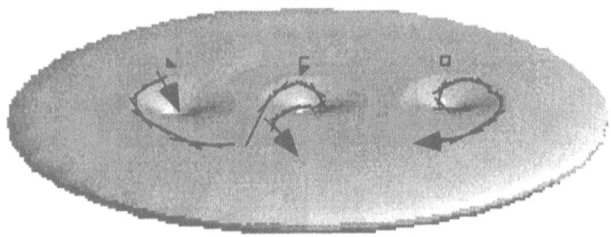

Figure 4

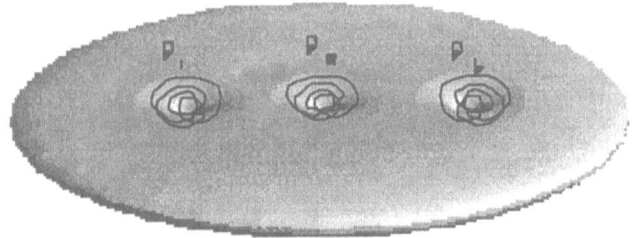

Figure 5

Some certitudes are more compelling than others. To visualize a *compelling cer titude*, we can take a perspective of the topography of <u>compellors</u>, i.e., attractors, repellors, and saddles, in a topology of certitude and visualize them as toroidal holes in a somewhat flattened tore such as shown in Figure 4. In general, these to pological holes may stand in any relationship to one another, may move about, may fluctuate in size and curvature, may combine with or obliterate one another, and may appear and disappear subject to the dynamics of the whole phenomena which they represent. In Figure 4 the hole at A behaves as an attractor, the hole at R be haves as a repellor, and the hole at S behave as a saddle (first attracting and then repelling). We could use such an image to visualize how one might be attracted from a plateau of neutrality or indecision or uncertainty toward a catchment of cer tainty such as A, perhaps by indirect and circuitous paths. In other situations one might be repulsed away from a certitude such as R or first attracted to then repelled from a certitude such as S. In a topology which has several compelling but distinct certitudes — such as C_1, C_2, and C_3 in Figure 5 — we might hold one or another or several at once, but with a degree of uncertainty which keeps our cognitions in fluc tuation in a kind of fractal limit cycle as regards each one. If compelling certitudes overlap as suggested by Figure 6, there can be mutually supplementary gradations of certitude regarding each one, hence a fuzziness of the logic of certainty and un certainty among them. At the same time there may be a "chaotic" trajectory of co gitation which interlinks them. Thus we can represent both "fuzziness" and "chaos" together without paradox in one image of a topology of certitude as below.

Finally let us consider *systemic certitude*. Certitude is, at the very least, systematic, since it is by definition orderly and repeatable. Suppose that we define a <u>closed certainty</u> to be that region which is entirely within a topological surface, assuming that we can ascertain where the bounding surface is at every instant, and define

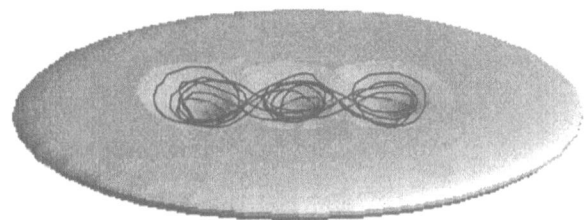

Figure 6

uncertainty to be everywhere else. To do this is quite consistent with the way in which sciences put their subject matter "under control," first setting it apart as in an environment-free laboratory, then taking it apart for analysis. Note that while a spherical topology captures a kind of static certainty, a toroidal topology embraces a kind of dynamic certainty. Clearly, by deformation of its surface, any part of an interior of certainty may be brought arbitrarily close to the external uncertainty. Furthermore, cavities may indent and protuberances may distend the topography of a surface in such a way as to permeate the region of certainty with uncertainty th roughout, though simply-connected paths of certainty may bypass the intruding uncertainties. In the case of a toroidal surface, however, one or more holes of un certainty may pass entirely through the certainty in such a way that simply-connected paths of certainty become multiply-connected networks instead.

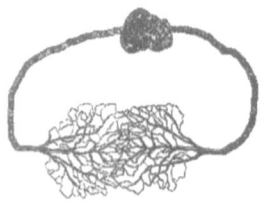

Figure 7

Conceived topologically, closed certainties can be seen to be organized, e.g., spherically in stasis or toroidally in dynamics. By the classification theorem [4], the toroidal topology offers the more generality, however, allowing not only for simple heterarchies of certainty but also for elaborate networks of certainty such as suggested by Figure 7. Here a circulating dynamical certainty is richly crumpled into a manifold (shown at the top) and broadly dispersed in a branching manifest (shown below).

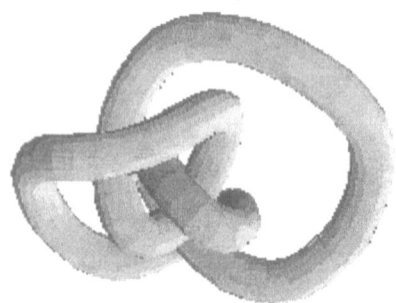

Figure 8

A toroidal topology of certainty can be further elaborated into knottedness, as suggested by Figure 8. A knot of certainty may impart yet a higher order to the consistency of certainty and to its interpenetration by uncertainties. Of course, we can combine the topological operations of brachiation and of knotting to produce certainties of any desired complexity which we care to imagine.

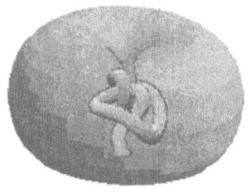

Figure 9

All of the closed systems of certainties mentioned heretofore are supplemented by everything else, i.e., the surrounding environment of uncertainties, so there is inevitably a <u>topology of uncertainty</u> which plays the role of ground opposite the figures of certainty which we have supposed. But uncertainty itself may form an interposing topology, as suggested in Figure 9 where the toroidal hole in a certain whole is knotted. Thus, while Figure 8 illustrates a certain complex knotted whole, Figure 9 illustrates a knotted hole, i.e., a complex of uncertainty.

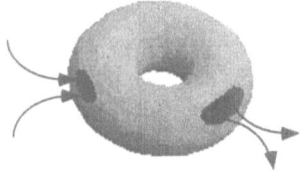

Figure 10

But all systemic certitude is not closed, i.e., certainty my be profused with uncertainties and vice versa. An <u>open certainty</u> might be represented topologically as in Figure 10 where a toroidal surface is punctured so as to allow uncertainty to flow in and certainty to flow out. This corresponds to the notion of an open system which intakes organized energy and diffuse information from its environment and outputs diffuse energy and orderly information to its environment ... and vice versa. Information input serves to organize and revise certitude, enabling us to make certain; an influx of uncertainty may dissipate existing internal certainties, making us uncertain. An outflow of certainty serves to introduce certain actualities into the uncertain environment; in this way a system of partial certainty can itself produce partiality and serve to organize an uncertain environment somewhat.

Certainty, after all, is all in the mind. Whatever the world of actualities may be, humans fabricate their certainties semiotically out of perception and interpretation [8]. thought establishes a standard against which other conceptions and perceptions may be repeatably measured. Moreover, we impose our purposes, i.e., intended ends, upon the uncertain world as <u>imaginary attractors</u> — be they rational, compelling or systemic — which offer the pseudo-certainty of equifinality achieved through deliberate causality [7]. Beneath the topography of a cognitive mindscape, then, we surely have a multiply-connected topology of compelling goals which first take shape as mental certainties and only become certified in reality insofar as they eventually become actualized.

4. Conclusion

Certainly there is much more to uncertainty than any single paper can begin to address. What can be asserted briefly, however, is that certitude is not to be found at large in nature but rather is for us to determine and make and grasp. Whatever may be the case in the world of actualities, people necessarily live in a construed world where certainty is created, re-created, and tested every day. In this paper we have identified the *rational*, the *compelling*, and the *systemic* certitudes as three possible artifices for coping with uncertainty. Each of these can be recognized in a topological perspective. The topology of our fabricated certainties is co-determined with the topology of everything else, i.e., with the topology of uncertainty. In all of this, the tore offers a rich paradigmatic image with which to visualize the heterarchical organization of the systems of concern. It helps us to understand how — in a world of change — the (spherical) nugget of actuality which we seek is not there to be discovered, since there is rather a hollow (toroidal) core at the center of every dynamical system which we can think of. The authors are uncertain what the reader will make of the topological approach to uncertainty proposed in this paper. We are partial to this approach, however, and we are certain that neither certainty nor uncertainty can be appreciated from less than a topological perspective.

References

1 Abraham, Ralph H., and Christopher D. Shaw, <u>Dynamics: The Geometry of Behavior</u>. New York: Addison-Wesley, 1992.
2 Ashby, W. Ross, <u>An Introduction to Cybernetics</u>. New York: John Wiley & Sons, 1966.
3 Casati, Roberto, and Achille Varzi, <u>Holes and Other Superficialities</u>. Cambridge: MIT Press, 1994.
4 Firby, P.A., and C.F. Gardiner, <u>Surface Topology</u> (Second Edition). London: Ellis Horwood Ltd., 1991.
5 Kosko, Bart, <u>Fuzzy Thinking</u>. London: HarperCollins, 1994.
6 McCulloch, Warren S., <u>Embodiments of Mind</u>. Cambridge: MIT Press, 1965.

7   McNeil, Donald H., "Re-framing Systemic Paradigms for the Art of Learning." Presented at the American Society for Cybernetics conference in Philadelphia, PA, in November 1993.

8   Merrell, Floyd, Semiosis in the PostModern Age. West Layfayette, Indiana: Purdue University Press, 1995.

9   Rosen, Steven M., Science, Paradox, and the Moebius Principle. Albany: SUNY Press, 1994.

10 von Foerster, Heinz, "Cybernetics of Epistemology." Proceedings of the 5th Congress of the Deutsche Gesellschaft fur Kybernetik, 1974.

11 Weinberg, Gerald M., An Introduction to General Systems Thinking. New York: John Wiley & Sons, 1975.

12 Young, Arthur M., The Geometry of Meaning. New York: Delacorte Press, 1976.

Part 3

Fuzzy System Design
for Engineering Applications

FUZZY CONTROLLER DESIGN FOR DIFFERENT APPLICATIONS: EVOLUTION, METHODS, AND PRACTICAL RECOMMENDATIONS

Leonid Reznik

Department of Electrical and Electronic Engineering
Victoria University of Technology
P.O. Box 14428 MCMC
Melbourne, VIC 8001
Australia
E-mail: leonreznik@vut.edu.au

Abstract. This paper attempts to classify different approaches applied in a fuzzy controller design until now and develop some general recommendations which could be useful in different applications ranging from engineering to social studies. The author's goal is to present not a mathematical theory but a procedure explaining some aspects of FC design. The design process is roughly divided into two stages: an initial choice of a controller structure and parameters and their further tuning. At the first stage the recommendations are given regarding the choice of the structure, scaling factors, rules and membership functions, though main attention is paid to membership functions and scaling factors. Different methodologies such as neural networks, genetic/evolutionary algorithms are considered at the second stage.

Keywords: fuzzy controller design, classification, parameter choice

1. Introduction

The history of a fuzzy controller design, originating from fuzzy sets and logic theory introduced by L. Zadeh in 1965, started as a practice with the pioneer experiments by I. Mamdani in 1974. Since then many successful design projects have been completed and a lot of applications have been reported in various areas. However, until now the design process was conducted in mainly intuitive, ad-hoc manner and results obtained sometimes looked rather 'mysterious', that even gave birth to different attempts to 'demystify' fuzzy control.

Investigating roots of this situation and fuzzy control itself the following reasons could be put forward, first of all:

1. Fuzzy control theory and especially fuzzy controller design theory are far from being completed and even not close to being developed. In 1990 C.-C.

Lee stated in his survey [9] that "there is no systematic procedure for the design of a fuzzy controller". In 1993 another author [2] asserted that "the design methodologies are in their infancy and still somewhat intuitive".

2. Fuzzy controller design occupies a 'boundary line' research field in which quite different approaches are applied, artificial intelligence (expert systems) and control engineering and optimisation theory, to name main of them. Combined these different methods could significantly enrich a fuzzy control methodology, bringing up new amazing results. However, behind the approaches could be seen different communities, traditionally applying various methodologies and criteria of their evaluation and having some lack of understanding and sometimes even some misunderstanding of each other methods.

On the stage of fuzzy control there are three main players, or better to say teams, trying to apply their traditional approaches and tools in FC design:

- mathematicians, both "pure" and "applied" ones, utilising methods of an optimisation theory, operation research, etc.,
- specialists in artificial intelligence, considering FC design as an application area of knowledge based systems,
- "conventional" control engineering theoretics expanding traditional control systems design approaches to fuzzy control.

Unfortunately all these teams do not play in unison causing confusion and frustration among designers community and braking down the development of a FC design technology.

This paper intends to analyse and classify different methods of FC design. Not attempting to develop a mathematical theory it aims at providing some practical recommendations how to design a FC and how to choose the right design method, first of all. The goal is to develop a more or less systematic procedure of FC design which can be used as a reference.

The design process is roughly divided into two stages: an initial choice of a controller structure and parameters and their further adjustment. At the first stage the recommendations are given regarding a choice of the structure, scaling factors, rules and membership functions, though the main attention is paid to membership functions and scaling factors. Different methodologies such as neural networks, genetic/evolutionary algorithms are considered at the second stage.

2. Design approaches classification

Basically all the approaches to FC design can be classified as follows:

1. expert systems approach,

2. control engineering approach,
3. intermediate approaches,
4. combined approaches and synthetic approaches.

The first approach originates from the methodology of expert systems. It is justified by a consideration of a FC as an expert system applied to control problem solving. In this approach fuzzy sets are applied to represent the knowledge or behaviour of a control practitioner (an application expert, an operator) who may be acting only on the subjective or intuitive knowledge. All the theoretical and practical methods of knowledge acquisition developed in artificial intelligence and other sciences are to be practiced here. One should note that by using linguistic variables fuzzy rules provide a natural framework for the human thinking and knowledge formulation. Many experts find that fuzzy control rules provide a convenient way to express their domain knowledge. So cooperation with the experts will be easier for a knowledge engineer. This approach was very popular in a design of pioneer FC's.

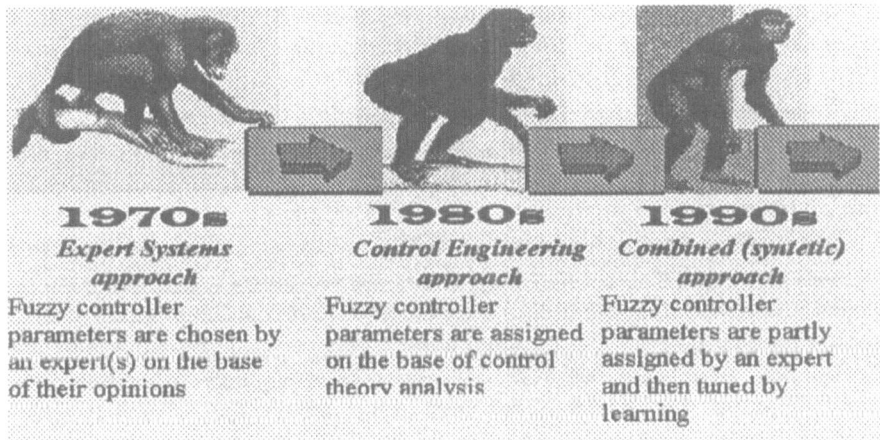

Figure 1. Fuzzy controller design methodology evolution

The supporters of the control engineering approach consider the first one as too subjective and prone to errors and try to make a choice based on some objective criteria. This approach proposes to design a FC by investigating how the FC stability and performance indicators depend upon different FC parameters. Thus this approach clearly incorporates the analysis of a FC as one of the important stages of design.

To evaluate a quality of a FC the criteria commonly used in control engineering practice are applied. As a performance indicator one can apply either of the following:

1. an integral criteria such as an integral of the absolute value of the error signal, an integral of the square of the error signal, etc.
2. one or a set of the parameters characterising the system response, e.g. the overshoot, the settle time, the response time, the steady-state accuracy.

The application of the same criteria facilitates a fair comparison of the conventional and FC's.

In a pure expert approach the choice of the structure, inputs, outputs and other parameters of a FC is the whole and solemn responsibility of the expert(s). Moreover, the supporters of this approach warn against further parameter modifications, pointing out that such adjustments can jeopardise the expert's instructions. Changing, for example, the scaling factors and/or membership functions may result in loosing original linguistic sense of a rules base. The experts may not recognise their rules after tuning and will not be able to formulate new rules. Generally speaking, here an expert system is designed. This expert system is specified for control applications and, after the design is completed, operates as a FC. In this approach any structure and set of the parameters of the FC can be chosen.

In the control engineering approach the feedback structure of the FC is commonly applied with the error signal chosen as one of the inputs. Here fuzzy PID-like (as well as PD-like, PI-like) controllers are extremely popular. The choice of the controller type determines other inputs and an output of the FC. The membership functions and scaling factors are selected on the base of their influence on the FC control surface, and rules are formulated considering the control trajectory.

Intermediate approaches suppose setting some of the parameters (e.g. membership functions) by the experts and fixing the others (e.g. rules) with the methods inherited from the control system design. Combined approaches include the initial choice of the FC structure and parameters made by the expert and further their adjustment performed with the control engineering methods. The development of these methods has lead to the application of models which computationally synthesise properties of expert production systems, neural networks, and fuzzy logic. The example of such methodology is ARTMAP [3] - a family of self-organising neural architectures that are capable of rapidly learning to recognise, test hypotheses, and predict consequences of analog or binary input patterns occurring in nonstationary time series.

Another area of a combined approach application has come from control engineering practice. In a typical for industry PID controller design, the controller parameters are determined initially and tuned after that manually to achieve a desired plant response. In this approach manual tuning can be replaced with a FC supervising a tuning process [4]. The resulting improvements in the system response are accomplished by making on-line adjustments to the parameters of the FC [17].

The simplified diagram of a historical development of the FC design methodology is presented in Fig. 1. It should be noted that an expert systems approach was very popular at the beginning though, it is being applied nowadays as well, of course, in a modified way. The example of such an application is given in [8] where a multiresolutional search scheme based on genetic algorithms is employed in a FC design.

All these approaches are equally right, and the goal of this paper is not to establish which one is more equal than others. From the theoretical artificial intelligence (AI) point of view fuzzy control can be observed as a small application part within a framework of approximate reasoning. However, from the practical point of view this small part looks covering a lion's share of all successful fuzzy technology implementations developed by now.

Considering the advantages of both (AI and control engineering) approaches one can conclude that the AI approach allows to capture in a FC design the vagueness of a human knowledge and express the design framework with natural languages. It leads to that feature of FC which becomes more and more important, especially in design applications: the design process of a FC becomes more understandable, looks less sophisticated and superficial to a human designer and becomes more attractive and threfore cheaper than a conventional one.

Control engineering approach allows to apply in a FC design traditional criteria and develop design methodologies to satisfy conventional design specifications including such parameters as e.g. overshoot, integral and/or steady-state error. Enhancing FC engineering methods with an ability to learn and a development of an adaptive FC design would significantly improve the quality of a FC, making it much more robust and expanding an area of possible applications.

One can argue which control methodology is better. It is very hard to find out a general answer, though, it becomes more and more obvious that advantages of FC should be looked for first of all in a design domain. FC methodology lets design capture a treasury of a human knowledge, express this additional information within a natural language framework and utilise it. This feature allows not only to absorb some extra data to increase the design quality but to make a design process more understandable and attractive to a human designer.

Although this paper attempts to incorporate all methods available, basically the following part includes analysis and design methodology typical for control engineering. Interesting enough, control engineering nowadays proposes methods of learning which can be applied in expert system design as well. So one can see here another way of an interaction for mutual benefits.

3. How to design a FC?

A FC is overparametrised. It means that there are too many parameters influencing its control surface. On the other hand, the same or similar effects can be reached by changing different parameters. This makes a comprehensive theory development extremely difficult and stimulate a "practical" way of design. A FC design process contains the same steps as any other process of practical design.

The practical way of a controller design is through prototyping and tuning. So the question in the title can be reformulated as two: *How to choose an initial controller model?* and *How to tune this model?*

One needs to choose initially the structure and parameters of a FC, test a model or a controller itself and change the structure and/or parameters based on the test results. One may see that an actual design process consists of choosing the controller structure and some parameters (a synthesis of the controller) and an evaluation of their influence on the controller stability and performance (an analysis of the controller).

The processes of the analysis and synthesis are interrelated and interdependable on each other. The process can be divided roughly into two steps: an initial choice of the structure and parameters and the following adjustment based on the analysis (Fig. 2). Because of a large number of parameters to be determined and an incompleteness of a design theory the first step in a fuzzy controller design is characterised by a high subjectivity degree, and as a result of that the second step may require a high effort to be implemented.

The complete set of the principal design parameters has been formulated in [9]. In practice, however, the things are different. The main reasons for this situation are that either even nowadays the literature contains no results representing the influence of some of them (e.g. definition of a fuzzy implication) on the FC performance, or a designer considers their determination as may be theoretically important but practically not essential. For example, a very good review of the fuzzy processing methods is available [see 21] but a practitioner commonly does not care about this choice and takes the method recommended in the design package. On the other hand, the same or similar effect can be reached by modifying different parameters of a FC.

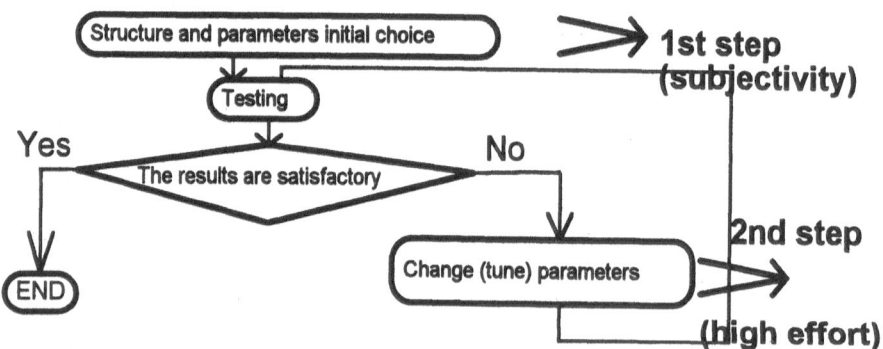

Figure 2. A fuzzy controller design process

So in practice, a FC designer usually needs to make a choice of:

1. a structure, inputs and outputs of a FC,
2. input and output scaling factors,
3. input and output membership functions,
4. rules.

4. How to choose the structure and parameters initially?

4.1. Type and Structure Choice

One of the main problems in a design of a FC as well as of any other controller is a concord of the required high performance under specified operating conditions and the desired possession of some other features, first of all stability and robustness. FC are proved to be rather robust to any changes in the environment, the plant and the controller itself. This feature was considered as one of the main advantages of the fuzzy control on the first (historically) stage of their development. As robustness we understand the ability to preserve or to avoid significant decay in the performance after some operating conditions have changed. To achieve this goal two basic ways are widely exploited:

1. an adjustment of the FC parameters after their initial choice and FC test - adaptive and learning controllers,
2. an application of special fuzzy control system structures, first of all hierarchical supervisory control structures.

The hierarchical rule-based controller usually consists of a simple upper-level "expert" controller, which is often called a supervisory controller [22] and the low level controller(s) (see Fig. 3). The switch between the levels can be realised

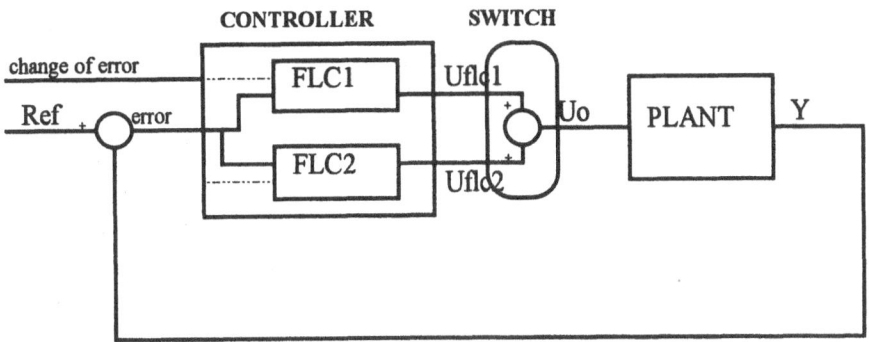

Figure 3. Hierarchical FC scheme

as a fuzzy controller as well. The application of a hierarchical structure lets distribute solving of different problems between the levels and apply various criteria in construction of different controllers. The top level controller should provide the approaching of the main goal of the system as the low level controllers should obtain the solutions of special problems.

The rule-based supervisory control has proven to be very effective in a number of applications, providing the required performance and stability of the whole system and comparing favourably to a variety of conventional techniques attempted to date. Unfortunately, except [15] the author knows no references comparing fuzzy hierarchical and adaptive controllers. One should note that in some applications low-level (executive) controllers are designed by conventional methods.

The low level controller(s) quite often can be conventional PID controllers (Fig. 4a). In this case a high level FC can realise a rules set used for a low level parameter adjustment, directly replacing an operator. Note that a FC can perform both on-line and off-line parameter change. It means that it can work at the same time as a conventional PID controller as well as before and after. In some cases FC can be applied only for preventing potentially dangerous situations in process control.

Another way of producing fuzzy - PID combinations can be done by putting a fuzzy and a PID controllers at the same level. Here both series [see, e.g. 1] and parallel [18] structures are considered. In a serial connection (fig. 4c) a FC develops an input signal for a PID. In most cases this FC replaces a human operator and is designed as a simple expert system. In a parallel connection (fig. 4b) a FC develops an extra control input which is applied to the plant in some combination with a PID control signal. These structures are good at nonlinear plant control and for the systems working under conditions of strong disturbances.

4.2. Fuzzy PID-like controllers

A PID controller is the most widely used in industry controller type. Unsurprisingly a lot of work has been done in design of fuzzy PID controllers (fuzzifying the parameters of a PID controller) and fuzzy PID-like controllers (employing the same set of the inputs and outputs as PID controllers). Some FC designers try to imitate the PID controller action [see 5, 18]. It has been proved that fuzzy controllers are capable of approximating any real continuous control function on a compact set to arbitrary accuracy. In particular, any given linear control can be achieved with a fuzzy controller for a given accuracy. [5] proposes methodology which enables the synthesis of a Sugeno or Mamdani type fuzzy controller precisely equivalent to a given PI controller. The main idea is to equate the output of the fuzzy controller with the output of the PI controller at some particular input values, called modal values. The rule base and the distribution of the membership functions can thus be deduced. The analytic expression of the output of the generated fuzzy controller is then established. For Sugeno-type fuzzy controllers, precise equivalence is directly obtained. For Mamdani-type fuzzy controllers, the defuzzification strategy and the inference operators have to be correctly chosen to provide linear interpolation between modal values. The usual inference operators satisfying the linearity requirement when using the center of gravity defuzzification method are proposed. This method can be used not only in design but in analysis of FC's. Nowadays there exist numerous examples of fuzzy PID-like (PI- and PD-like) controllers.

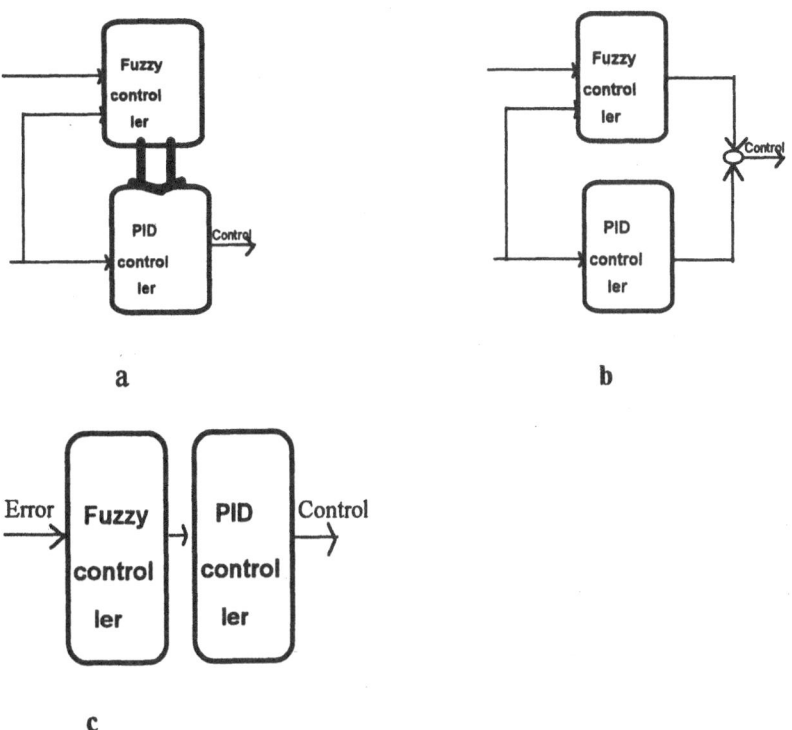

Figure 4. Fuzzy-PID control structures

4.3. Membership function selection

In the membership function choice one has to solve a few problems: how to choose general parameters like a number of classes (membership functions) to describe all the values of the linguistic variable on the Universe of discourse, the position of different membership functions on the Universe of discourse, the width of the membership functions, and concrete parameters like a shape of a particular membership function. In this paper the membership function selection is conducted on the base of the index which is called the whole overlap (WO) [16] which can be calculated according to the formula (1) [see Fig. 5]:

$$\int_x \text{Min} (\mu_1 (x), \mu_2 (x)) \qquad\qquad (1)$$
$$WO = \int_x \text{Max} (\mu_1 (x), \mu_2 (x))$$

where $\mu_A(x)$ and $\mu_B(x)$ are two adjacent membership functions.

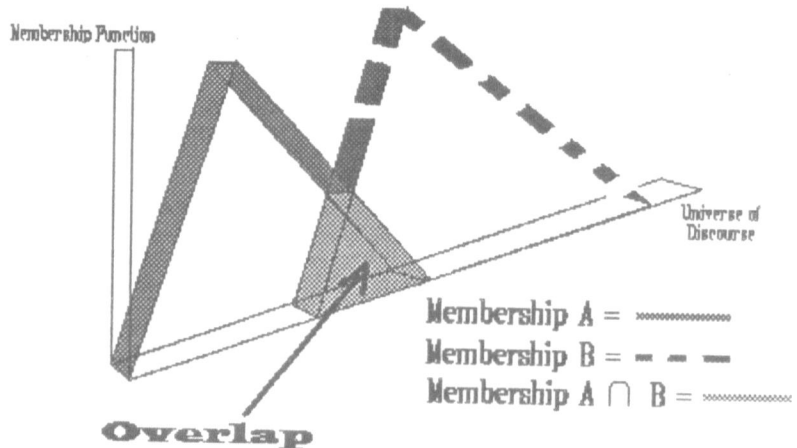

Figure 5. A whole overlap parameter evaluation

The research results [16,17] let us make the following assertions:

1. Initial choice of the membership function choice should provide the whole overlap parameter of about 12-14%,
2. Use of narrower membership functions results in the faster response (smaller response time),
3. Larger oscillation, overshoot and settling time appear when narrower membership functions are used,
4. Use of narrower membership functions produces the system with lower steady-state error but with a very narrow function the steady state can possibly not be reached at all,
5. The choice of the defuzzification method in many practical cases does not significantly influence the system performance characteristics,

6. The presence of small noise and disturbances generally keeps the statements made above valid

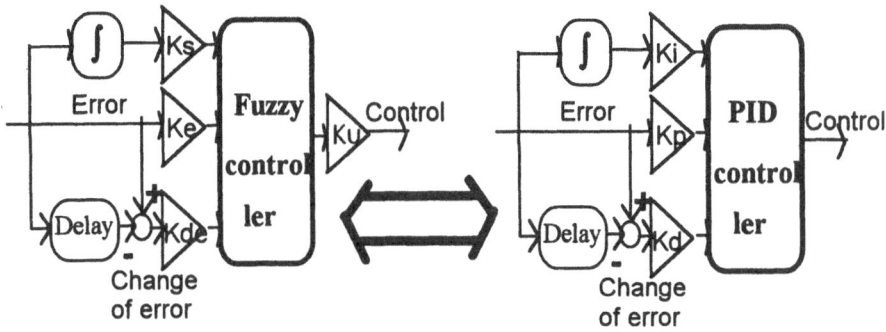

Figure 6. Similarity between scaling factors of a fuzzy controller and coefficients of a PID controller

If we compare an action of a fuzzy controller with the action of a conventional PID controller, we can state, that when we decrease a membership function width (decrease the WO ratio), we increase the differential part, and in the opposite case we emphasise an integration performance of the controller. So we can see some analogy between a membership functions choice and a PID-controller coefficients choice similar to that we will establish for a scaling factor choice later on.

4.4. Choice of the scaling factors

A choice and adjustment of the scaling factors is recommended to perform on the base of the similarity between FC scaling factors and PID coefficients (fig.6), which is referred to in [23]. Derived as a result the tuning actions for a PI-like FC scaling factors are given in Table 1. Other recommendations on an initial parameter choice are summarised in Table 2.

Table 1

Error Attribute	Tuning Action for		
	Kde	Kdu	Ke
Steady Divergence	Decrease	Decrease	Decrease
Overshoot / Oscillation	Decrease	Decrease	Decrease
Speed of Response	Increase	Increase	Decrease
Steady State Error	Decrease	Decr/Inc	Increase

5. How to tune (adjust) the parameters of a FC?

Some of the tuning methods assume an existence of the initial FC model and an availability of the plant model. However, most of the late design methods do not require any plant model at all. An example of such methods is given in [13] which proposes a complete design method for an online self-organising fuzzy logic controller without using any plant model. By mimicking the human learning process, the control algorithm finds the control rules of a system for which little knowledge has been available. In an expert approach, knowledge on the system supplied by an expert is required in developing control rules, however, the proposed fuzzy logic controller needs no expert in making control rules. Instead, rules are generated using the history of input-output pairs, and new inference and defuzzification methods are developed. The generated rules are stored in the fuzzy rule space and updated online by a self-organising procedure.

Table 2

Choice of the structure	Apply the hierarchical structure whenever there is any doubt in the stability of a fuzzy control system or in applications requiring high reliability
Choice of the inputs	The same as for a conventional control system
	The error and change_of_error (derivative) signals are often applied as the inputs for a fuzzy controller (fuzzy PID-like controller)
	Additional: choose the inputs regarding to which some control rules, expressing the dependence of the output on these inputs, can be easily formulated
Choice of the scaling factors	Initially choose the scaling factors to satisfy to the operational ranges (the universe of discourse) for the inputs and outputs, if they are known.
	Change the scaling factors to satisfy to the performance parameters given in the specifications on the base of recommendations provided
Choice of the number of the classes (membership functions)	There are several issues to consider when determining the number of membership functions and their overlap characteristics.
	The number of membership functions is quite often odd - generally, anywhere from 3 to 9.
	As a rule of thumb, the greater control required (i.e. the more sensitive the output should be to the input changes) the greater the membership function density in that input region

Choice of the membership functions	1) the expert approach - choose the membership functions determined by the expert(s) 2) the control engineering approach - 1) initially choose the width of the membership functions to provide the whole overlap (see section 4.4) about 12-14%, 2) in order to improve the steady-state error and the response time decrease the membership functions whole overlap, 3) in order to improve dynamic characteristics (oscillation, settling time, overshoot) increase the whole overlap, 4) the use of a fuzzy controller with wider membership functions and a large overlap can be recommended in the presence of large disturbances.
Choice of the rules	Main methods: 1) expert experience and knowledge, 2) operator's control actions learning, 3) fuzzy model of the process or object under control usage, 4) learning technique application. The whole rules set should be: - complete, - consistent, - continuous.
Choice of the defuzzification method	Choose the method according to the criteria The most widely used are: The Centre_of_Area and Middle_of_Maxima
Choice of the fuzzy reasoning method	Choose Mamdani method if: - the rules are expected to be formulated by a human expert Choose Sugeno method if: - computational efficiency and convenience in analysis are very important.
Choice of the t-norm and s-norm calculation method	The most widely used are: for t-norm Min or product operators, for s-norm Max or algebraic sum

Different techniques have been applied for a FC tuning (Fig.7). They can be divided into conventional and intelligent methods. In conventional methods classical approaches of the mathematical analysis are applied in searching for the parameter set optimising the performance criteria. In intelligent methods some heuristic procedures algorithms such as artificial neural networks (ANN), genetic (GA) and evolutionary (EA) algorithms are employed.

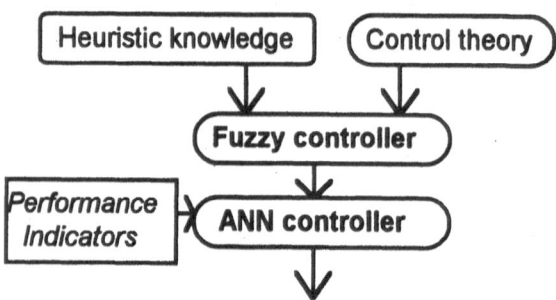

Figure 7. Combined structure for a FC

In an intelligent design fuzzy logic is utilised to incorporate the available knowledge into the controller design, and ANN and/or GA technology are applied to adaptively develop an optimal control strategy. The control system structure in this case can be presented as in Fig. 8 [11]. One should note that there exist another trend in combining FL and ANN technologies and creating new synergisms such as adaptive network based fuzzy inference systems (ANFIS) [7]. In this approach the controller design originates from the ANN framework.

Various fuzzy neural networks have been proposed to enhance the performance of ANN for control of complex dynamic plants with strong nonlinearity and high uncertainty. Nowadays there is a need for systematic techniques that takes the properties of the FL, ANN and GA into account in order to obtain fast convergence and to be able to tackle more complex control problems.

Recently ANN and GA were applied to tune one of the FC parameters. Modern methods tend to enable simultaneous determination of a few. To achieve this goal a combination of FL, ANN, GA, and conventional techniques is often employed. For instance,. [14] proposes the FC represented in the form of a neural network which can be trained using a GA. This enables the simultaneous determination of the membership functions for the fuzzy input variable, the quantisation levels for the output variable and the elements of the relation matrix of the FC.

In [6] a simultaneous design of FC rules and membership functions with GA is discussed. One has to note that these parameters are interrelated as the choice of membership functions determines in some aspect the choice of rules. On the other hand to achieve the same effect in many cases a designer is able to modify either the membership functions or the rules. Previous work using GA has focused on the development of rule sets or high performance membership functions; however, the interdependence between these two components suggests a simultaneous design procedure would be a more appropriate methodology.

When GA's have been used to develop both, it has been done serially, e.g., the des ign of the membership functions and then the use of them in the design of the rule set. This, however, means that the membership functions were optimised for the initial rule set and not the rule set designed subsequently. GA's are fully capable of creating complete fuzzy controllers given the equations of motion of the system, eliminating the need for human input in the design loop. [10] proposes the GA based algorithm which integrates even three design stages: the choice of the membership functions, the rules, and the rule consequent parameters.

1. Conventional methods:
    1.1. least-square method variations [19],
    1.2. gradient descent method variations [12]

2. Intelligent methods with applications of fuzzy logic, neural networks, and genetic algorithms
    2.1. tuning with fuzzy meta-rules,
    2.2 adjustment with neural networks,
    2.3 optimisation with genetic/evolutionary algorithms.

Figure 8. Tuning (adjustment) of fuzzy controller parameters classification

6. Conclusion

Summarising the review of the FC design methods it can be stated, that the most active issues these days are the specification of an intelligent control system design methodology based on a task-tcchnology mapping and the integration of specific technologies to obtain synergistic effects. The future development of fuzzy controllers and fuzzy controller design of more complicated control systems, making a control strategy more task-oriented rather than set-point and tracking following [24]. These controllers will incorporate different achievements from various fields of artificial and computational intelligence such as fuzzy logic, neural networks, genetic/evolutionary algorithms, expert systems.

References

[1] Bourmistrov A. and Reznik L. *Hybrid Guidance Control for a Self-Piloted Aircraft* Proceedings for the First International Discourse on Fuzzy Logic and the Management of Complexity, January 15-18, 1996, Sydney, Australia, vol.2, p. 155-159

[2] Brehm, T. and Rattan, K.S. *Hybrid fuzzy logic PID controller.* Proceedings of the IEEE 1993 National Aerospace and Electronics Conference. NAECON 1993 p. 807-13 vol.2, IEEE, New York, NY, USA, 1993

[3] Carpenter G. A. and Grossberg S. Learning, Categorization, Rule Formulation, and Prediction by Fuzzy Neural Networks In: Fuzzy logic and Neural Network Handbook/ed. C.H. Chen, McGraw-Hill, 1996

[4] Copeland R. P. and Rattan K. S. A Fuzzy logic Supervisor for PID Control of Unknown Systems. In Proceedings of IEEE Symposium on Intelligent Control, 16-18 August 1994, Columbus, OH, USA, pp 22 - 26

[5] Galichet, S. and Foulloy, L. *Fuzzy controllers: synthesis and equivalences* IEEE Transactions on Fuzzy Systems,Vol: 3 Iss: 2 p. 140-8, 1995

[6] Homaifar, A.and McCormick, E. *Simultaneous design of membership functions and rule sets for fuzzy controllers using genetic algorithms* IEEE Transactions on Fuzzy Systems Vol: 3 Iss: 2 p. 129-139, 1995

[7] Jang, J.-S.R. and Chuen-Tsai Sun *Neuro-fuzzy modeling and control* Proceedings of the IEEE Vol: 83 Iss: 3 p. 378-406, 1995

[8] Kim J. and Zeigler B.P. Hierarchical Distributed Genetic Algorithms: A Fuzzy Logic Controller Design Application IEEE Expert, June 1996, pp.76 - 84

[9] Lee, C.-C. *Fuzzy logic in control systems: Fuzzy logic controller* . IEEE Transactions on Systems, Man, and Cybernetics, vol. 20, No.2, p. 404 - 435, 1990

[10] Lee, M.A. and Takagi, H. *Integrating design stage of fuzzy systems using genetic algorithms* Second IEEE International Conference on Fuzzy Systems, p. 612-617 vol.1, IEEE, New York, NY, USA, 1993

[11] Menozzi, A. and Chow M.-Y. *A design methodology for an intelligent controller using fuzzy logic and artificial neural networks* Proceedings of the IECON '93. International Conference on Industrial Electronics, Control, and Instrumentation p. 408-13 vol.1, IEEE, New York, NY, USA, 1993

[12] Nomura H., Hayashi I., and Wakami N. *A Self-tuning Method of Fuzzy Control by Descent Method.* Proceedings of the 4th World Congress of

International Fuzzy Systems Association (IFSA'91), Brussels, 1991, p. 155 - 158

[13] Park Y-M., et. al. *A self-organizing fuzzy logic controller for dynamic systems using a fuzzy auto-regressive moving average (FARMA) model.* IEEE Transactions on Fuzzy Systems Vol: 3 Iss: 1 p. 7, 1995

[14] Pham, D.T. and Karaboga, D. *Design of neuromorphic fuzzy controllers* 1993 International Conference on Systems, Man and Cybernetics. Systems Engineering in the Service of Humans, p. 103-8 vol.4, IEEE, New York, NY, USA

[15] Reznik L. and O. Gnanayem *Hierarchical Versus Adaptive Fuzzy Logic Controllers: Design and Performance*, 2nd Australian and New Zealand Conference on Intelligent Information Systems, pp. 224 - 228 , Brisbane, November 29 - December 2, 1994

[16] Reznik L. and Shi J. *Fuzzy Controller Design From a Practitioner's Point of View: Membership Function Choice*, Australian Journal of Intelligent Information Processing Systems, 1995, vol.2, No.2, pp.38 - 46

[17] Reznik L. and Little A. *Fuzzy Controller Design From a Practitioner's Point of View: The Review of Methodologies*, Australian Journal of Intelligent Information Processing Systems, 1995, vol.2, No.4, pp.1 - 9

[18] Reznik L. and Stoica A. *Some Tricks in Fuzzy Controller Design*, Proceedings of the Australia and New Zealand Conference on Intelligent Information Systems, ANZIIS - 93, pp. 60 - 64, IEEE, Perth, Western Australia, 1993

[19] Takagi T. and Sugeno M. *Fuzzy identification of Systems and Its Applications to Modelling and Control.* IEEE transactions on Systems, Man, and Cybernetics, vol. SMC-15, No.1, 1985, p.116 - 132

[20] Tzafestas S. and Papanikolopoulos N.P. *Incremental Fuzzy Expert PID Control* IEEE Transactions on Industrial Electronics, vol.37, No.5, p. 365-371, 1990

[21] Tsukamoto, Y. *Some issues of reasoning in fuzzy control: principle, practice and perspective.* Proceedings. Sixth International Conference on Tools with Artificial Intelligence p. 192 - 196, IEEE Comput. Soc. Press, Los Alamitos, CA, USA, 1994

[22] Wang Li-Xin *A Supervisory Controller For Fuzzy Control Systems That Guarantees Stability.* IEEE Transactions on Automatic Control, vol.39, No. 9, pp. 1845 - 1847, 1994

[23] Yager R.R and Filev D.P. *Essentials of Fuzzy Modeling and Control.* John Wiley & Sons, 1994

[24] Zadeh, L.A. *Fuzzy logic: issues, contentions and perspectives* ICASSP-94. 1994 IEEE International Conference on Acoustics, Speech and Signal Processing, p. VI/183 vol.6, IEEE, New York, NY, USA, 1994

APPROXIMATION OF TRANSFER FUNCTIONS BY VARIOUS FUZZY CONTROLLERS

L.T. Kóczy* and D. Tikk

Department of Telecommunications and Telematics
Technical University of Budapest
H-1111 Budapest, Sztoczek u. 2
Hungary

E-mail: {koczy,tikk}@ttt-202.ttt.bme.hu

Summary. When Zadeh introduced the Compositional Rule of Inference, one of the most important aspects was that by this new technique of modelling it became possible to describe very complex systems in a way that enables drastic reduction of computational complexity (both in the space and time sense), as compared to previous symbolic methods of Artificial Intelligence that always neglected the subsymbolic structural information in the model. A computationally more effective version of the same idea was soon proposed by Mamdani, and so practical applications of fuzzy models and controllers were suddenly possible. The Mamdani (M-)model was followed several years later by a simplified technique, proposed by Sugeno (S-controller), a special case of the M-controller, using crisp output in the rules, and this was soon extended to the idea of Takagi–Sugeno-model that uses crisp outputs, however as functions valid for an element of the fuzzy partition of the input space.

The existence of these models started a philosophical question: Do very complex systems with no identifiable explicit models have a crisp transfer function at all that can be approximated (even if it is unknown)? The S- and TS-models seem to suggest an affirmative answer. The M-model, however, does not support the assumption. We think that both attitudes have a justification: some of the systems do have a theoretical crisp transfer function, but some of them do not. What is the solution of the problem in the latter case? There exists a fuzzy transfer function, that maps the fuzzy power set of the input universe of discourse onto the fuzzy power set of the output universe of discourse. We have recently discussed some aspects of this mapping, showing that all former models can be interpreted as special case of the general fuzzy model. (This model is also supported by recent considerations of Zadeh on a class of problems that cannot be discussed in terms of crisp transfer functions.)

This paper will add a few aspects to the ways of approximation both types of transfer functions, crisp anf fuzzy. The most important elements here are the evidence obtained by analysing the explicit functions of the three classical fuzzy controllers, the results available on the universal approximator property of fuzzy controllers, and the method of rule interpolation that enables the uniform handling of all models in question. Some advantageous properties of the interpolation techniques proposed by Kóczy and Hirota will be shown, especially stressing the stability of such interpolations.

In the concluding part some open questions will be hinted at, for further study.

Keywords: explicit functions of fuzzy controllers, universal approximator property, interpolation in fuzzy rule bases, Stability of the KH-interpolation method.

1. Introduction

Fuzzy control systems have proved their applicability in many areas. Their user-friendliness and transparency certainly belong to their main advantages, and these two enable developing and tuning such controllers easily, without knowing their exact mathematical description. Nevertheless, it is interesting to know, what mathematical functions hide behind a set of fuzzy rules and an inference machine. For practical purposes it is necessary to consider real, implementable fuzzy control systems with reasonably low computational complexity. In the following sections we discuss the problem of what types of functions are generated by realistic fuzzy control systems. We consider such practically important special cases, controllers having rules with triangular and trapezoidal membership functions, and crisp consequents.

Fuzzy control systems (FCS) consist of four main components: the input interface determining the degrees of matching (that is inherently connected with the rule base), the rule base itself, the inference engine that is the algorithm of determining the individual conclusions obtained from each rule by applying the degree of matching as a weight or delimiter (depending on the t-norm applied), and the defuzzification interface.

The inference engine using the rules in the base has both fuzzy input and output (the former being just a set of matching degrees, the latter a possibly nonconvex fuzzy set), the two interfaces connect the whole system with the external world through exclusively non fuzzy variables. The "black box" model of a FCS is identical with any (crisp) function generator.

Because of this, it is possible to construct the explicit formulae substituting the (ready) fuzzy control system fully.

2. Explicit formulae for fuzzy controllers with equidistant isosceles triangular rules

First results in this topic were presented by El Hajjaji and Rachid [7]. In their paper they considered only rule systems with the following properties: For every x the sum of the membership degrees of all antecedents for given x results in 1:

$$\sum_{i-1}^{r} \mu_{R_i}(x) = 1 \tag{1}$$

(the antecedents form a Ruspini-partition.)

Futhermore, both antecedents and consequents are identical isosceles triangular, equidistant (the common base is denoted by b), and so, for every x, either x is the core of one of the triangles, or x has positive membership in exactly two antecedents, A_i and A_{i+1}, so that because of (1)

$$\mu_i(x) + \mu_{i+1}(x) = 1$$

is always true. (The core or kernel of fuzzy set S is defined by $\text{core}(S) = \{x|\mu_S(x) = 1\}$.)

They determinined the defuzzified conclusion belonging to a crisp observation, by using the Mamdani method [19] with Center of Area defuzzification. Suppose that the degrees of matching for observation x^* are (d_1, d_2), where

$$d_1 = \mu_i(x^*) \geq d_2 = \mu_{i+1}(x^*), \tag{2}$$

then, the conclusion is

$$y^* = \left(i + \frac{1}{2}\right) b + \frac{(d_1 - 1)^2 - (d_2 - 1)^2}{2d_1 - d_1^2 + d_2} \times \frac{b}{2}, \tag{3}$$

Because $d_1 + d_2 = 1$, (3) can be rewritten in the form

$$y^* = \left(i + \frac{1}{2}\right) b + \frac{1 - 2d_1}{1 + d_1 - d_1^2} \times \frac{b}{2}. \tag{3a}$$

If x^* is substituted into y^*, we find that

$$y^* = c_1 i + c_1' + \frac{c_2 + x^*}{c_3 + c_4 x^* + c_5 x^{*2}},$$

where c_i are constants (derivable from the rules), and $i = [x^*/2b]$, $[\cdot]$ denoting the integer part (floor). It is important to note that c_4 and c_5 can never be 0. At the first sight, even the danger threatens that at some x^* the rational function is unbounded. However, because of the geometric and algebraic background of these formulae, such a possibility can be excluded. With further investigations, we can see that the function must be "almost monotone" in $[ib, (i+1)b]$. The first part represents a linear function, in terms of the rule subscript, the rational function part might be negligible if x^* is large, or might be close to the linear if x^* is far from the curved area. These facts reveal the true nature of fuzzy control systems implemented by isosceles equidistant triangular rules.

The limit behavior of such a system is that of a *crisp* expert control system, with intervals represented by their centers. Even for x^*-s with very small absolute value, the limit is $y^* = c_1 i + c_1''$, a somewhat different, but locally linear function. Statements on the limit behavior of the function can be found in [7].

At this point it is worth while comparing (3) (or (3a)) with the formula obtained when another very common defuzzification method is applied: the Center of Gravity method. The literature does not clearly distinguish these two defuzzification methods (see e.g. [5]), however, in this paper we shall observe the following: the essential difference is that the area, where both truncated consequents ($\min\{d_1, B_1\}$ and $\min\{d_2, B_2\}$) overlap, is considered with double weight (pulling the conclusion closer to the center of this area) in the COG, but with single weight only in the COA method. It is not necessary to suppose that (2) holds, as here the two consequents play a symmetrical role. The explicit formula of the conclusion is

$$y^{*\prime} = \left(i + \frac{1}{2}\right)b + \frac{(d_1 - 1)^2 - (d_2 - 1)^2}{2d_1 - d_1^2 + 2d_2 - d_2^2} \times \frac{b}{2}, \tag{4}$$

or, in another form

$$y^{*\prime} = \left(i + \frac{1}{2}\right)b + \frac{(d_1 - 1)^2 - (d_2 - 1)^2}{2 - (1 - d_1)^2 - (1 - d_2)^2} \times \frac{b}{2},$$

obviously a symmetrical expression in d_1 and d_2. (See [13, 14] for further details.) This can be again rewritten, since $d_1 + d_2 = 1$

$$y^{*\prime} = \left(i + \frac{1}{2}\right)b + \frac{d_1(1 - d_1)(1 - d_2)}{(1 + d_1 - d_1^2)(1 + d_2 - d_2^2)} \times \frac{b}{2}. \tag{4a}$$

The structure of this result is similar to that of (3):

$$y^{*\prime} = c_1 i + c_1' + \frac{c_2 + c_3 x^* + c_4 x^{*2}}{c_5 + c_6 x^* + c_7 x^{*2} + c_8 x^{*3}},$$

and also the rational function part has a similar behavior to the rational part in the former result. It is even more difficult to overview the function's general behavior as the denominator cannot be reduced in its degree, however, because of the geometric interpretations, the function is monotone in $[ib, (i + 1)b]$.

It is interesting to compare the error of approximating the COA-result by the COG-result (or the opposite). We can obtain easily the following result (the detailed computation can be found in [14]):

Statement 1. Applying the Mamdani-method for equidistant isosceles triangular rules, the maximal deviation between the COG and COA defuzzified control values is 2% of the consequent base length.

This result contributes to the explanation why the COG method has gained popularity compared to the COA method that needs more complicated formulae, and so, makes the runtime of the whole fuzzy control algorithm higher, seeing that the difference is not significant (at least in this special case, cf. [17]).

In the initial paper [7], also the formulae of the Larsen-style control [18] are discussed. For the same special case, the formula

$$y^* = \left(i + \frac{1}{2}\right)b + \frac{7d_2 d_1^2 - 7d_2 d_1^2 + 6d_2^3 - 6d_1^3}{6(d_1 + d_2)(2d_1^2 + 3d_1 d_2 + 2d_2^2)} \times b. \tag{5}$$

As it is supposed that $d_1 + d_2 = 1$, this formula can be simplified to

$$y^* = \left(i + \frac{1}{2}\right)b + \frac{7d_2 d_1^2 - 7d_2 d_1^2 + 6d_2^3 - 6d_1^3}{6(2d_1^2 + 3d_1 d_2 + 2d_2^2)} \times b. \tag{5a}$$

The structure of this formula is essentially similar to the previous ones, however, here, there is a real linear component (depending on x^*) obtained by executing the polynomial division in the rational function part. The whole function is monotone, so the remainder will be monotone, too.

Again, let us calculate the COG conclusion:

$$y^{*'} = \left(i + \frac{1}{2}\right)b + \frac{d_2 - d_1}{d_1 + d_2} \times \frac{b}{2}, \tag{6}$$

apparently considerably simpler than (5). Again, if $d_1 + d_2 = 1$ holds, this simplifies to

$$y^{*'} = \left(i + \frac{1}{2}\right)b + (d_1 - d_1) \times \frac{b}{2} = \left(i + \frac{1}{2}\right)b + (1 - 2d_1) \times \frac{b}{2}. \tag{6a}$$

As it can be seen especially from the latter form, this is linear in terms of x^* (as d_1 is linear itself).

Apparently, both the Mamdani- and the Larsen-controllers combined with both the COA or COG defuzzification methods realize rather similar type functions: Approximately piecewise linear in $[ib, (i+1)b]$ with a rational (polynomial fraction) "error" member that has a not too complicated behavior. The only convenient exception is the Larsen-method with COG defuzzification, where exactly piecewise linear functions are obtained.

We can determine again the error of approximating one formula by the other.

Statement 2. Applying the Larsen-method for equidistant isosceles triangular rules, the maximal deviation between the COG and COA defuzzified control values is 6% of the consequent base length.

Even though the COA defuzzification method seems to be more justifiable than the COG one, the very advantageous (simple) form of the Larsen method combined with COG deserves a more detailed examination. Let us transform (6a) by substituting d_1

$$y^{*'} = \frac{b}{a}x^* + (i - k)b \tag{7}$$

If $x^* = ka$ then $y^{*'} = ib$, and if $x^* = (k+1)a$ then $y^{*'} = (i+1)b$, as it is expected. Between these two values, the last method gives an exact linear approximation.

As a final conclusion of this section, we should state the following:

Statement 3. If a rule base is given, where all antecedents and consequents are equidistant isosceles triangular, the Mamdani- and Larsen-controllers combined with the COG and COA defuzzification techniques provide various approximately linear interpolations between the control points of the $y = f(x)$ function, determined by the corresponding core points of the rules. Among these four, only the Larsen-controller (applying algebraic t-norm for

the weighting of the consequents) with COG defuzzification offers an exactly
linear interpolation between two neighboring terms, according to (7).

The analysis of these methods leads to the recognition why the very simple equidistant isosceles triangular rule bases are rather popular among the appliers. From the four formulae it is clear, that from the mathematical point of view, it is reasonable to use the algebraic t-norm, and the COG defuzzification, in this case namely it is possible to estimate the maximum error of the fuzzy control system compared to the exact function approximated, as the linearity error of that function.

However, if real fuzzy control applications are discussed, it must be observed that the regularity of the terms and rules assumed in this approach can usually not be satisfied. Even, if the starting point is an equidistant term system, by tuning, the antecedents will change, and the consequents accordingly, as it was presented e.g. by Burkhardt and Bonissone [3]. The terms will be even more irregular if the rules are generated by qualitative modeling based on measured data as e.g. by Sugeno and Yasukawa [24]. Because of that, it is reasonable to extend the approach to more general formulae. As the computational complexity of the algorithm is considerably higher if the terms are not piecewise linear (even though some authors, especially in theoretical investigations, deal with truncated exponential, etc. shapes), the next study will be restricted to normal trapezoidal rules, as the overwhelming majority of the real applications use only such.

3. General explicit formulae for trapezoidal rules

If only the trapezoidal shape of the terms in the rules is examined (both in the antecedent and consequent parts), and the fact that for any x^* there are maximally two fixed antecedents, the degrees of matching for x^* are

$$d_1 = \frac{a - x^*}{b - a} + 1 \text{ and } d_2 = \frac{x - c}{d - c}$$

The left antecedent has its right core point in $(a, 1)$, its right support point in $(b, 0)$, and the right antecedent has its left support point in $(c, 0)$, and its left core point in $(d, 1)$; further on $a \leq c \leq x^* \leq b \leq d$. If this latter is not the case, one or both membership degrees will change to 1 or 0.

Next we discuss the consequents. Considering the simplicity of the COG method compared with the COA, and assuming from the previous results that the two in general do not deliver very much different conclusions, next we will calculate only the COG results.

Computing the corresponding areas and centers (cf. [14]) we can state the following:

Statement 3. Applying general trapezoidal rules with every observation firing maximally two rules only, if the Mamdani-method is adopted with COG defuzzification, the final explicit formula for the conclusion is

$$y^* = \left\{ \left[(s-p)d_1 - (s+r-q-p)\frac{d_1^2}{2} \right] \times \right.$$

$$\times \left[(1-d_1+d_1^2)(s^2-p^2) + (d_1-2d_1^2)(sr-pq) + d_1^2(r^2-q^2) \right] \div$$

$$\div 3\left[(2-d_1)(s-p) + d_1(r-q) \right] + \left[(w-t)d_2 - (w+v-u-t)\frac{d_2^2}{2} \right] \times$$

$$\times \left[(1-d_2+d_2^2)(w^2-t^2) + (d_2-2d_2^2)(wv-tu) + d_2^2(v^2-u^2) \right] \div$$

$$\left. \div 3\left[(2-d_2)(w-t) + d_1(v-u) \right] \right\} \div$$

$$\div \left[(s-p)d_1 - (s+r-q-p)\frac{d_1^2}{2} + (w-t)d_2 - (w+v-u-t)\frac{d_2^2}{2} \right]. \quad (8)$$

(supposing the four characteristic points (min of support, min of core, max of core, and max of support, respectively) of the left consequent are p, q, r, s, and of the right consequent are t, u, v, w).

This formula is rather complicated, and there is little hope for any simplification as the parameters are independent from each other. Nevertheless, from this general starting point we can derive, using simplifying conditions, the explicit formulae of all special cases.

For example, in practical fuzzy control, the simplifying conditions of $q = r$ and $u = v$ often apply and lead to some simplification of the formulae. Then we obtain:

Statement 4. Applying general triangular rules with every observation firing maximally two rules if the Mamdani-method is combined with COG defuzzification, the final explicit formula for the conclusion is

$$y^* = \left\{ \frac{s-p}{6} \left[\frac{2d_1-d_1^2}{2-d_1} + \frac{2d_2-d_2^2}{2-d_2} \right] \times \right.$$

$$\times \left[(1-d_+d_1^2)(s+p) + (d_1-2d_1^2)q \right] \div$$

$$\left. \div \left[(s-p)\left(d_1 - \frac{d_1^2}{2} \right) + (w-t)\left(d_2 - \frac{d_2^2}{2} \right) \right]. \quad (9)$$

The denotation is the same as at the previous statement, except here the three characteristic points (min of support, core, and max of support, respectively) of the left consequent are p, q, s, and of the right consequent are t, u, w.)

4. The Behavior of General Trapezoidal "Fuzzy Controllers"

Next we will analyze the advantages of using general trapezoidal rules in fuzzy reasoning in the style of generalized trapezoidal controllers (having even trapezoidal consequents).

Expression (8) is somewhat complicated, and it is not possible to simplify it in the general form. However, its structure can be seen better if it is expressed by using simpler denotations for the constants. Further on, d_1 and d_2 are expressed by x^*, so finally we obtain:

$$y^* = \frac{C_1 + C_2 x^* + C_3 x^{*2} + x^{*3}}{C_4 + C_5 x^* + C_6 x^{*2}} + C_7 + C_8 x^* + \frac{C_9 + x^*}{C_4' + C_5' x^* + C_6' x^{*2}}. \quad (10)$$

There are 6 constants occurring in (10), determined by the 12 parameters a, b, c, d; p, q, r, s; t, u, v w. The six (plus six) individual expressions $C_i = f_i(a, b, c, d; p, q, r, s; t, u, v, w)$ from (10) and the expressions for d_1 and d_2 are available with the authors.

Expression (9) is similar to the formulae obtained for the equidistant case, however, it has two essential differences: the polynomial (linear) part is not constant, i.e. even its dominant member does not depend only on the interval where the observation is located, but also on the exact value of the observation within the interval, and it contains a higher number of parameters. It would be easier to control the behavior of these formulae if due to the high number of free parameters, it were possible to eliminate the denominator, i.e. $C_6 = C_5 = 0$ (or $C_6' = C_5' = 0$) could be chosen, however, it is obvious from (8) that this is impossible, as

$$C_6' = \frac{s - p}{(b - a)^2} + \frac{w - t}{(d - c)^2}. \quad (11)$$

contains the sum of two nonnegative members, which can be 0 only, if both are 0, i.e. $s = p$, $w = t$ which would hold only for *crisp* conclusions. The basic structure of (9) is similar to (3), but for large x^*-s the formula behaves linearly for a given pair of consequents, allowing an approximately piecewise linear approximation of the control function. For very small x^*-s, the constant part is somewhat different, while in the "medium large" domain for x^*, the behavior of (9) seems to be rather unpredictable, and very different from linear. The complicatedness of this formula shows clearly the advantages of using fuzzy sets and rules instead of directly interpolating between the known characteristic points.

Finally let us examine the behavior of the formula in general briefly. For the detailed reasoning we refer to [14]. (8) and even (9) have a hardly overviewable structure. The nonlinear component has many degrees of freedom, and so, theoretically it is possible that the nonlinear component of y^* is nonmonotonic. But from as it is shown in the referred paper this case can be excluded, i.e. y^* behaves monotonicly.

Statement 5. With general trapezoidal rules, if an observation fires maximally two rules, and if the Mamdani-method, and COG defuzzification are applied the fuzzy control algorithm provides strictly monotone interpolation between two consequent core points.

The previous remarks will be true and meaningful for the observation in the interval $[c, b]$, where the output is determined by both rules. If for a given x^* there is only one such A_i that $d_i \neq 0$ (let us denote it by d_1), the situation is quite different, as y^* will be determined only by a single consequent.

5. The singleton-consequent Sugeno-controller

Let us turn to (11) now. We found that (10) might be reduced to a linear function only if the consequents are crisp singletons. Indeed, fuzzy controllers with crisp singleton consequents play a more and more important role. Such a controller is a special case of both the Mamdani- and Larsen-controllers, and the Takagi–Sugeno-controllers [25], and referred to in the literature as Sugeno-controller. Because of the importance of this special type of controller, next its explicit formulae will be determined.

If the same notation is used as in the previous two sections,

$$d_1 = \frac{a - x^*}{b - a} \text{ and } d_2 = \frac{x - c}{d - c},$$

hold again. The consequents of the rules are B_1 and B_2, in this case both are crisp singletons in Y. As normal and subnormal singletons have no geometric area, this time the center of gravity or, (what is equivalent to it in this special case), the center of area, is calculated by the weighted average of the two consequents.

Statement 6. With general trapezoidal rules in the antecedents, and crisp singletons in the consequents, (special Sugeno-controller) further on if every observation fires maximally two rules, the conclusion is given by

$$y^* = \frac{[(b - a)B_2 - (d - c)B_1]x^* + [b(d - c)b_1 - c(b - a)B_2]}{[b - a - (d - c)]x^* + (bd - 2bc + ac)} \tag{12}$$

We find that indeed, the quadratic part disappeared from the denominator (as it was stated in connection with (11)), but the result is nevertheless not linear. After executing the polynomial division, the structure becomes clear:

$$y^* = c_1 + \frac{c_2}{c_3 x^* + c_4}. \tag{13}$$

The behavior of such a superposed linear and hyperbolic pair has been investigated in [12], under certain circumstances its behavior is rather close to linear. However, if exact linear behavior is required, $c_3 = 0$ must be satisfied ($c_3 = b - a - (d - c)$). This condition is equivalent to the following:

$$d_1 + d_2 = \text{constant},$$

which is slightly more general than the case when the sum has to be exactly 1. The condition prescribes that all neighboring pairs of antecedents must have identical absolute value (but opposite sign) slopes turning toward each other. In case when the observation fires only a single rule, (16) simplifies to $y^* = B_1$, independently, whether x^* is in the core, or not. Summarizing the above observations, we have

$$y^* = \frac{B_2 - B_1}{b - c} + \frac{bB_1 - cB_2}{b - c}. \tag{14}$$

Suppose now that the antecedents are equidistant isosceles triangular ones. Then, (14) can be transcribed with the denotation used in Section 2

$$y^* = \frac{b}{a}x^* + (i - k)b.$$

This result is identical with (7) in Statement 2! As conclusion it can be stated that

Statement 7. The isosceles triangular Larsen-controller with COG defuzzification is identical in its behavior with the special (crisp consequent) Sugeno-controller.

Consequently, it is not reasonable to use any isosceles triangular Larsen-controller as the computational complexity involved with this one is considerably higher. The special Sugeno-controller is used often in practice, mainly because of its simplicity and piecewise linear behavior (see e.g. [15]).

In the next section, the general Takagi–Sugeno-controller will be investigated.

6. The general Takagi–Sugeno-controller

In the general form, the Takagi–Sugeno-controller has linear functions in the consequent part [20]. Suppose that every observation fires maximally two rules only. Let the ith consequent be

$$y = b_i x^* + B_i.$$

The conclusion is obtained by calculating the weighted average of the (maximally) two substitution values of the observation into the consequents. The final formula is

$$y^* = \frac{[(b - a)b_2 - (d - c)b_1]x^* + [b(d - c)b_1 - c(b - a)b_2]}{[b - a - (d - c)]x^* + (bd - 2bc + ac)}x^* +$$

$$+ \frac{[(b - a)B_2 - (d - c)B_1]x^* + [b(d - c)B_1 - c(b - a)B_2]}{[b - a - (d - c)]x^* + (bd - 2bc + ac)x^*}. \tag{15}$$

Statement 8. With general trapezoidal rules in the antecedents, and linear functions in the consequents, (general Takagi–Sugeno-controller) further on if one observation fires maximally two rules, the conclusion is given by (15).
The structure of this control function is

$$y^* = c_1 x^{*2} + c_2 x^* + c_3 + \frac{c_4}{c_5 + c_6 x^*}.$$

The above function has a different behavior from the Mamdani- or Larsen-controllers as its "main member" is parabolic. The non-polynomial member is hyperbolic here. Because of the parabolic part, there is no guarantee for the monotonicity of the function, obviously. When the non-polynomial member disappears, we can state the following:

Statement 9. The conclusion in the general Takagi–Sugeno-controller with maximally two fired rules is piecewise parabolic if and only if all "facing" antecedent flanks have identical absolute value (but opposite sign) slopes ($d_1 + d_2 = $ constant), and then the fuzzy controller realizes the the following:

$$y^* = \frac{b_2 - b_1}{b - 2a + c} x^{*2} +$$
$$+ \frac{ab_1 - cb_2 + B_2 - B_1 + (b - a)b_1}{b - 2a + c} x^* + \frac{aB_1 - cB_2 + (b - a)b_1}{b - 2a + c}. \quad (16)$$

The general (linear consequent) Takagi–Sugeno-controller realizes parabolic functions between two rules. However, the parameters occuring in (20) are not independent of the "next" member, as one consequent is always common. Suppose that a piecewise parabolic approximation is sought. The parabolic pieces should be adjacent, consequently, the supports of every other antecedent should be adjacent, too. In this case, the only free parameter for every new antecedent pair is the distance of the two cores (which must be necessarily single points, as the condition concerning the constant sum of the two weights prescribes it, i.e., the antecedents should be triangular), however, if a certain parabolic approximation is required, this distance will be automatically given by the intervals for which the individual parabolic pieces are valid. That means altogether $2n$ freely choosable parameters for n rules, while the parabolas themselves contain $3n$ prescribed parameters. From here it is obvious that it is not possible to construct piecewise parabolic approximations (with adjacent parabolic arcs) by the general TS-controller, and so, no arbitrary piecewise parabolic approximation can be realized.

Comparing these negative result with the results in connection with the Mamdani- and Larsen-controllers, further on, the special Sugeno-controller, it is obvious, why the general Takagi–Sugeno-controller is not used in the practice for general nonlinear systems.

There are many nonlinear functions in the practice which behave almost or exactly linearly within certain sub domains, while they change their behavior in a smooth way between such domains. The general TS-controller (with linear consequents) is a special tool to approximate such nonlinear functions, with the following conditions:

Statement 10. The conclusion in the general Takagi–Sugeno-controller with maximally two fired rules is alternatingly piecewise linear and parabolic, if the antecedents are trapezoidal, and the weights belonging to any observation sum up to 1. The linear pieces can be determined freely and if two subsequent linear pieces are $b_1 x + B_1$ and $b_2 x + B_2$, further on, the left boundary of the non-core zone between the two neighboring antecedents is a, and the length is d, the interconnecting parabolic piece is determined by

$$y^* = \frac{b_2 - b_1}{d} x^{*2} + \left(\frac{a(b_1 - b_2) + B_2 - B_1}{d} + b_1 \right) x^* + \left(\frac{B_1 - B_2}{d} a + B_1 \right).$$
(17)

Expresion (21) generates the average of the two linear functions. As a matter of course, if $b_1 = b_2$, and $B_1 = B_2$, then it reduces to the single linear function $y^* = b_1 x^* + B_1$.

All the concrete results in the previous sections referred to single input systems. The results can however easily be extended to multiple inputs as well.

7. Explicit function of multiple-input controller

Now we will examine the explicit output function of two-input controllers. This results can be generalized easily to determine the output crisp function of a multiple-input fuzzy controller.

In the previous sections the one dimensional fuzzy controller was discussed where any observation can give positive membership degrees for maximum two antecedents. Generalizing this state for two-input controllers we give that maximum four rules may fire an (x_1^*, x_2^*) observation. In the following we will investigate such rule bases where this maximal situation is reached. This condition, and supposing general trapezoidal shape for the antecedents with their four characteristic points $a_{i,j}$, $b_{i,j}$, $c_{i,j}$, $d_{i,j}$, can be interpreted as $a_{1,i} \leq c_{1,i+1} \leq b_{1,i} \leq d_{1,i+1}$ and $a_{2,i} \leq c_{2,i+1} \leq b_2, \leq d_{2,i+1}$. We concentrate to the region

$$c_{1,i+1} \leq x_1^* \leq b_{1,i}, \qquad c_{2,j} \leq x_2^* \leq d_{2,j+1},$$
(18)

this is one of the four subregions, where the observation (x_1^*, x_2^*) gives positive, non-trivial membership degree for the antecedents.

As in the former sections we calculate the explicit output function for Mamdani-controller with COG defuzzification. We obtain the formula

$$y^* = \frac{c_1 x_1^{*3} + c_2 x_2^{*3} + c_3 x_1^{*2} + c_4 x_2^{*2} + c_5 x_1^* + c_6 x_2^* + c_7}{c_8 x_1^{*2} + c_9 x_2^{*2} + c_{10} x_1^* + c_{11} x_2^* + c_{12}}$$
(19)

where the parameters c_k depend on the characteristic points of the membership functions. Similar results for Larsen-type controllers can be found in [26].

For the future, it seems to be reasonable to determine similar explicit functions for exponential antecedent and consequent rules, especially as they appear frequently in artificial neural networks and so the formal equivalence of such systems with fuzzy controllers might be directly investigated.

8. Fuzzy controllers as universal approximators

Considering the success of fuzzy controllers with non-linear plants, many authors have raised the question what is the reason of this versatility of the fuzzy controllers. In recent years an answer from the mathematical point of view was given: fuzzy control systems are universal approximators in the sense that it is possible to construct such rule based fuzzy controllers that approximate any (non-linear) function with arbitrary accuracy. Now we will examine the most important statement concerning the universal approximator property, and compare them with the results obtained by explicit formulae of the fuzzy controllers used in practice.

The first essential statement from mathematical point of view was given by Kosko [16]. His motivation is the idea that fuzzy rule bases are defining "patches" in $X \times Y$, which patches cover the hypothetical crisp fuction $y = f(x)$, and if the size of patches is getting smaller and smaller, i.e. the cover is refined, then the approximation of the fuction by the patches will be more and more accurate. His main theorem is the following:

Theorem (Kosko):. An additive fuzzy system uniformly approximates f : $x \rightarrow y$ if X is compact and f is continuous.

This theorem refers to the Mamdani-method. Its proof gives a clear idea of how the approximation works. The cardinal problem is the minimal distance if the centroids of two adjacent consequent sets: if they are denoted by y_i and y_{i+1}, and it is required that the approximation by the FCS is everywhere not worse then ε, then

$$|y_i - y_{i+1}| < \frac{\varepsilon}{2p - 1},$$

where p is the number of the maximal overlapping antecedents over x, in most practical cases 2, if the input is one dimensional. From here the number of rules in the base should be

$$|R| > \frac{|X|}{\varepsilon},$$

even in the one dimensional case. From this it is clear that the size if the rule base in arbitrary good approximation is *not bounded.*

At almost the same time, the results of Mendel and Wang [27, 28] have shown that fuzzy rule based systems with the Larsen-algorithm, however, and using rules with everywhere positive membership functions over all the input domain (in the latter paper, especially for exponential membership functions), can approximate with arbitrary accuracy.

Theorem (Wang): For any real continuous function g on the compact set $X \subset R^n$ and arbitrary positive ε there exists f (the fuzzy control function realized with Gaussian membership functions, Larsen-conclusion, and any centroid defuzzification method) that

$$\sup_{x \in X} |g(x) - f(x)| < \varepsilon.$$

Unfortunately, the same practical difficulty occurs here as in the case of Kosko's theorem: the number of the rules in the base is not bounded, in addition to that even the supports of the terms in the rules are not bounded (identical with the universe of discourse).

The above theorems were generalized by Nguyen and Kreinovich [21]:

Theorem (Nguyen–Kreinovich): The statement of Wang's theorem remains valid for arbitrary membership function type were every membership function can be generated from a given $\mu_0(x)$ such that $\mu(x) = \mu_0(ax + b)$ (a is non zero, b is an arbitrary real), and the support of μ_0 is non vanishing, further on, any t-norm and t-conorm is used for the construction of the conclusion (the extension of the Mamdani- and Larsen-method), finally arbitrary averaging defuzzification operator is used.

This very general theorem shows that the universal approximator property of fuzzy control is not a consequence of the choice of any particular norm, or defuzzification technique but rather that of the possible *density* of the rule base.

Castro [4] have shown the universal approximator property for wide classes of fuzzy controllers. The classes are controllers with:

- a prefixed type of membership functions (among others the most commonly used triangular or trapezoidal shaped),
- arbitrary t-norm used for the construction of the conclusion,
- center of area defuzzification.

Theorem (Castro): *Let* $f : X \subset R^n \rightarrow R$ *be a continuous function defined on a compact* X*. For each* $\varepsilon > 0$ *there exists a* $S_\varepsilon \in S$

$$\sup\{|f(x) - S_\varepsilon(x)| : x \in X\} \leq \varepsilon.$$

Here S *can be two general groups of fuzzy controllers: first one with fuzzy set (Mamdani-type), the second one with crisp consequent in the rules.*

Even this last theorem proves the universal approximator property for very wide and often used classes of fuzzy controllers it does not give neither the way of construction nor the size of this optimal fuzzy logic controller.

Even though these statement have theoretical importance, we shall show next that they *do not explain in reality success of fuzzy control.* The following statement is a corollary of the previous one's:

Statement 11. Let us define (crisp) expert control as follows: $X = \prod_{j=1}^{K} X_j$ be a compact domain that is partitioned by a net defined by x_{jk}. There is a set of rules in the form

$$R_i = \textbf{If } x_{ij} \in [x_{jk}, x_{jk+1}) \textbf{ then } y = b_i$$

Then any real continuous function g on the compact set $X \subset R^n$ and arbitrary positive ε there exists f obtained from R_i by applying any fuzzy control method and centroid defuzzification that

$$\sup_{x \in X} |g(x) - f(x)| < \varepsilon.$$

The proof of this statement is easy based on the previous ones (cf. [13, 14]).

From this statement it is obvious that the universal approximator property of fuzzy rule based control is but the consequence of its fuzziness. Any crisp expert control has the same universal approximator nature — however, as we have pointed it out, the necessary number of rules in the base is *unbounded*, as it is obviously in the case of applying crisp rules, and so the computational complexity of such a fuzzy control method is very high, which fact shows that these methods are unusable from practical point of view.

These proofs do not utilize the *interpolative* nature of fuzzy control which is the real advantage of it: It is possible to approximate acceptably well by a low number of rules, as it was shown in the previous sections, e.g. by linear pieces or alternating linear and quadratic sections.

There is one more interesting group of results concerning the universal approximator property of fuzzy control, the ones presented in [1], *the function generation by fitting antecedents*. These results refer to simplified Sugeno-controller having constants in the consequents. The theorems in referred paper show that for an arbitrary continuous function g on a compact domain it is possible to find a low number of rules with constant consequents so that the resulting function realized by the fuzzy control by the rule base is g. This surprisingly powerful statement hsa a weak point however: the membership functions of the antecedents most be constructed by a direct transformation of g.

It is interesting to mention that it is possible to realize any function within the given class by *only two rules*, however, then the convexity of the terms is not guaranteed, although convexity plays a central role in natural (e.g. linguistic) reasoning (cf. [8]). Obviously, this method also does not offer any reduction in the computational complexity as the non linearity in g is simply transferred into the membership functions.

9. Interpolation in fuzzy rule bases

In the previous sections we discussed systems where the observation known concerning the actual state of the system has matched with one or several rules in the model.

These systems are so-called *dense rule bases*, in the sense that any observation has non-zero degree of mathcing with at least one rule in the base. If the value of the minimal degree of mathing is α then the system is α-*covered* by the rules. If the knowledge is partial and therefore the union of supports of the rules does not cover the whole universe of discourse, the rule base will be *sparse*, which requires some other way of dealing with it which was proposed first by Kóczy and Hirota. (We can also get such rule bases when we want to reduce the complexity of a system by omitting rules containing redundant information [11].)

The rule bases containing gaps require completely new techniques of reasoning and control. In the following paragraphs, some general considerations will introduce a family of methods eliminating the difficulties caused by gaps while having lower complexity than the classical methods. The family of methods works well only if the system has some "nice" properties: it is not allowed to behave too unexpectedly at the areas where the model does not cover it. Luckily, in practice such a nice behaviour might be expected in most cases. The term for the class of systems where the following algorithms are applicable is *interpolative system*.

In order to treat sparse rule bases with observation in the gaps, the following extended concept of rule interpolation must be introduced. The starting ideas are the Extension Principle and Resolution Principle. The latter describes the decomposition of fuzzy sets to α-cuts:

$$F - \bigcup_{\alpha \in [0,1]} \alpha F_\alpha \tag{20}$$

(Here the union means maximum.) The former states that the solution of a problem for fuzzy sets can be found in the form of solving first for arbitrary α-cuts (N.B. these are crisp sets) and then extending the solution to the fuzzy case.

Every fuzzy set can be approximated by the family of approximations of its cuts. Although theoretically all infinite cuts should be treated separately, in most practical cases, if the membership function is piecewise linear, i.e. its shape is trapeziodal or triangular, it is often enough to calculate for only a few important or typical cuts (e.g. in the simplest case, the support and the core) [12, 22, 23].

Some condition must be fulfilled for the applicability of the KH-interpolation method (there exists such interpolation method which does not requires special conditions, cf. [2]): the fuzzy sets have to be normal and convex, and the state variables (included X_i and Y as well) must be bounded and gradual which guarantees that a full ordering in each of them exists. In this case a partial ordering can be introduced among the elements of X (i.e. among CNF sets) with the help of their α-cuts. If

$$\forall \alpha \in [0,1]: \ \inf\{F_\alpha\} \leq \inf\{G_\alpha\} \text{ and } \sup\{F_\alpha\} \leq \sup\{G_\alpha\}$$

then F and G are comparable, i.e. $F \prec G$.

Among comparable fuzzy sets there is a possibility to introduce a new concept of distance. For each significant α the two extremal points (or bounds) of the α-cuts are chosen and their pairwise distances defined as the lower and the upper fuzzy distance of the two cuts. The definition of these distances results in two families of distances. These can be represented by fuzzy families of distances as follows:

$$\mu_{d_L(F,G)}(z_i) = \sum_{\alpha \in [0,1]} \alpha / D(\inf\{A_\alpha\}, \inf\{B_\alpha\}) \tag{21}$$

$$\mu_{d_U(F,G)}(z_i) = \sum_{\alpha \in [0,1]} \alpha / D(\sup\{A_\alpha\}, \sup\{B_\alpha\}) \tag{22}$$

where $z_i \in Z$ is the variable representing the possible values of distances in X_i.

Using the concept of fuzzy distance, the closeness of two comparable fuzzy sets can be determined even if their supports are disjoint.

With the help of fuzzy distance, the classical methods of function approximation can be applied on the rule bases, even if they are sparse. Using the Resolution Principle, a rule base can be represented by a family of hyper-intervals in $X \times Y$, namely for every α, $A_{i\alpha}$, $B_{i\alpha}$ form a hyperinterval for CNF sets. The domain $A_{i\alpha} \times B_{i\alpha}$ can be unambiguously represented by the minimal and maximal points thus every rule base can be represented with an acceptable accuracy by $2l$ sets of points (l is the cardinality of breakpoint set), each containing r points (r is the number of the rules). These point sets represent the α-cuts of the fuzzy mapping $\mathcal{R}$ ($\mathcal{R}$ assigns a CNF set of Y to every CNF set of X) which is approximated by the rules. By the Extension Principle, instead of approximating the original fuzzy mapping, only its important (breakpoint level) cuts will be approximated. Hence the problem is reduced to a family of non-fuzzy approximation problems which can be solved with any of the classical function approximation methods like interpolation or extrapolation, etc.

The simplest of these methods is the linear interpolation of two rules for the area between their antecedents. This can be applied if the observation is located so that

$$A_{i1} \prec A^* \prec A_{i2} \text{ and } B_{i1} \prec B_{i2}.$$

Using the concept of fuzzy distance, the following fundamental equation of *linear interpolation* can be written, in accordance to the gradual semantic interpretation of fuzzy rules by [6].

$$d(A^*, A_{i1}) : d(A^*, A_{i2}) = d(B^*, B_{i1}) : d(B^*, B_{i2}). \tag{23}$$

After decomposing this equation to every $\alpha \in [0,1]$ or in breakpoints sets it can be solved for B_α^*. The following formulae are the solution for linear KH-controller:

$$\min\{B_\alpha^*\} = \frac{\frac{1}{d_L(A_\alpha^*, A_{i1\alpha})} \inf\{B_{i1\alpha}\} + \frac{1}{d_L(A_\alpha^*, A_{i2\alpha})} \inf\{B_{i2\alpha}\}}{\frac{1}{d_L(A_\alpha^*, A_{i1\alpha})} + \frac{1}{d_L(A_\alpha^*, A_{i2\alpha})}}, \tag{24}$$

$$\max\{B_\alpha^*\} = \frac{\frac{1}{d_U(A_\alpha^*, A_{i1\alpha})} \sup\{B_{i1\alpha}\} + \frac{1}{d_U(A_\alpha^*, A_{i2\alpha})} \sup\{B_{i2\alpha}\}}{\frac{1}{d_U(A_\alpha^*, A_{i1\alpha})} + \frac{1}{d_U(A_\alpha^*, A_{i2\alpha})}}, \tag{25}$$

For the two families of solutions to determine a fuzzy set B^*, it should satisfy $\min\{B_\alpha^*\} \le \max\{B_\alpha^*\}$ for every α [9].

The principle of interpolating two rules can be extended in many different ways. The most obvious extension of the interpolation of two rules is the interpolation of $2n$ rules (n and n flanking the observation in the sense of $\prec$) where pairs of flanking rules are considered and the farther the elements of the pair are located from the observation, the less weight the respective consequents play in the construction of the conclusion. The extended formulae (which will play a crucial role in the following sections) for this type of interpolation are obtained from the solution of the fundamental equation repeatedly for the pairs of points and by averaging the various solutions:

$$\min B_\alpha^* = \frac{\sum_{i=1}^{2n} \frac{1}{d_L(A_\alpha^*, A_{i\alpha})} \inf\{B_{i\alpha}\}}{\sum_{i=1}^{2n} \frac{1}{d_L(A_\alpha^*, A_{i\alpha})}} \tag{26}$$

$$\max B_\alpha^* = \frac{\sum_{i=1}^{2n} \frac{1}{d_U(A_\alpha^*, A_{i\alpha})} \sup\{B_{i\alpha}\}}{\sum_{i=1}^{2n} \frac{1}{d_U(A_\alpha^*, A_{i\alpha})}} \tag{27}$$

More details on this method can be found in [10].

10. Stability of the KH-interpolation method

In this section we proof the stability of the Kóczy–Hirota interpolation method, in the sense that it does not depend on the distribution of measurement points.

Let p_i ($i = 1, \ldots, k$) distinct points and let us consider the following approximation of the continuous function f (which is identical with expressions (26) and (27) rewritten in a somewhat more general way):

$$L_k(f, p) := \sum_{i=1}^{k} f(p_i^{(k)}) \frac{\frac{1}{|p-p_i|}}{\sum_{j=1}^{k} \frac{1}{|p-p_j|}}. \tag{28}$$

Here we interpreted the distance as the absolut value.

Statement 11. $L_k(f,p) \to f(p)$ *if f is continuous on the domain $X \subset R^n$ and the points $p_i^{(k)}$ become more dense uniformly on X in the sense that for an arbitrary subdomain $\omega \subset X$ for sufficiently great k (depending on ω; $k \geq k_0$) the number of the points $p_i^{(k)} \approx \frac{|\omega|}{|X|}$.*

Proof. Let us estimate the next difference

$$L_k(f,p) - f(p) = \sum_{i=1}^{k} [f(p_i) - f(p)] \frac{\frac{1}{|p-p_i|}}{\sum_{j=1}^{k} \frac{1}{|p-p_j|}} =$$

$$= \sum_{i,|p-p_i|\leq \delta} [f(p_i) - f(p)] \frac{\frac{1}{|p-p_i|}}{\sum_{j=1}^{k} \frac{1}{|p-p_j|}} +$$

$$+ \sum_{i,|p-p_i|> \delta} [f(p_i) - f(p)] \frac{\frac{1}{|p-p_i|}}{\sum_{j=1}^{k} \frac{1}{|p-p_j|}} =: \sum_1 + \sum_2$$

If δ is small enough then because of continuity of f

$$|f(p_i) - f(p)| < \varepsilon$$

for arbitrary $\varepsilon > 0$. δ can choosen on X when f is continuous independently from p and p_i. Thus

$$\sum_1 \leq \varepsilon \cdot \sum_{i=1}^{k} \frac{\frac{1}{|p-p_i|}}{\sum_{j=1}^{k} \frac{1}{|p-p_j|}} = \varepsilon$$

Now let us estimate $\sum_2$. Obviously

$$\sum_2 \leq 2\max_X |f| \cdot \sum_{i,|p-p_i|>\delta} \frac{\frac{1}{|p-p_i|}}{\sum_{j=1}^{k} \frac{1}{|p-p_j|}} \leq$$

$$\leq 2\max_X |f| \cdot \frac{\sum_{i,|p-p_i|>\delta} \frac{1}{|p-p_i|}}{\sum_{j,|p-p_j|>\delta} \frac{1}{|p-p_j|} + \sum_{j,|p-p_j|\leq\delta} \frac{1}{|p-p_j|}} \qquad (29)$$

Let us denote the sum $V := \sum_{j,|p-p_j|>\delta} \frac{1}{|p-p_j|}$. Hence

$$\sum_2 \leq 2\max_X |f| \cdot \frac{V}{V + \sum_{j,|p-p_j|\leq\delta} \frac{1}{|p-p_j|}}$$

Considering the case $n = 1$, $V = V(k) \sim \log(k)$. For estimating the sum in the denominator let us take the closest p_j in the δ-environment of p (for sufficiently great n such a p_j exists) keeping only this member in the sum, i.e.

$$\sum_{j,|p-p_j|\leq\delta} \frac{1}{|p-p_j|} > \min_{j,|p-p_j|<\delta} \frac{1}{|p-p_j|}.$$

Hence

$$\sum_2 \leq 2\max_X |f| \frac{\log k}{\log k + \frac{1}{k}} \sim 2\max_X |f| \frac{\log k}{\log k + k} \to 0 \qquad (k \to \infty)$$

which completes the proof for the case when $n = 1$.

Consider the case $n = 2$. Then (28) will have the form:

$$L_k(f,p) := \sum_{i=1}^{k} f(p_i^{(n)}) \frac{\frac{1}{|p-p_i|^2}}{\sum_{j=1}^{k} \frac{1}{|p-p_j|^2}}, \qquad (30)$$

i.e., the distance function changes to the Euclid-distance. This requires changing in the estimation (29):

$$\sum_2 \leq 2\max_X |f| \cdot \frac{\sum_{i,|p-p_i|>\delta} \frac{1}{|p-p_i|^2}}{\sum_{j,|p-p_j|>\delta} \frac{1}{|p-p_j|^2} + \sum_{j,|p-p_j|\leq\delta} \frac{1}{|p-p_j|^2}} \leq$$

$$\leq 2\max_X |f| \cdot \frac{V}{V + \sum_{j,|p-p_j|\leq\delta} \frac{1}{|p-p_j|^2}} \qquad (31)$$

similar to the former denotation $V = \sum_{i,|p-p_i|>\delta} \frac{1}{|p-p_i|^2}$, i.e., we sum for those control points which are farther from p then δ. Supposing uniform distribution for the control points for sufficiently great k it results in an equidistant grid on the plane. Let the distance of the grid points (control points) less then η:

$$|p_i - p_j| \leq \eta \quad \text{for all } i, j \neq i < k$$

The number of the grid points in a circle with radius R (choosing sufficiently great R for cover the whole domain of f) which are farther from the centre than δ can calculate as follows (cf. Fig. 1.):

$$\int_\delta^R \frac{2r\,dr}{\eta^2},$$

thus

$$V \approx \int_\delta^R \frac{2r\,dr}{\eta^2} \cdot \frac{1}{r^2} = \int_\delta^R \frac{2\,dr}{\eta^2} \cdot \frac{1}{r} \geq \frac{1}{\eta^2}$$

Let us turn to estimating $W := \sum_{j,|p-p_j|\leq\delta} \frac{1}{|p-p_j|^2}$. If the point p is equals to one of the grid points then this expression will be ∞, hence $\sum_2 < \varepsilon$. The maximal distance between the point p and the closest grid point is less than η.

$$\delta_0 := \max_p \min_{p_i} |p - p_i| < \eta.$$

Thus

$$W \approx \int_{\delta_0}^{\delta} \frac{2rdr}{\eta^2} \frac{1}{r^2} = \int_{\delta_0}^{\delta} \frac{2dr}{\eta^2} \frac{1}{r} = \frac{\ln \delta}{\eta^2} - \frac{\ln \delta_0}{\eta^2} = -\frac{1}{\eta^2} \ln \left(\frac{\delta_0}{\delta}\right)$$

where both $\ln \delta$ and $\ln \delta_0$ are big negative numbers. So $\ln \left(\frac{\delta_0}{\delta}\right)$ is negative, as well. Finally

$$\sum_2 \leq 2 \max_X |f| \cdot \frac{V}{V + W} = \frac{1/\eta^2}{1/\eta^2 - (1/\eta^2) \ln \delta} < \varepsilon$$

because we can choose sufficiently great δ for every η and ε. This completes the proof for $n = 2$.

11. Conclusions

In the previous sections we have discussed in some detail the explicit formulae of fuzzy control, comparing some methods in use, especially concentrating on triangular and trapezoidal rules that have a low computational complexity. After the presentation of some formulae already proposed in the literature, some more were introduced. It was shown that usually these formulae generate piecewise linear or close to piecewise linear approximations (monotone rational functions).

In the case of the TS-controller we have shown that it generates quadratic or quasi-quadratic splines, which are however parametrically interchained, and so the real applicability of that control method is in the cases where the control function consists of quasi-linear sections, and so, the interpolation sections will be generated automatically by the method.

In the second part of this paper the question of universal approximator property of fuzzy control was discussed. Even though a family of theorems exist that state that is possible to approximate an arbitrary continuous function of compact domain with arbitrary accuracy by a very wide class of fuzzy control, the real advantage of fuzzy control is not there, as these statements all refer to solutions where there is an immense computational complexity involved. We have shown that even crisp expert control approach satisfies the universal approximator nature while from pratical engineering applications point fuzzy control proved to be more popular and succesful. The real point of applying fuzzy systems lies in the reduction of nonlinearity and complexity of the real world systems. Instead of looking for very complicated and large size rule bases (that approximate very well) the target should be to find minimal acceptable realizations when observing a given objective function. Explicit formulae might help with evaluating any fuzzy control (or reasoning) system whether this objective function is satisfied.

In the third part of the paper the rule interpolation technique was discussed. This is applicable when the rule base is sparse. After the introduction of the essential idea of the KH-controllers, we deduced the formulae for the simplest cases, i.e. when using linear interpolation and considering two flanking rules in the computation of the conclusion. We derivated the extension of this case when $2n$ (n and n flanking) rules were taking into cosideration.

Finally the stability of the KH-controller was shown for the latter case. This offers another obvious reason for the fast success of fuzzy controllers especially if in the future similar stability can be proved for the Mamdani-controller, as well.

Acknowledgments

This research was supported by the National Scientific Research Foundation (OTKA), Grants No. T019671, T011600, by the Hungarian Ministry of Culture and Education (MKM), Grant No. 788, and by the Faculty of Economy and Commerce, the University of Trento.

References

1. P. Bauer, E.P. Klement, B. Moser and A. Leikermoser: Modeling of fuzzy functions by fuzzy controllers, in *Theoretical Aspects of Fuzzy Control* (H.T. Nguyen, M. Sugeno, R. Tong, R. Yager, Eds.), Wiley, Chichester, 1995.
2. P. Baranyi and T.D. Gedeon: Rule interpolation by spatial geometric representation, *IPMU'96*, Granada, 483–488, 1996.
3. D.G. Burkhardt and P.P. Bonissone: Automated fuzzy knowledge base generation and tuning, *IEEE Int. Conf. on Fuzzy Systems*, San Diego, 179–196, 1992.
4. J.L. Castro: Fuzzy logic controllers are universal appoximators, *IEEE Transac. in Systems Man and Cybernetics* **25**, 629–635, 1995.
5. D. Driankov, H. Hellendoorn and M. Reinfrank. *An Introduction to fuzzy control*, Springer, Berlin, 1993.
6. D. Dubois and H. Prade: Gradual rules in approximate reasoning, *Information Science* **61**, 103–122, 1992.
7. A. El Hajjaji and A. Rachid: Explicit formulas for fuzzy controller, *Fuzzy Sets and Systems* **62**, 135–141, 1994.
8. N. Foo: New perspective and old problems... *Soft Computing Symposium*, Sydney, 1994.
9. L.T. Kóczy and K. Hirota: Ordering, Distance and Closeness of Fuzzy Sets, *Fuzzy Sets and Systems* **60**, 281–293, 1993.
10. L.T. Kóczy and K. Hirota: Approximate reasoning by linear rule interpolation and general approximation, *Internat. J. Approx. Reason.* **9**, 197–225, 1993.
11. L.T. Kóczy and K. Hirota: Size reduction by interpolation in fuzzy rule bases, *IEEE Trans. on SMC.* **27**, 14–25, 1997.

12. L.T. Kóczy and Sz. Kovács: On the preservation of convexity and piece-wise linearity in linear fuzzy rule interpolation, *TR 93-94/402*, LIFE Chair of Fuzzy Theory, TIT, Tokyo, 23p, 1993.

13. L.T. Kóczy and D. Tikk: Approximation in rule bases, *IPMU '96*, Granada, 489–494, 1996.

14. L.T. Kóczy and M. Sugeno: Explicit Functions of Fuzzy Control Systems, *International Journal of Uncertainty, Fuzziness and Knowledge-Based Systems* **4**, 515–535, 1996.

15. J.T.K. Koo: Design of stable adaptive fuzzy control, M. Phil. Thesis, Div. of Info. Engineering, The Chinese University of Hong Kong, 1994.

16. B. Kosko: Fuzzy systems as universal approximators, *FUZZ-IEEE '92*, San Diego, 1153–1162, 1992.

17. C.P. Kwong: Fuzzy inference without membership function, Manuscript. Div. of Info. Engineering, The Chinese University of Hong Kong, 1993.

18. P.M. Larsen: Industrial application of fuzzy logic control, *Int. J. of Man Machine Studies* **12**, 3–10, 1980.

19. E. H. Mamdani and S. Assilian: An experiment in linguistic synthesis with a fuzzy logic controller, *Int. J. Man-Machine Studies* **7**, 1–13, 1975.

20. M. Mizumoto: Improvement of fuzzy controls (IV), *6th Fuzzy System Symposium*, Tokyo, 1990, 9–13 (in Japanese), 1990.

21. H.T. Nguyen and V. Kreinovich: On approximations of controls by fuzzy systems, Technical Report TR 92-93/302, LIFE Chair of Fuzzy Theory, Tokyo Inst. of Technology, 1992.

22. Y. Shi and M. Mizumoto: On Koczy's interpolative reasoning method in sparse rule bases, *10th Fuzzy Systems Symposium*, Osaka, 211–224, 1994.

23. Y. Shi and M. Mizumoto: Some considerations on Koczy's interpolative reasoning method, *FUZZ-IEEE/IFES '95*, Yokohama, 2117–2122, 1995.

24. M. Sugeno and T. Yasukawa: A fuzzy-logic-based approach to qualitative modeling, *IEEE Trans. on FS* **1**, 7–31, 1993.

25. T. Takagi and M. Sugeno: Fuzzy identification of systems and its applications to modeling and control, *IEEE Trans. on Syst., Man and Cybernetics* **15**, 116–132, 1985.

26. J. Varga and L. T. Kóczy: Explicit formulae of two-input fuzzy control, *BUSEFAL* **63** (Été), 58–66, 1995.

27. L.X. Wang and J. Mendel: Generating fuzzy rules from numerical data with applications. *TR USC-SIPI #169*, Signal and Image Processing Institute, Univ. of Southern California, 1991.

28. L.X. Wang: Fuzzy systems are universal approximators, *FUZZ-IEEE '92*, San Diego, 1163–1169, 1992.

FUZZY PERFORMANCE INDICATORS FOR THE CONTROL OF MANUFACTURING PROCESSES

L. Berrah [1,3], **G. Mauris**[2], **L. Foulloy**[2] and **A. Haurat**[1]

[1]Laboratoire de Logiciels pour la Productique LLP-CESALP
[2]Laboratoire d'Automatique et de MicroInformatique Industrielle LAMII-CESALP
Université de Savoie
41, Av. de la Plaine
F-74016 Annecy
France
[3]IPI - ENSGI - INPG
46, Av. Félix Viallet
F-38031 Grenoble Cedex
France
Email: {berrah,mauris,foulloy,haurat}@esia.univ-savoie.fr

Abstract. This study shows one use of the fuzzy subset theory with the aim of taking into account the qualitative, the imprecise and/or uncertain knowledge in a manufacturing context. The focus is more particularly on the evaluation aspects of the industrial processes and activities. This evaluation concerns the results of the different activities or processes' enactments and consists in comparing them to the assigned objectives. One means to effect this comparison is given by performance indicators. Usual indicators only treat precise and numerical data, while the objectives can be imprecisely, subjectively or gradually defined and the measures uncertainly expressed. By using the fuzzy, the possibilistic and the probabilistic formalisms, the proposed indicators deal with both numerical and symbolic data, and provide either a performance measure or a performance evaluation. Moreover, always in contrast with the usual evaluation, the proposed indicators evaluate a relative performance, with regard to the objective on the one hand, and to the real conditions of the considered enactments on the other hand. While the absolute performance is related to the efficiency measure, the relative performance is useful for the control of the processes. The evaluation of these performances is also based on some fuzzy extensions of general concepts of correspondence, proximity, distance...These ideas are applied to some problems encountered in one ski production process. This production involves first an assembly activity whose performance is driven by many parameters; and at the end of the production, a control quality activity which is performed by human operators, on the basis of subjective features.

Keywords: performance indicators - fuzzy set comparisons - fuzzy rule-based aggregation, manufacturing process.

1. Introduction - Context

To improve their performance in an "affluent society", with its wealth of goods and services on the one hand, and the critical consumers on the other hand, the manufacturing companies have to deal with a strong competition between manufacturable comparable products. If they want to survive, they must increase customer satisfaction, by quickly reacting to their changing demands. They also have to continuously improve the quality and the delivery of the products, and to reduce costs. In this context, traditional management systems have shown their limits, since they are based on the *one cost system*, which only considers financial performances. To remedy the inadequacies of this system, companies turn towards a balanced representation of both financial and operational measures. The latter complement the former by taking non-financial aspects into account, such as performance of the industrial process, userfriendliness of a CAM (Computer Aided Manufacturing) system,reliability of the manufacturing equipment.[1]

In the conventional system (taylorian system), measures were only used as a way to *verify* the productivity of the resources and also to assess the workmanship. The performance structure today is essentially a support system for *control, diagnosis* and *monitoring*. Performance indicators are a means to this end: they provide a tool to compare current results with pre-set objectives, and to measure the extent of any drift. In this sense, one definition of the performance indicator is: *a variable indicating the effectiveness and/or efficiency of a part or whole of the process or system against a given norm/target or plan* [2].

At an operational level, performance indicators are used for the evaluation of the activities of the industrial processes. For each activity, the comparisons of the results obtained with the assigned objectives provide the reached performances. Furthermore, the performance indicators *indicate quantitatively how the performance is quantitatively going*. An insufficient performance can be traced by looking at some parameters which *drive* the performance process, and so called *performance drivers*. In contrast to financial reports, the structure of performance indicators indicates quickly how sound the considered activites or processes are. Hence, this structure constitutes one essential element for a more reactive analysis. Whenever one drift is detected by any performance driver, supervisors can immediately react, rather than passively wait for the subsequently financial information [3].

Besides, the reached performance results from the comparison of the measures with the objectives. We adopt first the traditional definition of the performance, as being the *absolute* result of the analysis of the measure according to the assigned objective. Thus, a relative performance is proposed, which compares the measure to a *contextualized* objective. Such a contextualized performance takes into account the effective conditions of the execution of the considered activity or process, and also any malfunctioning or drift eventually detected by a performance driver. The relative performance gives a more realistic local appreciation.

Furthermore, the historical development of measurement means in general shows that the use of the indicators is still limited to the measurement of precise numerical variables. Indicators are conceived to provide objective data, describing a situation from the strict quantitative point of view. In many manufacturing situations, the precise numerical representation is not relevant, since the variables involved in the performance evaluation, essentially the objectives and the measures, are not always of a numerical or a precise nature. Indeed, the objectives can be:

- precise but with a flexible satisfaction *(the delivery of products within ten days, knowing that the satisfaction function decreases all the more as the delivery delay is far from ten days)*,
- imprecisely defined *(delivery within a few days)*,
- associated to subjective entities as *the cleanliness of a ski.*

Concerning the measures, they can be:

- eventually pervaded by errors, when they are performed by physical sensors,
- described as linguistic characterization when they are related to subjective entities, and thus effected by human operators, *(e.g. a dirty ski)*.

Knowing thus that the objectives and measures can be vague or subjective, imprecise or uncertain, their handling requires more than a classic analytic framework. To this end, we propose to use the fuzzy subset, the possibility and the probability formalisms [4] [5] [6]. Furthermore, many mathematical concepts used for the comparison have been defined or extended in order to handle fuzzy data, such as the correspondence, the distance, the proximity... [7] [8] [9]. Some of these approaches are used here for the evaluation of both the absolute and relative performances, in the case of numerical fuzzy subsets.

Performance indicators are used everywhere, at different levels of the manufacturing processes. At a management level, only global pieces of information are needed, such as e.g. the global quality of the products, the production costs... These "global" performances constitute a scorecard, used as a strategic decision support. The global performances aggregate many "elementary" ones. This aggregation requires on the one hand a homogeneous expression of the elementary performances, and on the other hand a relation between them. The latter is often difficult to express in analytical or numerical ways. Fuzzy rules are considered here to aggregate the basic performances; and also to determine the validity of human information, which is based on some

228

parameters such as skill...

Some of these propositions are applied first to one problem of ski production at the Skis Dynastar S.A. company. This production involves one activity related to the assembly (moulding) of the elements that constitute the skis. The performance of this assembly activity depends on many drivers such as the quality of the raw materials, the moulding tools, the elements to assemble, the know-how of the operators who assemble the skis... All these criteria are taken into account by the so called "Contrôle_Utile" indicator (which could translate as "relevant control" indicator) which evaluates the global assembly of the skis.

Another activity is considered in this study, related to the quality control which is performed at the end of the production cycle of the skis. This control is effected by human operators, with regard to the functional and aesthetic features of the skis. The first problem submitted by the company concerns the global evaluation of the quality of the skis, knowing that the involved features are not easily quantifiable, such as e.g. the proportion of the streaks on a sole. The second problem is related to the validity of the information returned by the operators. The proposed solutions are based on the use of rules deduced from human expertise.

2. The indicators as an illustration of the performance concept

Through the evaluation framework they propose, performance indicators have essentially to show whether actions for improvement have the desired results and to what extent. In this study, they are used for the particular evaluation of the industrial process activities. Knowing that a process can be seen as *a set of partially ordered activities intended to reach a goal* [10], its control is naturally based on the results of the different activities, with regard to the objectives respectively assigned to each of them.

Therefore, more than a sensor, a performance indicator performs an evaluation of the measure of the enactment activity, according to the assigned objective. The performance elaboration can be described through the following stages:

- the expression of the objective to realize,
- the acquisition of the measure,
- the elaboration of the performance.

2.1. The expression of the objectives

The objectives are expressed from customers'[1] requirements, and appear in

---

[1]Each element in a process acts as a "supplier" to the next one and a "customer" to the previous one.

different aspects, which are presented here by means of examples concerning the assembly of skis.

Example 1: *The production supervisor defines a total number of exactly 1000 skis to be produce.d In this case the objective is said to be precise.*

Example 2: *Suppose that the production supervisor wants to introduce some flexibility concerning the satisfaction of an objective of assembly of a total number of 1000 skis. The objective is still precise but the closer to 1000 skis, the higher the supervisor satisfaction. This satisfaction function could be associated to a fuzzy subset as shown in Fig.1. The objective is said to be flexible (see after in section 2.4.1.1 a more detailed discussion) and could be considered here as "around 1000 skis".*

Example 3: *Now, the supervisor's constraints can be less strict, leading to an objective defined with some imprecision, for example, "between 850 and 1250 skis". Such an objective could be represented by the numerical discrete fuzzy subset ([850.. 950...1050... 1250]) (see Fig. 2).*

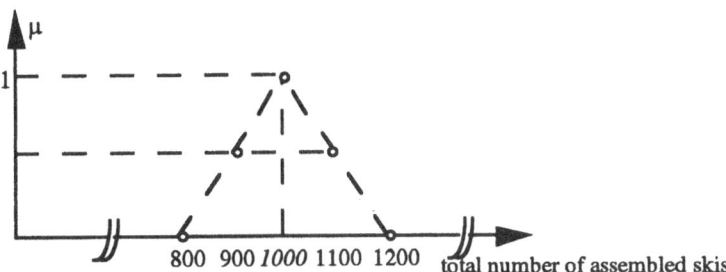

Figure 1. A fuzzy representation of a flexible objective

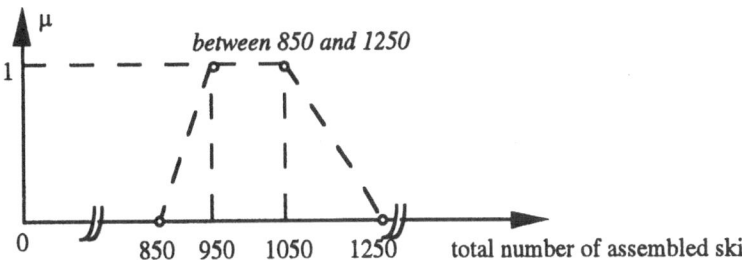

Figure 2. A fuzzy numeric representation of an imprecise objective

Example 4: *Another situation could be imagined, where the supervisor expresses*

*the objective in natural language, for example "a small number of rejected skis".
In this case, the objective is represented by a symbolic expression. This objective
can be numerically described by a fuzzy subset (called the meaning of the term)
which takes into account the vagueness of the considered term "small". The
meanings of such terms are generally provided by the expert estimations
according to their know-how of the analysed process.*

Example 5: *Finally the objective can be related to a subjective entity, such as
the cleanliness of the heel of a ski, which is rarely described in a numerical
way. The objective can be expressed e.g. by {1/clean} which is a precise symbolic
objective, or by e.g. {0.8/dirty, 0.2/fairly_clean, 0/clean}(the degrees could be
obtained by a numeric transformation of linguistic labels), which is an imprecise
symbolic objective.*

The different described situations are summarized Fig. 3.

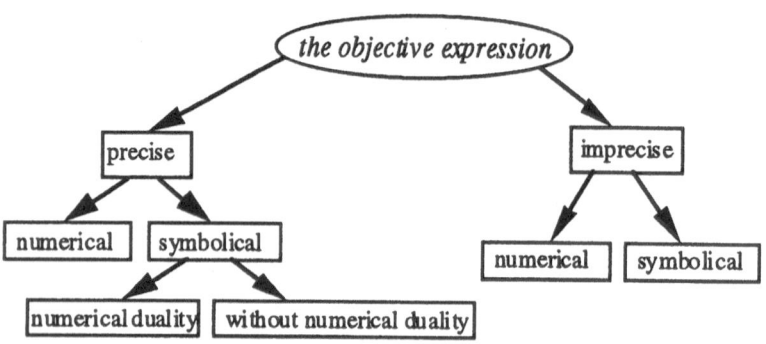

Figure 3. The different aspects of a manufacturing objective

2.2. The expression of the measure

The measures coming from the physical sensors or human operators are often
expressed by precise numerical values (*e.g. 1100 skis assembled today*), but in
reality they are pervaded with errors. Concerning the physical sensors, many
methods which deal with the measurement errors are proposed [11]. For this
case, we consider here two approaches: the most frequently used one in our
context, based on the probability theory, and the second one based on the
possibility theory. This second approach allows a homogeneous representation of
measures and objectives in the fuzzy framework.

Example 6: *Consider again the objective related to the number of assembled
skis (Example 1). The effective number is measured by an operator who gives his
measure as "approximatively 1100 skis". This kind of information could be*

represented by a uniform (or eventually other forms) probability density function (Fig. 4).

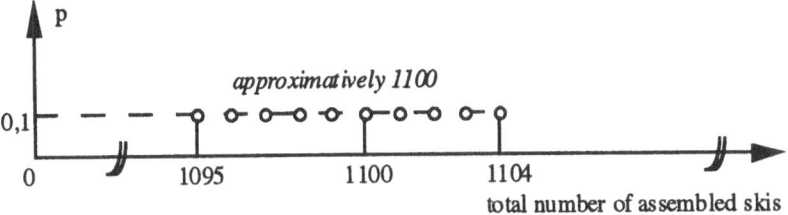

Figure 4. A probabilistic characterization of a measure

As pointed by Dubois [12], this kind of qualitative information coming from operators could be advantageously represented by a possibility distribution (Fig.5), because the operators provide several nested intervals with associated levels of confidence (from which the possibility distribution is deduced) more easily than a function of density of probability.

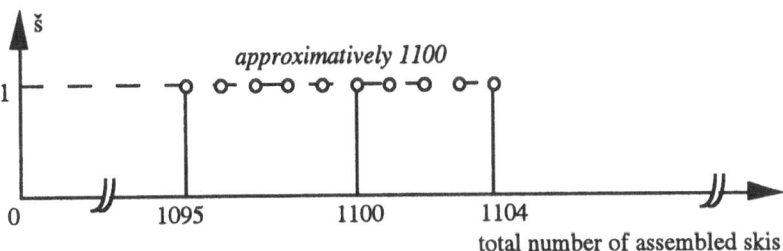

Figure 5. A possibilistic characterization of a measure

Another approach consists in characterizing the measure on the pre-defined terms L of the objective, i.e. doing a symbolic measurement, which could be returned by fuzzy sensors [13]. One manner to proceed is to deduce a degree of description ($\mu_{D(x)}$) for every term L from its meaning ($\mu_{M(L)}$) by [14]:

$$\mu_{D(x)}(L) = \mu_{M(L)}(x) \qquad (x \in \mathcal{N}; \ L \in L) \ (1).$$

If the measure is given with an imprecision represented by a fuzzy subset $\mu_E(x)$, a lower and an upper descriptions D^- and D^+ could be respectively computed by [15]:

$$\mu_{D^+(E)}(L) = \perp_{x \in E} T(\mu_{M(L)}(x), \mu_E(x)) \tag{2}$$

$$\mu_{D^-(E)}(L) = T_{x \in E} \perp (\mu_{M(L)}(x), 1 - \mu_E(x)) \tag{3}$$

Example 7: *Consider the feature related to the "sole_blows", which illustrates the number of blows on the sole of a ski. The fuzzy meanings of the terms used for describing the number of blows are given by:*

*M(low) = {1/0, 1/1, 1/2, 0.66/3, **0.33**/4, 0/5, 0/...},*
*M(medium) = {0/0, 0/1, 0/2, 0.33/3, **0.66**/4, 1/5, 0.66/6, 0/...},*
*M(high) = {0/0, 0/1, 0/2, 0/3, **0**/4, 0/5, 0.33/6, .0.66/7, 1/8, 1/...}.*

The transformation from a numerical measure x into a symbolic one is given by using (1); 4 blows are described as {0.33/low; 0.66/medium; 0/high}.

Example 8: *Suppose now that the operator gives an approximative measure of the number of blows as around 5, which is modelled as the fuzzy discrete interval* $E = \{0/..., 0/3, 0.5/4, 1/5, 0.5/6, 0/7, 0/...\}$. *Using* (2) *and* (3), ($\perp$ =max, T =min), *we obtain a characterization of the sole blows between* D^- =*{0/low; 0.66/medium; 0/high} and* D^+ =*{0.33/low; 1/medium; 0.33/high}.*

Finally, the linguistic characterization of information can also be obtained directly according to the operator's estimation.

Example 9: *With regard to the feature "heel", the operator evaluates the cleanliness of the heel of a ski and gives for example the following characterization: {0/dirty; 0.20/fairly clean; 0.80/clean}.*
The different expressions of the measure are summarized Fig.6.

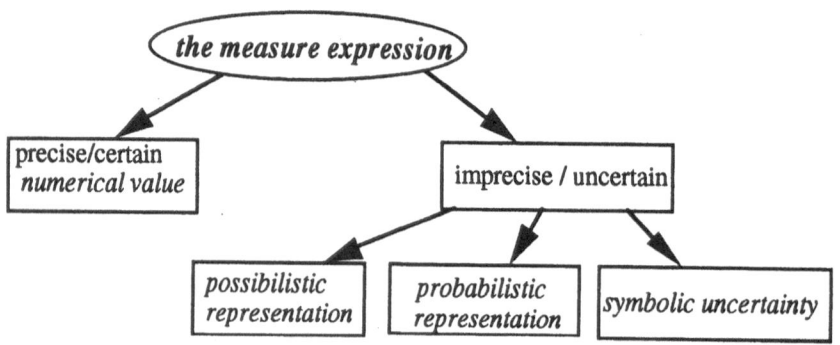

Figure 6. The different representations of a manufacturing measure

2.3. Performance Measure

From a general point of view, specifying the performance is done by so-called performance indicators which reflect the activity of an entity with respect to a given objective.

The problem which is addressed in this section concerns performance indicators which return information concerning *a single entity* and which can be used to implement *local control actions* for improving the *performance of this single entity*. Since the returned information is local, several definitions may be considered leading to define the performance on any set. Such a performance elaboration will be called *performance measures*.

Definition 1: Let O be the universe of definition of the objectives, and M be the universe of definition of the measures. A performance measure for a given entity is defined by the mapping P_m,

$$P_m: O \times M \to E$$
$$(o, m) \mapsto P = P_m(o, m)$$

where E can be any set.

Example 10: *Let us consider a production whose objective is to produce* 1000 *skis per day (the objective is precise and numerical). If the measure is* 800 *skis for the considered day (the measure is also precise and numerical), several performance measures can be considered:*

$$P_1 = P_{m1}(o, m) = m / o = 0.8$$
$$P_2 = P_{m2}(o, m) = o - m = -200$$

In the first case E is included in the set of rationals, in the second one in the set of signed integers.

Each performance measure has its own meaning and generally allows a quick analysis of the entity. For example, P_1 (resp. P_2) indicates that the current activity is below the objective if the result is below 1 (resp. negative).

Example 11: *Let us now consider the case of a cleaning machine whose objective is to produce "clean" skis (the objective is a precise and symbolic one). Let the measured activity be subjectively appreciated by a human being on a set of terms T_m={dirty, quite_dirty, quite_clean, clean, fairly_clean, very_clean}. Let the performance be defined on a set of terms T_p={very_bad, bad, fair, good, very_good}. A rule-based approach can be used to solve that case. The measure is symbolic and can be precise or imprecise (i.e. represented by a fuzzy subset of T_m). Let us define the following set of rules for the set of two objectives*

To={clean , very_clean}.

				Tm			
		dirty	q_dirty	q_clean	clean	f_clean	v_good
To	clean	v_bad	bad	fair	very_bad	bad	v_good
	v_clean	v_bad	v_bad	bad	bad	fair	v_good

Figure 7. Set of rules for symbolic performance measures

Let o=clean and m=fairly clean, then $P_3=P_{m3}(o, m)=\{good\}$ (here $E=T_p$).

Now let us use a fuzzy representation because the measure is symbolic and imprecise. Let o=1/clean, and m={0.8/clean, 0.2/fairly_clean}, according to Zadeh's compositional rule of inference [14], the performance is: $P_4(o,m)=P_{m4}(o,m)=\{0.8/good\}$, here E is included in the set of the fuzzy subsets of T_p.

2.4. Performance Evaluation

The problem is now slightly different. It concerns performance indicators which return *performance evaluation,* i.e. information concerning *a single entity* and which can be used to implement *global control actions* to improve the *performance of a set of entities*.

Example 12: *Let us consider a soccer team. Performance measures will be used to define the performance of each player in order to improve his own performance. Performance evaluations will be used to define the performance of each player in order to improve the team performance.*

Since the control actions are global, the returned performances must be homogeneous, thus leading to define them on a unique reference set. In order to deal with the fuzzy representation of the objectives and the measures, we propose to define this reference set as the unit interval [0,1], leading to the following definition.

Definition 2: Let O be the universe of definition of the objectives, and M be the universe of definition of the measures. A performance evaluation for a given entity is defined by the mapping P_e,

$$P_e:O \times M \to [0, 1]$$
$$(o,m) \mapsto P = P_e(o,m)$$

where :

1. means that the entity is not performant at all (i.e., the objective is not reached),
2. means that the entity is fully performant (i.e. the objective is fully satisfied),
3. means that the entity is partially performant (i.e. the higher P is, the more satisfied the objective is).

In the following, several examples are presented in order to illustrate this concept without being exhaustive.

2.4.1. Precise numerical objectives

2.4.1.1 Precise measures

Let us reconsider the example 1 of a ski production where m=800 and o=1000. The simplest performance evaluation which can be considered is:

$$P = Pe(o, m) = \begin{cases} 1 \; if \; o = m \\ 0 \; otherwise \end{cases}$$

Obviously this performance evaluation is not very useful due to the crisp answer provided. In general one may want to introduce some flexibility in the satisfaction of the objective.

A first approach consists in defining the performance by means of a function defined on [0,1]as in [16].

Example 13: *Let o=1000, m=1000 then Pe(o,m)=1, for o=1000 and m=800* $Pe(o, m) = exp(-|o - m| / o) = 0.135$

Another approach consists in providing a fuzzy subset O(o) of M only for a given objective o. The grade of membership expresses how the objective is satisfied for a given measure. In this case, we have:

$$Pe(o, m) = \mu_{O_{(o)}}(m) \tag{4}$$

Since the membership function depends on the objective o, and the semantics expresses some flexibility as regards its satisfaction, it is called a flexible objective by Zadeh in [17]. However, note that $\mu_{O_{(o)}}(m) \neq \mu_{\{o\}}(m)$, (except for o=m) where the latter is the membership function of the precise objective o.

Example 14: *Knowing the flexible objective "around 1000 skis" as described in Fig.1, suppose that the effective number is precisely measured as 1100 skis, then,* *by using (4), Pe(around 1000, 1100) = 0. 5 .*

2.4.1.2 Imprecise measures

Another situation must be considered when the measure is given with uncertainty/imprecision represented by a fuzzy subset μ_M. The latter could be propagated to the evaluated performance, which this time will not be precisely expressed. Equation (4) could be extended and computed by Zadeh's extension principle:

$$\mu_{Pe(o,M)}(m) = \begin{cases} \sup_u \mu_M(u) \ if \ m = Pe(o,u) \\ 0 \ if \ \forall u \ m \neq Pe(o, u) \end{cases} \tag{5}$$

Thus, the evaluated performance is a fuzzy subset of $[0,1]$. Because of the computation complexity of this method, when μ_M can be represented as a possibility distribution (i.e. a normalized fuzzy subset), Dubois & al. [18] have proposed two scalar degrees which summarize this fuzzy number. The possibility $\Pi(o; M)$ and the necessity $N(o; M)$ degrees are given by:

$$\Pi(o; M) = \sup_u min(Pe(o, u), \mu_M(u))$$
$$N(o; M) = \inf_u max(Pe(o, u), 1 - \mu_M(u)) \tag{6}$$

Moreover, they also point out that any modal value τ (i.e. $\mu_{Pe(o,M)}(\tau) = 1$) could be a scalar representation of Pe since τ is in the interval $[N(o; M), \Pi(o; M)]$. Note that in the summarization operation of Pe by scalar values, information is lost, which could lead to the same performance for different measures.

Example 15: *Let us consider the flexible objective given in example 2 "around 1000 skis", and let the measure be as defined in Fig. 5. It results in:*

$N(around 1000, approximatively 1100) = 0.48$
$\Pi(around 1000, approximatively 1100) = 0. 52$.

These results can be interpreted as, for the uncertain number "approximatively 1100 skis ", with regard to the flexible objective "around 1000 *skis*", the performance is between 0.48 and 0.52 .

When the uncertainty is handled by the probabilistic formalism, the equation (4) can no longer be applied. But a scalar evaluation of the performance can be obtained thanks to the concept of the probability of fuzzy events given by Eq. 7 for the discrete case [19]:

$$Pe(o, M) = \sum_{i=1}^{n} p(x_i) . \mu_O(x_i), \ x_i \in M \tag{7}$$

Example 16: *For the same flexible objective and the probability distribution given in Fig.4, we obtain* $Pe(around\,1000, approximatively\,1100) = 0.50$.

If supervisors want more indications when $Pe = 0$, it is possible to evaluate the bad measures with regard to a more tolerant flexible objective $O'_{(o)}$ deduced from the basic one $O_{(o)}$ as described in section 3.2.2. It would also be possible to compute normalized distances between the measures and a representative value of $O_{(o)}$(e.g. the center of gravity).

2.4.2. Imprecise numerical objectives

When the objectives are numerical and imprecise, the performance evaluations are less immediate. Therefore, the evaluation should be defined as a kind of "correspondence" between the measure (denoted M) and the objective (denoted O). This concept could be implemented by so-called "matching indices", which give degrees between 0 and 1 to qualify the correspondence, when both the objective and the measure are represented by fuzzy subsets. Note that here, we consider that such matching indices are not using a relation (e.g. a distance) on the universe of definition of the fuzzy subsets considered, but are only using the membership degrees.

A lot of such indices are encountered in the literature, based on intersection, inclusion, equality operations [20][7]. The problem of the choice of such indices according to the applications under consideration will not be considered here. But to illustrate our approach, among many others, we have considered the intersection operator proposed in [20] for the three following examples which are related to Fig.8:

$$I(O, M) = \max_{i} (min(\mu_O(x_i), \mu_M(x_i))) \tag{8}$$

Example 17: *Consider now the imprecise objective "between 850 and 1250 skis". Suppose that the measure indicates precisely 1075 skis. We obtain then* $Pe = 0.87$.

Example 18: *For the same objective, suppose now that the measure is described by the fuzzy number "approximatively 1075 skis".We obtain then* $Pe = 0.88$.

Example 19: *Suppose now that the effective number of good assembled skis is in the nearness of 750. Naturally,* $Pe = 0$.

238

Here also a supplementary parameter is required to distinguish the different situations where $Pe = 0$. A simple method consists in computing the minimal distance normalized to 1 between the centers of gravity of O and M *(here $d = (1025 - 750)/1025 = 0.27$).*

A more elaborate method consists in extending the scalar distances between ordinary sets (e.g. the Hausdorff distance) to scalar distances between fuzzy ones, by using the alpha-cuts. More sophisticated methods could also be used for determining fuzzy distances if required (see [8] for example).

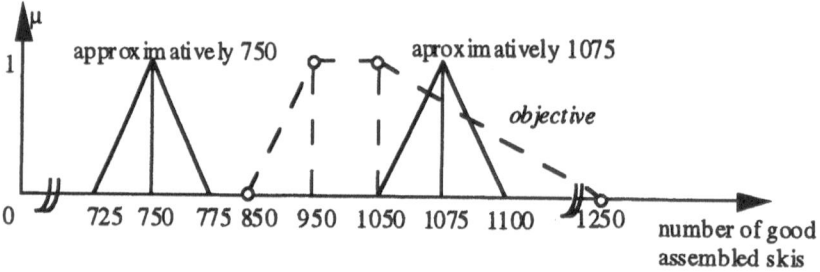

Figure 8. Imprecise objective and uncertain measures

When the uncertainty is handled by the probabilistic formalism, the evaluation is more delicate. How indeed can we compare two pieces of information which are not homogeneously handled? Some studies have been proposed for transforming a probability distribution into a possibility one by means of the evidence theory [21].

2.4.3. Symbolic objectives

When the objectives are defined by symbolic expressions, different situations are encountered. If the symbolic expressions have a numeric duality, that is the meanings of the terms are defined, the methods described above could be applied. When the objective is relative to a purely qualitative entity, i.e. defined by a symbolic characterization without a numeric duality, and when the measure is thus defined by another symbolic characterization on the same terms, the comparison of these two symbolic fuzzy subsets is a problem little considered in the literature.

A particularity is that often these fuzzy symbolic subsets are not normalized. This aspect must be taken into account in the choice of the matching indices referred to in 2.4.2.

3. How to use the evaluated performance for the manufacturing process control ?

3.1. The expression of the validity

We have seen that the manufacturing activity measures are provided either by physical sensors or operators. Up to now, we have proposed an evaluation which takes into account the eventual uncertainty related to the acquired measures. Another information concerning more generally the "validity" of the measures can be useful for defining the control actions. Indeed, even if the validity is weak, the supervisor will ask for a new measurement which better represents the reality, rather than wrongly evaluate the analysed activity. In contrast, if the validity is strong and the evaluated performance is bad, a diagnosis must be performed.

The proposed validity degree is defined in this study as being the result of the aggregation of some parameters intervening in the measurement, such as:

- the relevance of the parameter of measurement chosen by the operator to do his assessment. This choice can be linked on one hand to his knowledge of the problem, and on another hand to any technical problem (unavailability of measuring instruments...),
- the skill of the operator, i.e. his knowledge and his know-how about the analysed product. (From one model to an other, the measurement criteria vary.)

Example 20: *Concerning the ski quality control (see 4.2.), it has been noted in practice that, with regard to a given characteristic, the detection and the assessment of anomalies varied from an operator to an other. This is caused by the fact that the assessment criteria are not objectively measurable, but rather depend on the know-how of the operator. The main consequence of the subjectivity of operators' decisions is the eventual rejection of some satisfactory skis, or on the opposite, the acceptance of others which are not satisfactory.*

One solution would be the adjusting of the operators' assessment by a validity degree deduced from a rule-based aggregation of adapted parameters, by using a similar approach to the one presented in paragraph 4.3.

3.2. From an absolute performance to a relative one

3.2.1. Context

Up to now, we have proposed only an a posteriori performance evaluation of the activities. This kind of information is interesting for a strategic management of the production. However it is less relevant at the operational level, where local actions must be effected in a more reactive way. For example, the "Contrôle_Utile" performance indicator is an evaluation which helps the production supervisor to manage the global production. But this evaluation is given once the activities have been performed. This does not lead to a reactive control and so to an improvement of the activities. Indeed, the aim of the indicators' structure is to support a more reactive control, by rapidly detecting anomalies or drifts during the activity or process enactment. In this sense, performance drivers are put into place. They are particularly related to the parameters which have an influence on the result of the enactment. Whenever a drift is signalized, on the one hand corrective actions must be driven. On the other hand, the initial assigned objective must be revised, the reason of the noticed drift must be given, and the experience about the considered analysis or activity must be taken into account. The result of the activity enactment has to be then appreciated with regard to this modified objective, thus giving a local performance evaluation.

In this sense, one problem submitted by the Skis Dynastar company concerns the analysis of the impact of the elementary performances (of the drivers) on the global performance related to the quantity of assembled skis. The moulding of the skis depends on many drivers such as the quality of the elements to assemble and the available process tools... If any drift is detected concerning one of these drivers, the "Contrôle_Utile" evaluation will be affected. To adjust the evaluation and get a more realistic assessment, a relative performance has been elaborated. It takes into account the detected drift by modifying the objectives, thus allowing a more relevant local management of the activity execution, since it is not pervaded by the drifts.

3.2.2. The elaboration of the relative performance

One problem in the evaluation of the relative performance is the transformation of the assigned objective into a "contextualized" one. Two alternatives are possible, whether the repercussions of the detected drift on the global performance are well known or not:

- when the repercussions are well known, the activity supervisor can directly define the new objective according to his know-how,
- when the repercussions are imprecisely known, one manner is to use one method which supports the elaboration of the contextualized objective, such as the modification of the basic objective by a proximity relation [9].

Example 20: *Let us consider again the flexible objective given in example 2 Suppose that some defects are detected in the new raw material (e.g. the glue) used in the upper part manufacturing, (no resistance to the moulding conditions; heat and pressure). This implies a delay in the delivery of the planned lot of* *upper parts, and subsequently a delay in the assembly of the fixed quantity, thus leading to assign a new objective for the considered period. To estimate then the number of skis that can be effectively assembled, the experts consider some aspects as the available elements (other models, stocks...), the delivery of the manufactured upper parts, the machine loading... Thus, they define a new objective O' "almost around 1000 skis", which is more tolerant.*

O' can be defined by the intermediary of a proximity relation $\mu_{proximity}(n_1, n_2)$ between two numbers of skis n_1, n_2. It is computed using Eq.(9):

$$\mu_{O'}(n_2) = \sup_{n_1} \min(\mu_O(n_1), \mu_{proximity}(n_1, n_2)) \tag{9}$$

In Fig. 9, O' is plotted using:

$$\mu_{proximity}(n_1, n_2) = \max\left(0, 1 - \frac{|n_1 - n_2|}{100}\right) \tag{10}$$

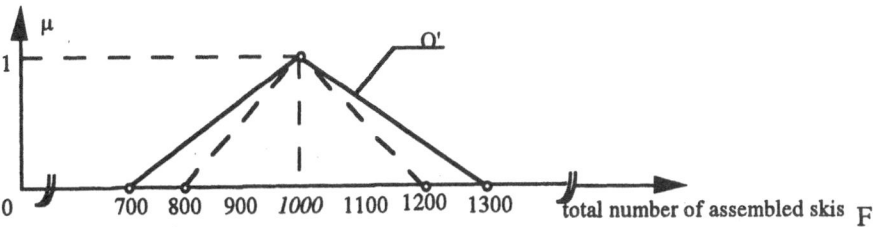

Figure 9. Modification of a flexible objective

The support of O' is larger than the support of O, thus leading to a better decision aid when the measure is outside [800..1000..1200].

Once the contextualized objective is defined, the relative performance is elaborated in the same manner as the absolute one.

4. From a basic evaluation to a synthetic one

To have a global assessment of the production process analysed, it is often necessary to aggregate the elementary assessments which are associated to the different activities of the process. Concerning the aggregation operation, many procedures described in the literature are distinguished, in accordance with the nature of the basic data and the links between them:

- if an analytical relationship between the numerical elementary data is known, the global one is simply deduced from it,

- if the elementary data under consideration are of a numerical type, and if the relation between them is not precisely known, but expresses some specificities of the application considered, such as compromise, tolerance... different numerical fuzzy operators could be used [22][23][24],
- if the aggregation relation is given by a set of rules linking the symbolic expressions of the elementary indicators and the global one, a fuzzy symbolic representation of such human knowledge is available. This section emphasizes the latter aspect.

4.1. Formal framework of fuzzy linguistic rules

Generally, most of the rules are expressed in the following manner: if V_1 is A_i then W is B_j, as e.g."*if the upper part of the ski is unstuck then the ski is not satisfactory*".

In fact, the rule represents a correspondence between the V_1 variable and the W one. Note that the premise and the conclusion variables could be of a numerical or symbolic type (*the cleanliness of a heel is of a symbolic type, while the number of blows on a sole is of a numerical one*). Here, we will denote y_1 the characterization of the variable V_1, i.e. y_1 is either a number or a symbol, and z the characterization of the variable W.

As proposed by Zadeh, the if... then... meta-implication can be viewed as a fuzzy relation represented by $\mu_{\Gamma_1}(y_1, z)$, which gives the relationship degree between the values y_1 and z. One can consider a conjunctive understanding of this rule. In this case the fuzzy relation Γ_1 is given by:

$$\mu_{\Gamma_1}(y_1, z) = T(\mu_{V_{1isAi}}(y_1) \, T \, \mu_{WisBj}(z)) \tag{11}$$

where T is a fuzzy conjunction operator (such as the min operator) [23].

Let us now consider a set of rules. If we use a conjunctive view of each rule, as we will do later, for a set of rules, the global relation Γ can be defined by an upper bound of each elementary relation Γ_i [10], i.e. $\mu_{\Gamma}(y_1, z) = \perp_i \mu_{\Gamma_i}(y_1, z)$, with $\perp$ a fuzzy disjunction operator (such as the max operator) [23].

Once the formal representation of the relation induced by a set of rules has been defined, the remaining stage is the computation of the image F of a fuzzy subset E of V_1. It is obtained by the Zadeh's compositional rule of inference [14]:

$$\mu_F(z) = \mu_{f(E)}(z) = \perp_{y_1} (\mu_E(y_1) \, T \, \mu_{\Gamma}(y_1, z)) \tag{12}$$

More complex rules can also be considered such as if V_1 is A_i and V_2 is A_j then W is B_k, as for example "*if the upper part of the ski is not unstuck and the heel is clean then the ski is satisfactory".* (12) becomes then:

$$\mu_F(z) = \underset{y_1,y_2}{\perp} [(\mu_{E_1}(y_1) T_1 \mu_{E_2}(y_2)) T \mu_\Gamma(y_1, y_2, z)] \tag{13}$$

Here, $\mu_F(z)$ could be seen as the result of the aggregation of $\mu_{E_1}(y_1)$ and $\mu_{E_2}(y_2)$. However, to compute $\mu_F(z)$, we must know $\mu_{E_1}(y_1), \mu_{E_2}(y_2)$, and $\mu_\Gamma(y_1, y_2, z)$. For our application, $\mu_{E_1}(y_1)$ and $\mu_{E_2}(y_2)$ are provided by elementary indicators, (e.g. the evaluation of the upper part of the ski or of the heel cleanliness). $\mu_\Gamma(y_1, y_2, z)$ is deduced from rules, such as the one described above. Furthermore, four aspects for the semantics of the rules could be envisioned, based either on a symbolic or a numerical view of the premises and the conclusions. Hereafter, one use of a symbolic rules representation is described, applied to one problem about the ski quality control.

4.2. The ski quality control

The ski quality control is situated at the end of the production cycle of skis before the activity of polishing. The control is performed by human operators on the basis of criteria relative to the functional and aesthetic features of skis (Fig. 10). The Dynastar company is very demanding about the quality of its products. A ski will not be judged satisfactory unless it meets strictly all the requirements. It is apparent that all these characteristics do not have the same weight in the control. Those concerning the ski sliding function have to be absolutely satisfied, while those relative to aesthetic aspects only play a part in the quality classification.

Ski Quality	Control
aesthetic features	*functional features*
screen	plating
polishing_ski_tip	sole
heel	sanding
varnish	upper_aspect
...	...

Figure 10. Functional and aesthetic features of a ski quality control.

4.3. Symbolic aggregation of aesthetic features for ski quality control

We consider here the aggregation of the aesthetic features in order to assess the aesthetic performance. These features being not referred to a numerical universe, the semantics of rules based on a symbolic view of the premises and the conclusion is used [25].

Consider the feature both related to the "heel", and the "ski_tip_polishing". The aesthetic conformity of the skis is evaluated from these two features. Let the variables be:

V_1 = *the heel_cleanliness,* V_2 = *the ski_tip_polishing and* W = *the aesthetic performance.*

$L(V_1)$ ={*dirty; fairly_clean, clean*},$L(V_2$) ={*not polished, fairly_polished, polished*} *and* $L(W)$ ={*not satisfactory, doubtful, satisfactory*}.

4.3.1. Definition of the symbolic relation

According to his know-how, the operator expresses rules of the following type: *"if the heel is clean and the tip polished then the ski is satisfactory".* According to the operator's know-how, Fig. 11 shows the table of the rules defined for these two features, from which Γ is deduced. For example, the last line and the last column represent the rule *"if the heel is clean and the tip polished then the ski is satisfactory",* that is mathematically written as: μ_Γ (*clean, polished, satisfactory*) = 1 .

		tip		
		not polished	fairly polished	polished
	dirty	*not satisf.*	*not satisf.*	*not satisf.*
Heel	fairly clean	*not satisf.*	*Doubtful*	*doubtful*
	clean	*not satisf.*	*Doubtful*	*satisf.*
		Aesthetic	**Performance**	

Figure 11. Aesthetic performance

4.3.2. Inference

Suppose now that the elementary performances have given the following assessments:

{ *0 /dirty; 0.2/fairly clean; 0.8/clean* },
{ *0 /not polished; 0.90/fairly polished; 0.10/polished* }

If we take $T(a, b) = a.b$ and $\perp(a,b) = \min(a+b,1)$, using the set of rules of Fig. 11, and the preceeding input descriptions, it results from Eq.(13): F={*0/not satisfactory; 0.92/doubtful; 0.08/satisfactory*}, which is a quite natural result since the ski is not really polished.

5. A performance indicator model for a software implementation

In application, the ideas developed before are formalized by means of a performance indicator model integrating on three facets (Fig. 12):

- the expression of the objective to realize,
- the acquisition and the evaluation of the measure provided by physical sensors or human operators,
- the appreciation of the performance evaluated before according to the context of the process or the activity enactment.

The object techniques have been chosen for the modelling of the performance indicator concept. One benefit of these techniques is assuredly their ability to consider three aspects of a system: the information aspect which introduces the data structure, the functional aspect which concerns the semantics of object activities, and the behaviour aspect which is interested in temporal models of system evolutions.

In this approach, the performance indicator is seen as an object composed of three objects, with regard to the previous facets: objective, evaluation, appreciation. The data exchanged between objects are also represented by objects, such as for example performance data (number of skis being assembled...) from which the objective and evaluation objects inherit.

The implementation has been realized with the OMT method [26] in a "Windows" environment. One interest for such a software is to constitute a support for manufacturing companies which want to put into place and memorize performance indicators for the control of their production in an homogeneous way.

6. Conclusion

This chapter has been concerned with one means for the evaluation of a manufacturing process performance. The considered approach is based on indicators which constitute nowadays an essential support for control, diagnosis and monitoring.

Usual indicators are designed to only handle precise and numerical data, while the variables involved in a manufacturing context, such as the objectives and the measures, can be expressed in a linguistic way, or respectively in an imprecise or an uncertain ones. One benefit of using fuzzy approaches is actually in the representation and the handling of such vague or subjective data.

In this sense, thanks to the fuzzy formalism, the objectives can be treated as directly expressed by supervisors, with flexibility, imprecision or subjectivity, without any restriction established by a crisp numerical description.

Furthermore, the eventual imprecision, incompleteness and uncertainty of the operators' measures can be easily handled by a possibility framework.

Moreover, the fuzzy comparison concepts are useful for the evaluation of the performance. Many methods are proposed, according to the nature of both the objectives and the measures to compare, and leading to scalar or fuzzy-value performances.

Besides, one particular interest of the use of the fuzzy theory for the manufacturing diagnosis is in the possibility of relaxing the initial objective by defining a more tolerant one, thus leading to one evaluation of a relative performance, which better illustrates the local effective execution of the activities.

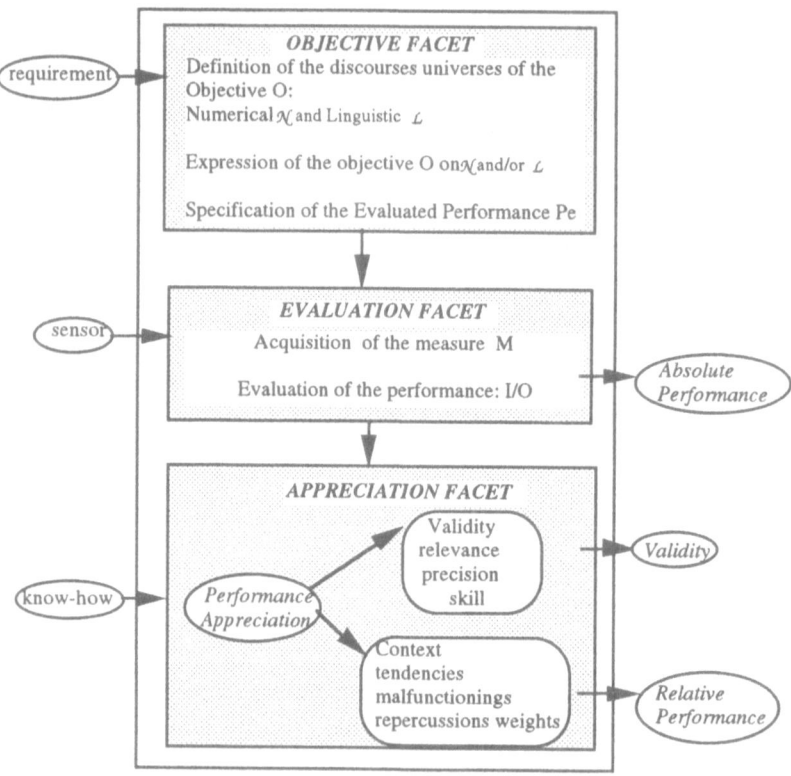

Figure 12. A performance indicator model

Another appropriate information concerns the validity of the measures returned by the operators, which has been defined in this study as the result of the aggregation of some subjective parameters involved in the measurement. Fuzzy aggregation mechanisms are also helpful to elaborate global evaluations from elementary ones, offering thus a scorecard used as a management decision support system.

Acknowledgements

This work is supported by the "Rhône-Alpes Region" of France, within the frame work of the MOPIC project [27].

References

[1] R. S. Kaplan, D.P. Norton, "The balanced scorecard - measures that drive performance", Harvard Business Review, January - February, 1992, pp. 71-79.

[2] L. Fortuin, "Performance indicators - why, where and how?", Europ. J. of Operational Research, 34, 1988, pp. 11-20.

[3] C. Berliner, J. Brimson (edited by), "Cost management for today's advanced manufacturing. The CAM-i conceptual design", Harvard Business school Press 1988.

[4] L. A. Zadeh, "Fuzzy sets", Information and Control, Vol. 8, 1965, pp. 338 - 353.

[5] L. A. Zadeh, "Fuzzy sets as a basis for a theory of possibility", Fuzzy Sets and Systems, Vol. l, n°l, 1978, pp. 3-28.

[6] M. Kendall, A. Stuart, The advanced theory of statistics, Ed. Griffin and Co., 1977.

[7] T. W. Liao, Z. Zhang, "A review of similarity measures for fuzzy systems", FUZZIEEE'96, New Orleans, USA, September 1996, pp. 930-935.

[8] C. Bertoluzza, N. Coral, A. Salas, "On a new class of distances between fuzzy numbers", Mathware and Soft Computing, 2, 1995, pp. 71-84.

[9] D. Dubois, H. Prade, C. testemale, "Fuzzy pattern matching with extended capabilities, proximity notions, importance assessments, random sets", Conf. of the North American Fuzzy Information Processing society (NAFIP 86), New Orleans, USA, June 1-4, 1986, pp. 125-139.

[10] W.S. Humphrey, P.H. Feiler, "Software process development and enactment: concepts and definitions", Tech. Rep. SEI -92-TR-4. Pittsburgh: Software Engineering Institute, Carnegie Mellon University, 1992, dans "Process Modeling", (B. Curtis, M.I. Kellner, J. Over) - Communications of the ACM, Volume 35, n°9, Septembre 1992, pp. 75-90.

[11] G. Shafer, "A mathematical theory of evidence", Princeton, Univ. Press Princeton, USA, 1976.

[12] D. Dubois, H. Fargier, H. Prade, "Fuzzy constraints in job-shop scheduling", J. of Intelligent Manufacturing, 6, 1995, pp. 215-234.

[13] G. Mauris, E. Benoit, L. Foulloy, "Fuzzy symbolic sensors: from concepts to applications", Measurement, 12, 1994, pp. 357-384.

[14] L. A. Zadeh, "Quantitative fuzzy semantics", Information Sciences, 3, 1971, pp. 159-176.

[15] L. Foulloy, S. Galichet, "Typology of fuzzy controllers", in Theoretical Aspects of Fuzzy Control (H. T. Nguyen, M. Sugeno, R. Tang, R. R. Yager Eds), Wiley, 1995, pp. 65-90.

[16] L. A. Zadeh, "Similarity relations and fuzzy orderings", Information Sciences, 3, 1971, pp. 177 - 200.

[17] R. Bellmann, L.A. Zadeh, "Decision-making in a fuzzy environment", Management Science, 17, 1970, B-141 -B-164.

[18] D. Dubois, H. Prade, "Weighted fuzzy pattern matching", Fuzzy Sets and Systems, 28, 1988, pp. 313-331.

[19] L. A. Zadeh, "Probability measures of fuzzy events", J. Math. Anal.& Applicat., 23, 1968, pp.421 -427.

[20] D. Dubois, H. Prade, "A unifying view of comparison indices", in fuzzy sets and possibility theory recent advances, R.R. Yager Ed, Pergamon press, 1982, pp. 3-13.

[21] D. Dubois, H. Prade, S. Sandri, "On possibility / probability transformations", in Fuzzy logic, R. Lowen and M. Roubens Eds, 1993, pp. 103-112.

[22] D. Dubois, H. Prade, "A review of fuzzy set aggregation", Information Sciences, 36, 1985, pp. 85-121.

[23] R. R. Yager "Connectives and quantifiers in fuzzy sets", Fuzzy Sets and Systems, 40, 1991, pp. 39 - 75.

[24] M. Grabisch, "Fuzzy integral in multi-criteria decision making", Fuzzy Sets and Systems, 69, 1995, pp. 279-298.

[25] L. Berrah, G. Mauris, A. Haurat, L. Foulloy, "A fuzzy approach for the validation of performance assessment in ski quality control", IPMU 96, Granada, Spain July 1996, pp. 623-628.

[26] Rumbaugh J., Blaha M., Premerlani W., Eddy F., Lorensen W; Object-oriented modeling and design, Prentice Hall Ed.

[27] MOPIC, Modélisation de la performance pour le pilotage à court terme d'un système de production, Research Project on Rhône-Alpes, 1994, France.

ADVANCED NEURO-FUZZY ENGINEERING FOR BUILDING INTELLIGENT ADAPTIVE INFORMATION SYSTEMS

Nikola K Kasabov

Department of Information Science
University of Otago
P.O.Box 56
Dunedin
New Zealand
E-mail: nkasabov@otago.ac.nz

Abstract. Intelligent adaptive information systems are systems which can automatically adapt their structure and behaviour in order to react better to a dynamically changing environment, and to provide knowledge which explains their behaviour. Fuzzy neural networks have features which make them useful for building such systems, namely: fast adaptive learning, good generalisation, good explanation facilities in form of fuzzy rules, abilities to accommodate both data and existing knowledge about the problem under consideration. This paper introduces a model of fuzzy neural networks, called FuNN, and a general methodology for building adaptive, intelligent multi-modular FuNN-based systems. The use of this methodology for building intelligent adaptive speech interfaces to databases and for adaptive control and adaptive time-series prediction has been given as case study problems.

Keywords: adaptive systems; intelligent systems; fuzzy neural networks; speech interfaces; adaptive control

1. Introduction

Intelligent adaptive information systems are systems which can automatically adapt their structure and behaviour in order to react better to a dynamically changing environment, and to provide knowledge which explains their behaviour. Developing such systems is crucial for the future use of computer systems in almost all areas of engineering and social sciences. For example, speech interfaces, which can adapt to new speakers on-line would produce near to 100% recognition rate in a speaker independent mode of operation. Such interfaces can be successfully used to search and access data from huge databases, or in real time operations when time is a crucial component. Adaptive control systems, which automatically change to dynamically changing parameters of the controlled objects, or adaptive prediction systems which learn continuously the behaviour of the predicted processes, are crucial in many areas of control. Systems which adapt and predict chaotic time-series are still difficult to develop with the use of the current AI technologies. Combining fuzzy logic and connectionist techniques is a promising approach to attempt this challenging task and the task of building adaptive systems in general. This is the focus of this paper.

Fuzzy neural networks (FNN) have been suggested and applied by several authors, among them Yamakawa and Uchino [2], Uchikawa and Furuhashi [3,5], Gupta [6], Kasabov [1,4], and others. These FNN have been successfully used for learning and tuning fuzzy rules and solving classification, prediction and control problems. Some recent publications suggest methods for training FNN in order to adjust them to new or dynamically changing data and situations [7-12].

This paper introduces in section 2 a new architecture of FNN, called FuNN. In section 3, a general methodology for building modular FuNN-based adaptive intelligent systems is presented. Section 4 investigates a possible application of this methodology for building adaptive intelligent speech interfaces to databases. Section 5 illustrates the methodology on adaptive prediction of chaotic time-series

2. FuNN - A Fuzzy Neural Network Model for Adaptive Learning and Monitoring of Knowledge

2.1. The Architecture of FuNN

The FuNN model is designed to be used in a distributed environment for facilitating learning from data, fuzzy rules extraction, fuzzy rules insertion, using both data and rules in one system, approximate reasoning, adaptation (adaptive learning in a dynamically changing environment), experimenting with different adaptation strategies. FuNN uses a multi-layer perceptron (MLP) network and a modified backpropagation training algorithm. The general FuNN architecture consists of 5 layers as shown in Fig. 1. It is adaptable FNN where the membership functions of the fuzzy predicates, as well as the fuzzy rules inserted before training or adaptation, may adapt and change according to new data. Here, a brief description of the FuNN architecture is given.

The input layer represents the input variables. The condition element layer performs fuzzification where triangular membership functions are used with centers attached as connection weights to the inputting connections. The triangular membership functions are in principle not symmetrical. Any input value would belong to maximum two membership functions with a degrees different from zero which degrees always sum up to one. There are no "bias" connections. The layer is potentially expandable during the adaptation phase with more nodes representing more membership functions for the input variables. Simple activation functions are used in the condition element nodes which functions perform fuzzification.

In the rule layer, one node represents one fuzzy rule. The layer is expandable, ie more nodes can be added to represent more existing rules or potential rules to be learned during training. The activation function is the logistic function with a variable gain coefficient g (a default value of 1 is used; values larger than 5 will make it close to the hard limited thresholding function). The semantic meaning of the activation of a node is that it represents the degree to which input data match the antecedent part of the fuzzy rule represented by this node.

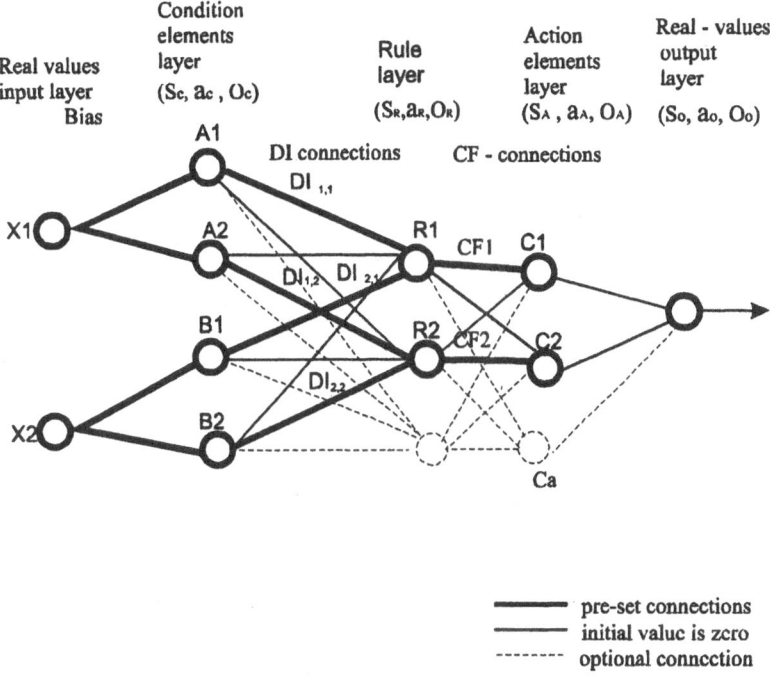

Figure 1. A FuNN structure for two initial fuzzy rules: R_1: IF x_1 is A_1 $(DI_{1,1})$ and x_2 is B_1 $(DI_{2,1})$ THEN y is C_1 (CF_1); R_2: IF x_1 is A_2 $(DI_{1,2})$ and x2 is B_2 $(DI_{2,2})$ THEN y is C_2 (CF_2), where DIs are degrees of importance attached to the condition elements and CFs are confidence factors attached to the consequent parts of the rules

In the action element layer a node represents a fuzzy label from the fuzzy quantisation space of an output variable, eg. small. medium, large. The activation of the node represents the degree to which this membership function is supported by all fuzzy rules together, so this is the level to which the membership function for this label is 'cut' according to the rules and current facts. The connections from the rule layer to the action element layer represent contextually the confidence factors of the corresponding rules when inferring fuzzy output values. The activation function for the nodes of this layer is the logistic function with a variable gain factor (as in the previous layer).

a)   The output layer performs a modified center of gravity defuzzification. Singletons representing centers of triangular membership functions, as it was the case of the input variables, are attached to the connections from the action to the output layer. Linear activation function is used. For example, small, medium and large are represented as 0, 0.5 and 1.0 as connection weights from the output range of [0,1] if normalised outputs are considered. Adapting the output membership functions would mean moving the centers, when the requirement for membership degrees to sum up to one, is always kept. More than one output variable can be used in a FuNN structure and the different output variables can have different number of membership functions.

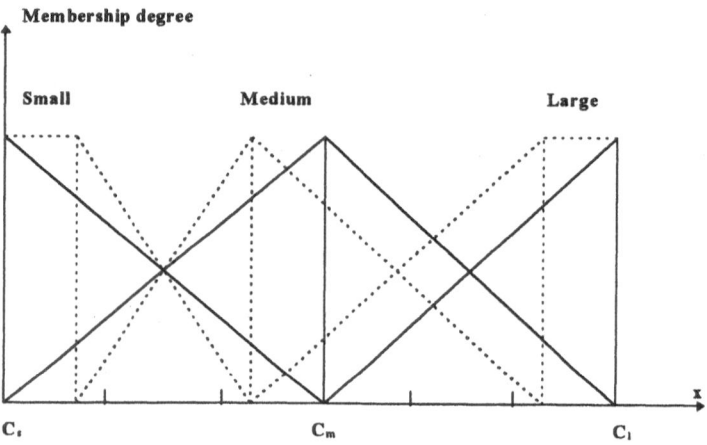

Figure 2. Initial membership functions (solid lines) of a variable x (either input, or output) represented in a FuNN, and the membership functions after adaptation (dotted lines). The boundaries, to which each centre can move but not cross, are also indicated.

There are three versions of the FuNN according to the mode of training and adaptation:

- a fixed version: the membership functions of the input and the output variables do not change; a modified backpropagation algorithm is used for this purpose;
- adaptive version with an extended backpropagation algorithm; in this mode all the connection weights change during training. The connection weights representing membership functions change in a way that the centers of the membership functions move preserving the requirement that each value from

the corresponding domain of an input or output variable belongs to not more than two fuzzy values to degrees different from 0 and the degrees sum up to 1. There are constraining bands in which centres are allowed to move so that the centers can not overlap;

- adaptive version with the use of a genetic algorithm (GA); a GA is used for adapting the membership functions. The three modes above are presented in [11];
- adapting the B-spline of order 2, triangular membership functions (MF) in a FuNN structure is graphically illustrated in Fig.2.

2.2. Experimenting with different methods for training, adaptation, rules insertion and rules extraction in FuNN

The FuNN (Fig.1) has a flexible architecture which allows for different training and adaptation strategies to be tested before the most suitable is selected for a certain application. For the initialisation procedure, uniformly distributed triangular MF can be used as initial values for the input fuzzy sets. If initial set of rules is available, it can be used for initialisation of the FuNN structure (the rule layer).

Training the FuNN can be accomplished either for the inner three layers, in which case the system adapts its fuzzy rules but does not adapt the membership functions, or for the five layers, in which case the system adapts both the rules and the membership functions. The connections in the fuzzification and defuzzification layers are "frozen" in the former case and they are subject to change in the latter case. The network can be trained with the use of different methods, one of them being the method of training and zeroing based on the backpropagation algorithm [9].

Adaptation can be performed in a FuNN system by further training the system with new data. Depending on the size of the new data two general learning strategies can be distinguished [12]:

- *aggressive (fast)* adaptation - a section of new data is used for further training and adaptation without using any of the old, previously used data;
- *conservative (slow)* adaptation - a new data is added to a portion of the old data and training is performed.

In both cases above the system adapts its previous rules and membership functions to the new data.

Different methods for rules extraction are applicable on FuNN. One of them, called REFuNN, is published in [1,4]. It is based on a simple idea of thresholding connection weights according to a pre-set threshold. The weights, which are above the threshold are represented as condition, or action elements in the extracted fuzzy rules with their corresponding weights representing degrees of importance and confidence factors.

254

Another algorithm for interpreting a FuNN structure in terms of aggregated fuzzy rules is presented in [11,12]. Each rule node is represented as one fuzzy rule. The strongest connection from a condition element node to the rule node, along with the neighbouring condition element nodes, are represented in the corresponding rule. The connection weights of these connections are interpreted as degrees of importance attached to the corresponding condition elements. One rule has in general as many consequent elements as the number of the nodes in the action element layer, each action (class) inferred with a certainty degree (confidence factor) defined by the connection weights from this rule node to that class node. The procedure of extracting such type of rules is called here masking of the FuNN [11,12]. Evidential fuzzy reasoning is applicable to this type of fuzzy rules [11,12].

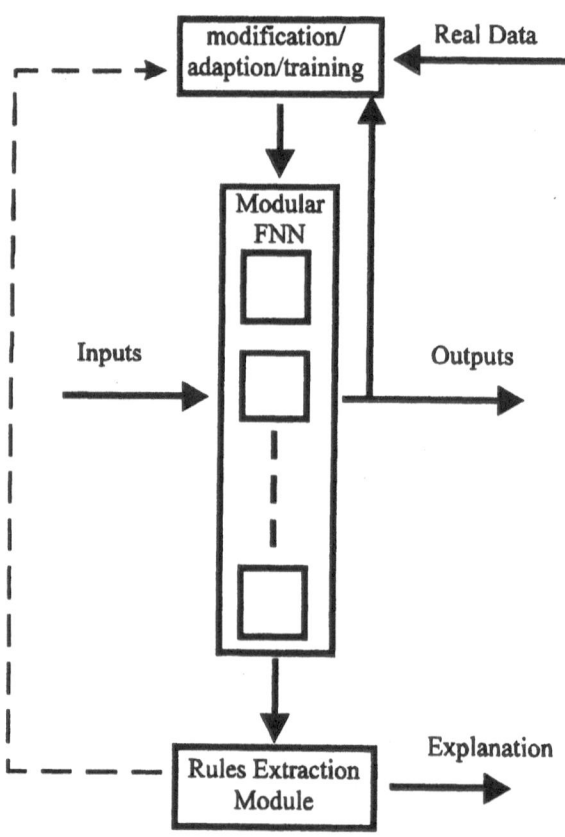

Figure 3. A general architecture of a multi-modular adaptive, intelligent FuNN-based system

3. A Methodology for Building Adaptive FuNN-based Multi-Modular Systems

Here, multi-modular systems which consist of several FuNN modules, a rule extraction module and a module for adaptation, are proposed. A general architecture is given in Fig. 3. Such systems can efficiently be used for classification tasks when one FuNN is trained to classify examples of one particular class. This structure allows for adaptive individual tuning of each of the class FuNN units according to the performance of this class unit during the operation of the whole system. Only the FuNN units which do not operate sufficiently well need to be tuned and not all of them. This makes adaptive tuning in a real time possible and efficient. For example, initially, all the modules may be trained with identical data for a small number of epochs. After that, each of the units may be tuned using specific variant of the general learning and adaptation techniques [1].

Extracting rules and explaining the behaviour of each of the units at each moment of their operation, as well as of the whole system, is accomplished by using a rule extraction module as shown in Fig. 3. The rules extracted may be used if necessary for a better adaptation in the adaptation module.

There are several hybrid environments published and made available for advanced neuro-fuzzy engineering, which environments make possible experimenting with the above models. One of them, FuzzyCOPE [1,13], has been designed to automate the whole process of building adaptive, intelligent neuro-fuzzy information systems. It has modules for the following functions:

- data fuzzification, data manipulation, visualisation and analysis;
- standard neural networks training;
- fuzzy-neural networks training;
- rules insertion and extraction;
- adaptation through modified backpropagation or through using a genetic algorithm;
- fuzzy reasoning over extracted fuzzy rules;
- functional links between several modules in one system in a way, that outputs from one module are used as inputs to other modules.

The system is available on the WWW from the home page:

http://divcom.otago.ac.nz: 800/COM/INFOSCI/KEL/software/

4. Building real-time adaptive spoken language interfaces to databases

A general block diagram of a spoken language interface system to databases is given in Fig.4. It consists of the following blocks [1]:

256

- Speech recognition and language modelling blocks.
- Similarity-based query block. This module does approximate reasoning over user's query and allows for vague, fuzzy queries to be used.
- Knowledge acquisition block. This module performs rules extraction from raw data. The module can be used for explanation of underlying rules in a database.
- Answer formation block. This module produces the answer to the user and performs a dialogue at any phase of the information retrieval. It has a speech synthesis and a text generation sub-modules.

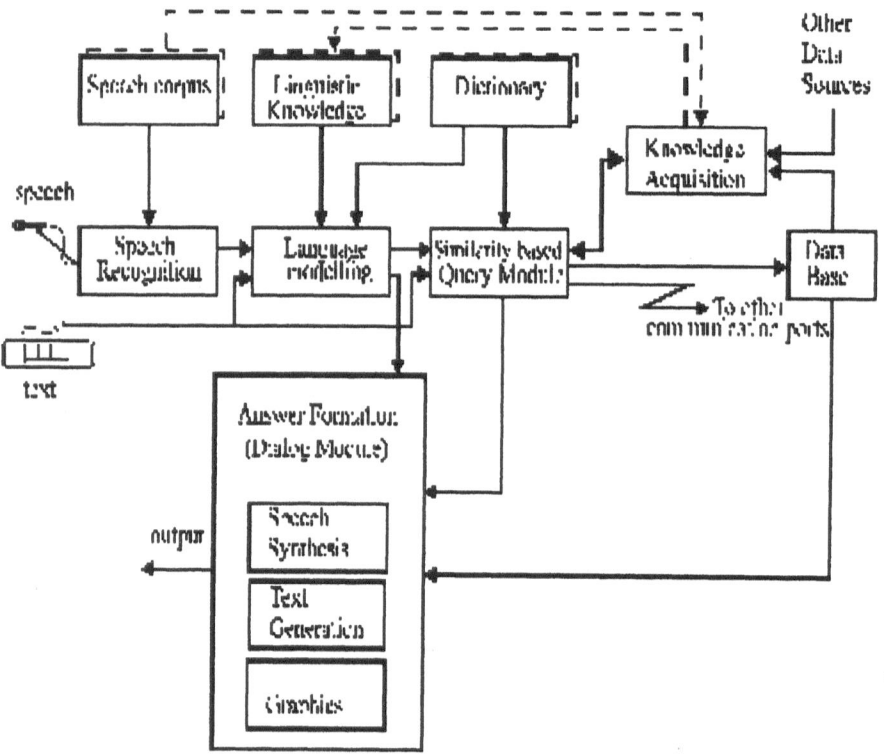

Figure 4. A general architecture of an intelligent speech interface to a database

Building adaptive phoneme-based speech recognition systems is discussed in [1]. Standard multilayer perceptrons are used there for each of the phonemes in English. Instead of a three layer perceptron, a single FuNN can be used for classification/

recognition of each of the phonemes as it was described in a general model presented in section 3.

A single FuNN can be adapted in a real time if necessary if the spoken word (phrase) pronounced by a (new) speaker is not correctly recognised. The adaptation feed-back module will investigate which particular phoneme (phonemes) causes the problem and will additionally train the corresponding FuNN unit with the new data for a small number of epochs. The adaptation process will repeat until all the wrongly recognised phonemes are recognised correctly. As FuNNs train fast and are robust to catastrophic forgetting, this process can be accomplished in a real time.

Figure 5 shows how the RMS error decreases with the number of generations in an adaptation process with the use of a genetic algorithm, on data for the phoneme /e/ as part of a word pronounced by a new speaker. Initially a FuNN was trained with data for /e/ from other speakers (Otago Speech Data Corpus) [1].

The initial value of RMS from Fig.5 is the error calculated in the /e/ FuNN when the speaker spoke the word for the first time. Speech corpus on New Zealand English has been used for initial training of the FuNN units in several experiments conducted in the Knowledge Engineering Research Laboratory in the Department of Information Science, University of Otago.

The corpus is available from the WWW:

http://divcom.otago.ac.nz: 800/COM/INFOSCI/KEL/software/

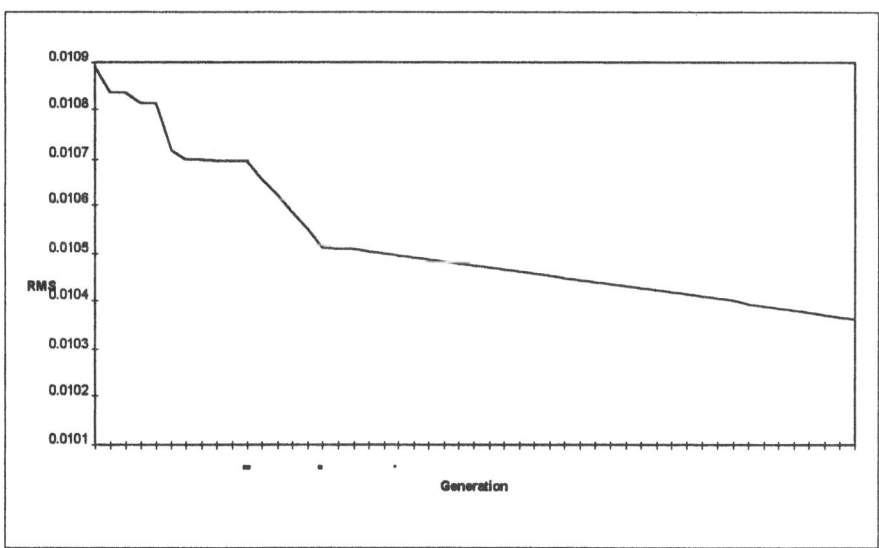

Figure 5. The decrease of RMS with the increase of generations for phoneme /e/ after 100 epochs of training with FuNN in an adaptation mode

5. Adaptive intelligent systems for chaotic time-series prediction and control

Predicting chaotic time series with changing dynamics is extremely difficult process. Such time series can be observed in many application areas, some of them having been explored with the participation of the author, namely:

- process control of sewage treatment plant;
- financial decision making based on time series data;
- data mining and knowledge extraction from databases;
- fruit-growth prediction in horticulture;
- image recognition and classification of environmental and medical images;
- oestrus prediction for artificial insemination of cows;

Here the gas-furnace bench-mark chaotic time series is used to show how a FuNN can be used for adaptive learning. The gas-furnace data was first used by Box and Jenkins (1976) and then by many researchers in the area of neuro-fuzzy engineering. This data set provides a nice example of time-series prediction and, having been used in many data analysis and signal processing studies, is well benchmarked allowing for comparisons between the current technique and alternatives. The data set consists of 292 consecutive values of methane at a time moment (t-4), and the carbon dioxide CO_2 produced in a furnace at a time moment (t-1) as input variables, with the produced CO_2 at the moment (t) as an output variable. The following steps were repeated in a cycle after the whole data set was segmented in consecutive segments: (1) train a FuNN system on a current data segment; (2) test its ability to generalise on the future data segments; (3) extract rules which explain the behaviour of the time series so far; (4) test the validity of the rules through fuzzy inference on the same future data segments.

```
MEMBERSHIP FUNCTIONS

INPUT Input1: {0.0052863 , 0.237243 , 0.504048 , 0.751044 , 0.999987}
 >  A: { 1, 0, 0, 0, 0}
 >  B: { 0, 1, 0, 0, 0}
 >  C: { 0, 0, 1, 0, 0}
 >  D: { 0, 0, 0, 1, 0}
 >  E: { 0, 0, 0, 0, 1}
INPUT Input2: {0.0019096 , 0.218233 , 0.486948 , 0.727425 , 0.957124}
 >  A: { 1, 0, 0, 0, 0}
 >  B: { 0, 1, 0, 0, 0}
 >  C: { 0, 0, 1, 0, 0}
 >  D: { 0, 0, 0, 1, 0}
 >  E: { 0, 0, 0, 0, 1}
OUTPUT Output1: {0.000124257 , 0.168399 , 0.500664 , 0.811863 , 0.999536}
 >  A: { 1, 0, 0, 0, 0}
```

> B: { 0, 1, 0, 0, 0}
> C: { 0, 0, 1, 0, 0}
> D: { 0, 0, 0, 1, 0}
> E: { 0, 0, 0, 0, 1}

RULES

if <Input1 is not D 0.126392> and <Input1 is E 0.116577> and <Input2 is A 0.184112> and <Input2 is B 0.178631>
 then <Output1 is A 0.687749> and <Output1 is B 0.00587244> and
  <Output1 is C 0.117495> and <Output1 is not D 0.97957> and
  <Output1 is not E 0.594911>
else
if <Input1 is D 0.0667426> and <Input1 is E 0.147035> and <Input2 is D 0.103517> and <Input2 is E 0.337239>
 then <Output1 is not A 0.173416> and <Output1 is not B 0.918238> and
  <Output1 is not C 0.685181> and <Output1 is D 0.274782> and
  <Output1 is E 0.570629>
else
if <Input1 is A 0.268865> and <Input1 is B 0.157016> and <Input2 is C 0.160078> and   <Input2 is D 0.338455> and <Input2 is not E 0.0757806>
 then <Output1 is not A 1> and <Output1 is not B 0.596618> and
  <Output1 is C 0.450101> and <Output1 is D 0.674112> and
  <Output1 is not E 0.524792>
else
if <Input1 is B 0.0503882> and <Input1 is C 0.107062> and <Input1 is D 0.0555892> and   <Input2 is not A 0.0275438> and <Input2 is B 0.212702> and <Input2 is not C 0.0185434>
 then <Output1 is not A 0.382772> and <Output1 is B 0.800337> and
  <Output1 is not C 0.414382> and <Output1 is not D 0.712416> and
  <Output1 is not E 0.276774>

Figure 6. Extracted fuzzy rules from a trained FuNN on the first two data sections from the gas furnace chaotic time-series data.

Here a FuNN is set with five membership functions for the input and output variables and 4 rule nodes in the rule layer of the FuNN architecture. The whole data set is divided into 4 consecutive sections (sub-sets) on equal time intervals. The experiments were performed with the use the conservative (slow) training strategy as follows: the FuNN was trained for 100 epochs on the first data segment; then the second data segment was added to the first one, and the FuNN was additionally trained for 100 more epochs, fully adaptive, with the use the modified backpropagation algorithm for adaptation. Aggregated rules were extracted form the trained FuNN as shown in Fig.6. The FuNN was tested (recalled) on the whole data set (Fig.7a) and the extracted rules was tested on the whole data set too, with an use of a fuzzy reasoning method [12] as shown in Fig.7b. The results show that the more a FuNN system is adaptively trained on new data segments, the better it adapts.

(a)

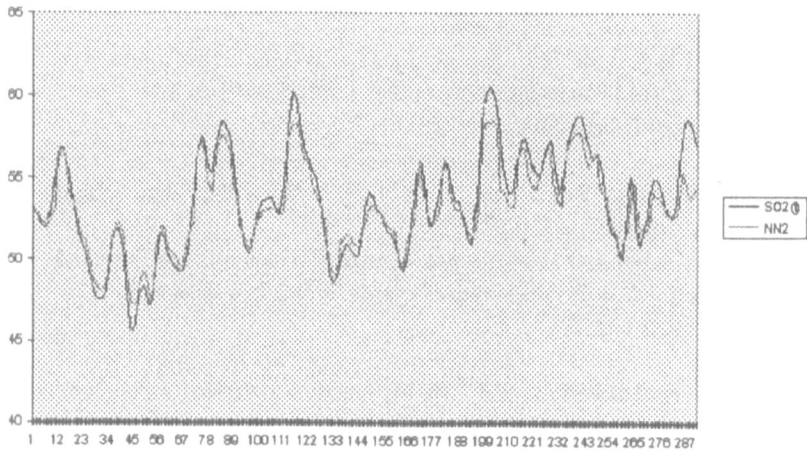

(b)

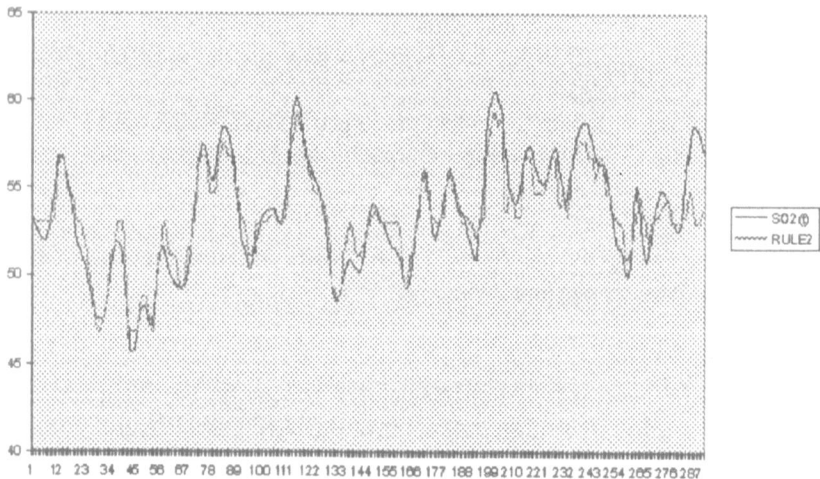

Figure 7. Testing the generalisation and the adaptation of a trained FuNN when the first half of the gas-furnace data is used for adaptive training; (b) Fuzzy reasoning over extracted from FuNN fuzzy rules (shown in Fig.6)

6. Conclusions

The paper discusses a neuro-fuzzy engineering methodology for building adaptive, intelligent information systems. The approach uses fuzzy neural networks and their main characteristics: fast training; good explanation facilities; means for rules insertion, rules extraction and rules refinement; means for adaptation; robustness. This approach has been illustrated on intelligent adaptive speech interfaces to databases and on chaotic time series prediction problems. Directions for further research are: developing methods for a better control of the adaptation and forgetting phenomena in FuNN; developing new methods for knowledge acquisition from FuNN; applications in modelling brain damage and brain recovery; modelling more physiologically plausible methods for speech recognition.

Acknowledgment

This research is partially supported by a research grant UOO 606 funded by the Public Good Science Fund of the Foundation of Research Science and Technology (FRST) in New Zealand. The following graduate students from the Laboratory for Knowledge Engineering and Computational Intelligence, Department of Information Science, University of Otago, took part in the implementation of the FuNN simulator and the FuzzyCOPE environment: J Kim, A.Gray, M Watts, B Hallett, F Zhang.

References

[1] N. Kasabov, Foundations of Neural Networks, Fuzzy Systems and Knowledge Engineering, The MIT Pres, CA, MA, 1996
[2] T.Yamakawa, H.Kusanagi, E.Uchino and T.Miki, "A new Effective Algorithm for Neo Fuzzy Neuron Model", in: Proceedings of Fifth IFSA World Congress, (1993) 1017-1020.
[3] T.Hashiyama, T.Furuhashi, Y.Uchikawa, "A Decision Making Model Using a Fuzzy Neural Network", in: Proceedings of the 2nd International Conference on Fuzzy Logic & Neural Networks, Iizuka, Japan, (1992) 1057-1060.
[4] N. Kasabov, "Learning fuzzy rules and approximate reasoning in neuro-fuzzy hybrid systems", Fuzzy Sets and Systems,1996,
[5] T. Furuhashi, Hasegawa, T., Horikawa S., Uchikawa, Y., An Adaptive Fuzzy Controller Using Fuzzy Neural Networks, in:Proceedings of Fifth IFSA World Congress (1993) 769- 772.
[6] M.M. Gupta, D.H. Rao, On the principles of fuzzy neural networks, Fuzzy Sets and Systems, 61 (1) (1994) 1-18.
[7] M.Brown and C.Harris, Neurofuzzy Adaptive Modelling and Control, Prentice Hall, 1994
[8] N. Kasabov, Adaptable Neuro Production Systems, to appear in Neurocomputing, Elsevier Science Publ. B.V.

[9] N.Kasabov, "Investigating the adaptation and forgetting in fuzzy neural networks by using the method of training and zeroing", in: Proceedings of the International Conference on Neural Networks ICNN'96, IEEE Press, Washington DC, June 3-6, 1996

[10] N.Kasabov, Hybrid Connectionist Fuzzy Production Systems - Towards Building Comprehensive AI, Intelligent Automation and Soft Computing, vol.1, No.4, 351-360, 1995

[11] N.Kasabov et al, FuNN - A fuzzy neural network architecture for adaptive learning and knowledge acquisition in multi-modular distributed environments, Information Sciences: Applications, Prentice Hall, 1996, to appear

[12] N. Kasabov, Adaptive Learning in Modular Fuzzy Neural Networks, in: Proceedings of International Conference on Neuro Information Processing ICONIP'96, Springer Verlag, 1996, to appear

LINGUISTIC VARIABLES IN SURVEYS OF INTENTIONS TO PURCHASE ENERGY EFFICIENT EQUIPMENT

W. Mielczarski, G. Michalik and M. E. Khan

Department of Electrical and Computer Systems Engineering
Monash University
Clayton, VIC 3168
Australia
E-mail: wlad.mielczarski@eng.monash.edu.au

Abstract. The chapter presents the use of linguistic terms to forecast purchase probabilities calculated from surveys of customer intentions. A short review of publications dealing with linguistic terms, their meanings and assigned subjective probabilities introduces to the survey analysis. An application of fuzzy filters to the analysis of customer responses results in a universal tool which provides continuous subjective probabilities of equipment purchase. Simulations carried out for two types of fuzzy filters, three sets of linguistic variables, and two distributions of customer responses show that this method leads to continuous cumulative probability of purchase which is insensitive to the number of linguistic terms and assigned probabilities. This new method can be considered as the generalization of the Juster survey approach.

Keywords: linguistic terms, forecasting, subjective probabilities, fuzzy filters, purchase intentions, customer behavior, the Juster survey.

1. Introduction

Electricity supply companies have to plan the future energy demand by forecasting the penetration of new energy efficient technologies that can significantly affect patterns of energy demand [Gellings, 1992]. One of the methods is to use customers' purchase intentions to forecast sale of new energy efficient equipment [Comerford and Gellings, 1982]. Customer intentions of purchase the new equipment can be investigated in many ways. A well-established method is to use linguistic terms and assigned subjective probabilities of purchase in surveys of prospective buyers proposed by Juster [Juster, 1966]

$$ \text{Forecast} = \sum_{i=1}^{N} n_i p_i \, / \sum_{i=1}^{N} n_i $$

where: n_i - number of declarations for assigned probability of purchase

p_i - probability of purchase assigned to particular linguistic terms and
N - number of linguistic terms.

Since then the method has been used by many researchers, finding applications in marketing studies of customer behaviors as well as in prediction voting behaviors [Hook and Gendall, 1993]. This approach arises two basic problems: what should be the number of linguistic terms and the subjective probability assigned to the particular declaration. These questions have stimulated research which deals with the meaning of particular linguistic terms, their vagueness and the selection of descriptions for particular linguistic descriptions.

The linguistic terms have been considered from different points of view. Meanings of particular terms have been of interest of experimental and general psychology. The statistical sciences deal with assigned probabilities to particular variables and the marketing sciences investigate the application of linguistic terms to customer surveys. Linguistic terms are of interest of fuzzy logic as stated by Zadeh [Zadeh, 1996], "fuzzy logic = computing with words" with significant contribution of Dubois and Prade, [Dubois and Prade, 1986, 1989, 1994].

This chapter presents a simulation study focusing on the determination of the optimal number of linguistic variables and assigning of most appropriate probabilities for various distributions of customer responses. The study employs two categories of fuzzy filters (delta and trapezoidal shaped) with two spans over the subjective probabilities assigned to the linguistic terms in order to investigate the sensitivity of the final results on the filter shapes and spans as well as a number of linguistic terms and values of the subjective probabilities.

The chapter demonstrates a general methodology which can be used in any type of forecasting based on customer surveys. This method can be a tool for the analysis of linguistic terms, assigned subjective probabilities and sensitivity studies.

The organization of the chapter is as follows. Section 2 provides an introduction to the research and applications of linguistic terms. This short introduction does not pretend to be a comprehensive overview of publications dealing with linguistic variables. It only focuses on indicating some important problems selected for the purpose of this chapter. Section 3 presents the simulation study carried out for two types of fuzzy filters, three sets of linguistic variables and two distributions of customer responses. The calculations aim at the cumulative probability of purchase investigating differences for various options of linguistic variables, fuzzy filters and distributions of responses. Conclusions discuss shortly the results obtained.

2. Investigating Customers' Intentions by Surveys

The basic idea behind surveys of customer anticipations is that customer purchases, in particular, for such items as houses, cars and appliances, are subjected to fluctuations that are to some degree independent of movements in observable financial variables such as incomes, assets, income change and so on, [Juster, 1966].

Fluctuations in these postponable types of expenditures are presumed to be foreshadowed by changes in anticipatory variables that reflect consumer optimism or pessimism. It relates in particular to purchasing intentions of new energy efficient equipment where customers' decisions are not only functions of possible benefits calculated using, for example, a simple pay-back period, but also customer anticipation of their future positions which may allow to participate in energy conservation programs leading to the improvement of environmental conditions.

Initially, intention surveys tended to split customers into two categories: intenders and non-intenders. In such a way it is convenient to express the purchase rate for the population identified by 'y' as a weighted average of the purchase rates of intenders 'r' and non-intenders 's' [Juster, 1966]. The purchase rate of the entire sample is determined as

$$y = [p \times r + (1 - p)s]N_s \tag{1}$$

where p is proportion of both groups and N_s is the number of surveyed customers. When results obtained from a representative sample are applied to a large customers' population, the following purchase rate 'x' is given as:

$$x = [p \times r + (1 - p)s]N_s \times R_p \tag{2}$$

where R_p is the ratio of the surveyed sample to the total population.

Customers are distributed between two groups with the following purchase rates:

$$P_{intender} = p \times r \times N_S \times R_p \tag{3}$$

$$P_{non\text{-}intender} = (1 - p) \times s \times N_S \times R_p \tag{4}$$

It is quite clear that this segmentation reduces the complexity of the problem, since it creates a gap between two distinct groups. However, in reality, there can be many steps between these two options.

If we suppose that all households regard a specific question about buying intention as having a cut-off probability of C_i, (as shown in Fig. 1), it can be observed that the fraction of 'p' of the sample is noted as intenders, while the fraction of '1 - p' as non-intenders.

The intenders have a mean purchase probability of 'r', while the non-intenders mean purchase probability is 's'. The sample as a whole has a mean value of 'x' [Juster, 1966].

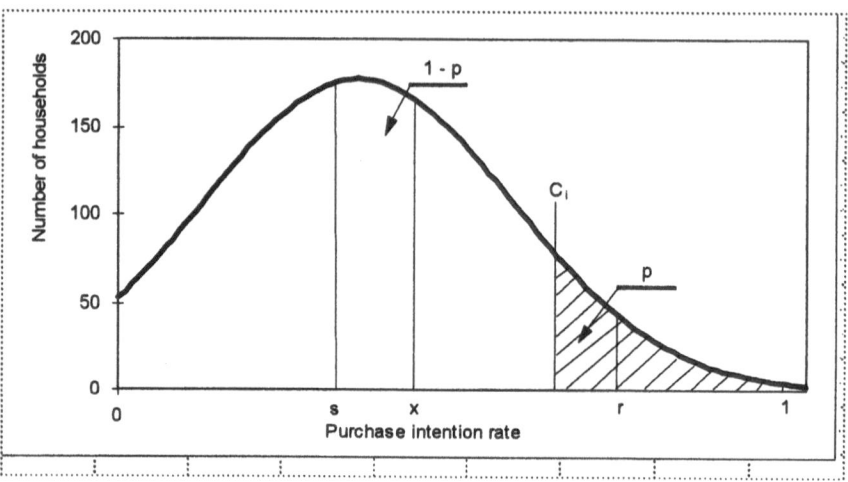

Figure 1. Distribution of intenders and non-intenders

The proportion of intenders can be calculated as follows:

$$p = \frac{1}{N_s} \int_{c_i}^{1} f(IN)d(IN) \tag{5}$$

The proportion of non-intenders is determined as

$$1 - p = \frac{1}{N_s} \int_{0}^{c_i} f(IN)d(IN) \tag{6}$$

where the function $f(IN)$ is the distribution of customer intentions. The number of customers for a given intention 'IN' is computed as $n_{IN} = f(IN)$. The total number of customers surveyed is given by

$$N_S = \int_{0}^{1} f(IN)d(IN) \tag{7}$$

The cut-off dividing customers between intenders and non-intenders is deferred prior and represents the same value for all customers. However, the cut-off probability associated with a specific question varies among households, and the probability distributions for intenders and non-intenders can overlap to some extend [Juster, 1966]. This conclusion leads to the concept of a membership function which can express the uncertainty of customers' declarations allowing to be a member of the intenders' family and a member of the non-intenders' family at the same time.

2.1. Purchase probability

There have been several attempts to measure consumer purchase probability using customer surveys and assigning customer declarations to the fixed levels of the purchase probability. One of the first surveys, a pilot test conducted by the US Bureau of the Census in November 1963, was carried out on a non-random sample of customers in one of Detroit suburbs (Detroit experiment). Another study (QSI experiment), conducted by the US Census Bureau in July 1964, was based on a random sample drawn from the 16,000 odd households [Juster, 1966]. The levels of probability assigned to the customers' answers expressed in linguistic variables are shown in Table 1.

It has been proved that the surveys with eleven levels of probability can express customer intentions more precisely than a two-pole division into intenders and non-intenders [Juster, 1966].

Linguistic variables with the fixed levels of probabilities can be applied to forecasting of customer intentions of purchasing energy efficient equipment like solar water heaters, energy-efficient refrigerators, lamps, or water heaters equipped with heat pumps. The forecast of the penetration of energy-efficient appliances is very difficult since, in many cases, a simple pay-back period calculated for these appliances is longer than 7 years. Therefore, the purchase intentions can not be only supported by an economical analysis but they strongly depend on customer's optimism.

	Detroit experiment		QSI experiment	
No	Answers	Probabilit y scale	Answers	Probability scale
1	absolutely certain	10	certain, practically certain	9.9/10
2	almost certain	9	almost sure	9/10
3	much better than even chance	8	very probable	8/10
4	somewhat better than even chance	7	probable	7/10
5	slightly better than even chance	6	good possibility	6/10
6	about even chance (50/50)	5	fairly good possibility	5/10
7	slightly less than even chance	4	fair possibility	4/10

8	somewhat less than even chance	3	some possibility	3/10
9	much less than even chance	2	slight possibility	2/10
10	almost no chance	1	very slight possibility	1/10
11	absolutely no chance	0	no chance, almost no chance	0/10

Table 1. Levels of probabilities for linguistic terms [Juster, 1966]

2.2. Linguistic scales

Accepting the Juster's scale as an 11-point probability scale to estimate the mean purchase probability, the problem to consider is the construction of a questionnaire with the linguistic or numerical descriptions. The initial research has suggested that the linguistic descriptions should be eliminated leaving only the numerical probabilities [Juster, 1966]. In another study Isherwood and Pickering [Isherwood and Pickering, 1974] have reported a successful application of the Juster scale with non-verbal descriptions other than *no chance* at 0 and *completely certain* at 10. In that study questionnaires have been collected by interviewers providing the opportunity to explain customers the meanings of the numerical values in the Juster scale. However, the cost of surveys has significantly increased.

The value of the Juster scale, as has been reported in the research, would be enhanced significantly if the self-completed questionnaires were as accurate as face-to-face interviews. The version of the Juster scale with the verbal probability descriptions has produced slightly more accurate predictions of the purchase behavior for a range of durable and non-durable products than the version containing only the purchase probabilities [Gendall, Esslemont and Day, 1991].

2.3. Environmental effects

People's attitudes to buying products strongly depend on the type of commodities. Goods relating to environmental situations are treated as items with heavy modal contents [Irwin, 1994]. In decisions on purchasing new energy efficient equipment the environmental issues can cause the increase in buying even if market prices are higher than comparable products, and a simple pay-back period is longer than 3 years. Therefore, people declarations on purchasing the new energy-efficient equipment will usually have a higher probability than the purchase declarations relating to other durable goods. However, consumer pessimism or optimism can have a strong impact on purchasing decisions. These show that the information on customers' intentions of purchasing new equipment can be very vague and ambiguous, and the fixed levels of probabilities can lead to significant errors in forecasts.

2.4. Mood effects

A person's mood can directly affect the judgment of the uncertainty of a future event. Happy people (optimistic) report higher probabilities for positive events and lower probabilities for negative events than sad people (pessimistic) [Wright, 1992]. This indicates that the forecast in which subjective probabilities are assigned to the peoples' declarations can be affected by the mood of respondents.

2.5. Subjective probability and verbal expressions

There are two approaches to the interpretation of probabilities: the objective and the subjective approach. The former states that the probability is the relative frequency of some event when an experiment is repeated a large number of times. This interpretation seems to be unrealistic, since the events evaluated cannot be treated as replicated experiments. The subjective interpretation of probability states that the probability is a measure of one's personal beliefs in a particular outcome of an experiment [Savage, 1954]. Subjective probabilities must satisfy the basic axioms of a probability theory given by:

$$0 < P(E_i) \leq 1 \text{ and } \sum P(E_i) = 1 \qquad (8)$$

where E_i is an event in the sample space of the experiment.

The subjective probabilities can be incorporated into statistical decision analysis applying the Bayes Law [Sullivan and Claycombe, 1977].

2.6. Representation of vague uncertainties

A probability estimation in the subjective probability approach suffers when information is vague, ambiguous, or indirect. Many people prefer to use expressions such as *probable*, *slight chance*, etc., which make information coming from surveys ambiguous because the meanings of linguistic expressions may vary among respondents.

There are three reasons for using imprecise linguistic expressions instead of numerical expressions. First, the information on which the forecasts or evaluations are based is itself not sufficiently precise. Second, the transition of this information into a numerical form can suggest to the user of information a level of precision and confidence that is inappropriate. Third, many people feel that they can understand and respond to information more precisely when it is in a linguistic rather than numerical form [Budescu and Wallsten, 1987]. About 65% of the surveyed respondents generally prefer to communicate their uncertainty verbally rather than numerically [Budescu and Wallsten, 1990].

The study carried out to compare decisions where forecasters communicate survey outcomes to decision makers either verbally or numerically has showed significant similarities between verbal and numerical bids: higher bids for gains than for losses, and similar decision time [Budescu and Wallsten, 1990]. However, there are well-documented differences between verbal and numerical expressions of probabilities that mean bids in response to verbal and numerical descriptions of common underlying probabilities tend to be different [Budescu and Wallsten, 1985].

2.7. Selection of verbal probabilities

One of the main problems with the use of linguistic variables is the selection of the number of expressions and their order in a list. The validity of using the verbal expressions can be compromised if the meaning of the words or phrases depends on contextual factors. There are several possible types of context effects. First, a phrase's immediate neighbors in a list can affect its meaning; thus *rarely* may mean something different if positioned between *very unlikely* and *absolutely impossible* than if its neighbors are *good chance* and *slightly less than half the time* [Hamm, 1991]. Second, presenting expressions in the sequential order exploits the immediate neighbor effects: it makes the meaning subjects assign less variable and closer to the conventional meaning of the phases. The use of the ordered list enables the subjects to select phrases that are more accurate answers for the word problems. In addition, the use of their own interpretations rather than the common meaning usually makes their answers more accurate [Hamm, 1991]. Next, the number of linguistic expressions used in the survey should be large enough allowing customers to present their intentions precisely. On the other hand, this number should be as small as possible to avoid the large overlaps leading to ambiguity in customers' declarations [Michalik, Khan and Mielczarski, 1996].

2.8. Application of verbal expressions

An example of the use of verbal expressions can be customers' surveys which have been used to collect the data on buying attitudes by the Survey Research Center, University of Michigan. These surveys have been conducted monthly since 1978 and quarterly before that [Curtin, 1982]. The specific question on buying attitudes was: 'Generally speaking, do you think now is a good time or bad time to buy a house?' An index of buying attitudes is constructed as follows [Dua Smyth, 1995].

$$\text{Index} = 100 \times (good + 0.5 \times uncertain) / (good + bad + uncertain) \qquad (9)$$

This index is a linear transformation of the traditional balance score that can be written as:

$$\text{Balance} = 100 \times (good - bad) / (good + bad + uncertain) \qquad (10)$$

The two indexes are related as follows: Balance = $2 \times$ (Index - 50). The study has suggested that the additional data may not have an independent influence on the home scale. This is an example of three verbal expressions applicable to calculations of customer intentions. Similarly, household survey data have been used for forecasting car expenditures. The attitudinal questions have been measured on a three-point scale of the form: *Better*, *The Same* and *Worse*, and the balance index based on the difference between the proportion of responses to the alternative *Better* and *Worse* has been calculated. The attitudinal indices have been based on estimated proportions (%) of households in the population for the given response categories [Jonsson and Argen, 1994]. The intentions have been measured on an 11-point likelihood scale. The study has found that among attitudinal indices those based on questions about households' own financial situation perform best regardless of an evaluation approach.

2.9. Measuring verbal probability

When customers' purchase intentions are expressed using linguistic terms, the problem arises how to measure linguistic variables, and which numerical values reflect most precisely linguistic terms. Another problem to solve is the determination of the number of linguistic variables required to cover precisely the entire range of probabilities avoiding large overlaps between linguistic variables which may result in respondent's confusion and lead to incorrect interpretation of their intentions.

Regan, Mosteller and Youtz [Regan, Mosteller and Youtz, 1989] have investigated the meanings of 18 verbal probability expressions studying frequency distributions of what a single number best represents each expression as well as word-to-number and number-to-word acceptability functions. They have investigated complimentary symmetry among expressions with *likely* and *unlikely* showing that *likely* is closer to 50% position than *unlikely*. The study has provided several pairs of synonymies for which a correlation coefficient was higher than $r = 0.834$. Seventeen of the correlation coefficients (11%) surpassed this criterion. *Almost impossible* was nearly synonymous with *very improbable* which was nearly synonymous with *very unlikely*. Also *medium chance* was nearly synonymous with *even chance*. By the more stringer criterion $r \geq 0.922$, the expression pairs (7%) were examples of synonymy. It has been found that 15 of the investigated expressions were very good at representation of extremes of the 0 - 100% range, and two expressions were good for the middle. There were no appropriate expressions to cover probabilities in the range of 30 - 35%.

2.10. Quantifying probabilistic expressions

A comprehensive study, done by Mosteller and Youtz with 8 comments and the rejoinder, was published in Statistical Science [Mosteller and Youtz 1990]. They have presented the overview of 20 different studies showing tabulated numeral averages of opinions on quantitative meanings of 52 probabilistic expressions selected from an initial list of 300. These studies have been based on questionnaires of population with various occupations such as: undergraduate and postgraduate students, medical workers, and science writers.

It has been observed that the extreme expressions like *always* and *never* have small variations as measured by the interquartile range. The 11 expressions covere median probabilities ranged between 33 and 67% on the scale from 0 to 100%. The value of 50% is anchoring these responses, especially with *even chance* and *as often as not*. Linguistic modifiers like *very* and *not* lead to three chooses of expressions where:

a) *very* reduces the probability that is originally less than 50%;
b) *very* increases the probability that is originally greater than 50%;
c) *not* changes an expressions from originally greater than 50% to less than 50%.

Mathematically it can be described in the following way. If 'x' is the initial value, 'x_{mod}' the modified value, and 'k' the multiplier, then the estimated relations for the three classes are [Mosteller, Youtz, 1990]:

a) Expression where *very* reduced the estimate: $x_{very} = k\,x$
b) Expression where *very* increases the estimate: $100 - x_{very} = k\,(100 - x)$
c) Expression of negations *not, in-, im-, un-*: $x_{not} = k\,(100 - x)$

This study has confirmed stable positions of the extreme probabilities such as 98% or 99% for *always* and *certain* and 2% or 1% for *impossible* and *never*. The probability offers (with its modifiers) a good spread from about 5% to about 90%. Precisely, 50% has been obtained by *even chance* and *as often as not*. The similar value expressing more uncertainty would be delivered by the moderate probability.

Kadane [Kadane, 1990] has discussed the possible applications of verbal expression standardization pointing that it is not important that each possible word is represented and that each state of uncertainty should have its representation. He has also proposed to reduce the number of verbal expressions from 52 used in Mosteller and Youtz' study to a dozen or less. Analyzing the similarities in the verbal expressions, Kadne has proposed 11 verbal expressions and their ranges of probabilities, to which we will return later.

Although Kadane, Mosteller and Youtz have not mentioned the Juster scale in their discussion, the comparison done by us shows that 11 verbal expressions and ranges of probabilities presented by Kadane [Kadane, 1990] and discussed by Mosteller and Youtz [Mosteller and Youtz, 1990] seem to be very close to the verbal expressions and average probabilistic values of those proposed by Juster [Juster, 1966], as shown in Table 2.

However verbal expressions used by Kadane, Mosteller and Youtz are different from the expressions used by Juster (as the former relate more to a frequency of events, while the later present possibilities/probability of purchasing intentions), similarities between these expressions, average values of probabilities, and probability intervals can be noticeable.

2.11. Vague meanings and membership functions

When verbal expressions are used in customers' surveys there is a need for tools to measure the vague meanings of linguistic variables. One of the several possible methods is to calculate probabilities as the ratio of YES answers to the total number of answers when objects with adjectives such as small, large are presented [Hersh, Caramazza and Brownell, 1979]. Another method to determine the membership functions is the direct scaling in which subjects are rated on a scale from definitely in the concept to definitely not in the concept [Oden 1976 and Zysno, 1981].

A different approach which utilizes a modified pair comparison method for measuring the vague meanings of probability terms has been proposed by Wallsten, Budescu, Rapoport, Zwich and Forsyth [Wallsten and others, 1980]. This comprehensive study, based on two empirical experiments, has provided a family of membership functions for 10 verbal expressions: *almost certain*, *probable*, *likely*, *good chance*, *possible*, *tossup*, *unlikely*, *improbable*, *doubtful*, and *almost impossible*. The best coincidence has been obtained for three variables: *almost certain*, *almost impossible*, and *tossup*. Larger diversities have been noted for the remaining phrases resulting in both monotonic and single-peaked functions. Furthermore, in these cases same-shaped functions do not take on similar values, so that none of the remaining terms provided precisely the same functions for any 2 subjects. Although this study has not resulted in a precise range of the membership functions for the given linguistic expressions but it has provided an approximate range of the membership values and suggested the shapes of these functions.

Kadane and Mosteller Intervals			Juster Scale	
Verbal Expressions	*Kadane Intervals*	*Mosteller Intervals*	*Verbal Expressions*	*Average Probabilit y*
Almost never	0.00-0.05	0.01-0.05	No chance, almost no chance	0.01
Seldom	0.05-0.15	0.07-0.18	Very slight possibility	0.1
Infrequent	0.15-0.25	0.10-0.23	Slight possibility	0.2

Sometimes	0.25-0.35	0.18-0.35	Some possibility	0.3
Less than an even chance	0.35-0.45	0.40-0.45	Fair possible	0.4
Even chance	0.45-0.55	0.50^- - 0.50^+	Fairly good possibility	0.5
More often than not	0.55-0.65	0.57-0.60	Good possibility	0.6
Often	0.65-0.75	0.65-0.75	Probable	0.7
High probability	0.75-0.85	0.77-0.87	Very probable	0.8
Very high probability	0.85-0.95	0.90-0.95	Almost sure	0.9
Certain	0.95-1.00	0.99-1.00	Certain	0.99

Table 2. Probabilities of verbal expressions

2.12. Fuzzy set preference model for consumer choice

An application of fuzzy set models to prediction of a customer choice has been presented by Turksen and Willson [Turksen and Willson, 1994]. In this study fuzzy measurements of consumer preferences have been examined. A comparison test of predictive validity using identical data from 199 subjects has showed significant improvements in the rank order of prediction validity in comparison with a non-fuzzy approach. Seven linguistic terms have provided 7 anchored subjective evaluations with 3 set sizes considered as 7, 15, and 29-element options. In calculations of membership functions, weighting coefficients w_i have been assumed to be a crisp attribute with the important weight (1 - 7). Although the simplified membership functions, such as discrete functions with 7, 15 or 29 points, have been implemented, the prediction improvement of 83% has been attributed to the use of fuzzy set definitions for subject ratings with the firs 2 or 6-ordered ranking predictions on the average [Turksen, Willson, 1994].

2.13. Fuzziness versus probability

There have been hundreds of papers written on the topic of fuzziness versus probability. Zadeh as the first writer has discussed models that have had elements of both fuzziness and probability. On the other hand, there have been opinions that uncertainty is rather of a statistical than a fuzzy nature, and the Bayesians approach can solve adequately the problem of uncertainty. For example, Laviolette and Seaman [Laviolette and Seaman, 1992] presenting the Bayesian view have claimed that the probability provides better modelling than fuzzy mathematics can do. There have been many publications contradicting this opinion. Their authors have indicated a number of areas where uncertainty cannot be adequately expressed by a probabilistic approach [for example Kosho, 1994 and Bezdek,1994]. In particular, Zadeh stated "Probablility theory and fuzzy logic are complimentary rather than competitive" [Zadeh 1995].

2.14. Forecasting customers intentions as fuzzy probabilistic problem

In our opinion, the problem discussed in this chapter has both a probabilistic and a fuzzy nature. Data acquisition from a selected sample of customers or calculations of customer intentions for a large population with probabilities resulting from the survey analysis has a probability nature. On the other hand, data are collected by asking customers to provide responses on verbal expressions so the meanings of these expressions by an individual have rather a fuzzy than a statistical nature. Thus both approaches should be employed to deliver adequate forecasts based on surveys of customer intentions.

3. Analyzing uncertain data from customers' surveys

The problem of the meaning of verbal expressions has still being discussed revealing various aspects of their adequate representation. In mean time, there is a strong need for tools allowing to accommodate human knowledge or human intentions to decision making processes in engineering and management. In particular, we are interested in the long-term forecasting of new energy efficient technologies. Such a forecast can be only constructed by a survey of customer intentions and a proper analysis of the data collected.

Published to the date research on the quantification of the probabilistic expressions may have some applications to our task, but it suffers from a number of drawbacks. Many tests have been carried out using groups of undergraduate students, postgraduate students and science writers. Surveying students reduces costs of research, but they cannot be accepted as representative groups for a large population. In particular, the recognition of the words' meanings and the differentiation of these meanings is a part of the university education or a part of professional skills in the case of science writers. Most customers, as possible objects of surveying, do not have the tertiary education so their perception of verbal expressions can be significantly different.

In our opinion, the large number of verbal expressions does not necessary lead to precise answers. Moreover, it can sometimes create confusing situations which may result in large forecasting errors. An increase of the number of linguistic variables can create unjustified confidence of forecasters in the data collected, while in reality it could cause hidden, unrecognized errors.

A solution proposed is to reduce the number of variables used in the survey and focus affords on data analysis and investigation how uncertainty in the input data is transferred through analysis processes into the output data, and how it may affect final decisions.

In order to do it we have undertaken a simulation study for two shapes of fuzzy filters (delta and trapezoid filter) with two spans over subjective probabilities assigned to linguistic terms. Simulations have been carried out for two distributions of customers' responses (uniform and Gamma distribution). Customers' intentions collected in a survey in linguistic terms are not assigned to "a priori" assumed levels as if it was in the Juster survey but to fuzzy filters which span subjective probability ranges over each linguistic term. Fuzzy filters represent membership functions expressing a degree of a purchase probability. It provides a flexible tool to deal with uncertainty of customers' expressions as well as with changing intentions in a period between declarations and purchase.

From a group of customers who declared for example "*certain*" purchase some may change their minds moving to groups with lower probability of purchase. It can be said that a customer declaration gives him the highest membership of the group declared but uncertainty causes that he is also a member of other groups with a lower level of membership. A fuzzy membership concept is the best way to deal with uncertainty of human declarations.

3.1. Fuzzy membership functions

Two types of fuzzy membership functions are taken into account: the delta-shaped and the trapezoidal-shaped functions. Each membership function has two options: a narrow probability range and a wide probability range. 'Narrow' means a span of ±0.25 of the membership function from an average value, while 'Wide' denotes a span of ±0.5 from the mean value. These lead to four options, as shown in Table 3.

Probability range	Shapes of the membership functions	
	Delta	Trapezoidal
Narrow	DN	TN
Wide	DW	TW

Table 3. Four options of the membership functions

As there is no common agreement on the number of verbal expressions required, ranges of probabilities, and descriptive words, we proposed three scenarios with 9, 7, and 5 verbal expressions. To avoid a detailed discussion what are the best words for a particular verbal expression, we coded these expressions using the notation V9_n, V7_n, and V5_n where the first part of expression denotes a scenario, while 'n' is a description of linguistic variables starting from the right side of the probability interval [0,1]. Tables 4, 5 and 6 present the probability ranges for the narrow delta option showing average values and possible verbal expressions.

The option with five verbal expressions is constructed in a special way. Of the five linguistic variables two represent extremes: high intention of the purchase (Max) and no intention (Min). The third variable (Middle) is settled above 50% (or 0.5p.u.) of a purchase chance (*even chance*) being an anchor for the 4th and 5th variables resulting from the modification by the use of two modifiers *NOT* (Mid_Min) and *SOME* (Mid_Max). The former reduces the assigned probability of the anchor variable to a new value of 0.3, while the latter increases the assigned anchor probability to a new value equal to 0.7. These two modifiers provide the symmetrical probabilities, as shown in Table 6. The membership functions for the 5th -variable scenario are shown in Figure 2.

Variable Notations	Probability Ranges	Average Values	Possible Verbal Expressions
V9_1	1 - 0.65	0.9	Certain, virtually certain
V9_2	1 - 0.55	0.8	We are convinced
V9_3	0.95 - 0.45	0.7	Very probable
V9_4	0.85 -0.35	0.6	We believe, better than even chance
V9_5	0.75 - 0.25	0.5	Probable, even chance, about even chance
V9_6	0.65 - 0.15	0.4	Less than even chance, fair possible
V9_7	0.55 - 0.05	0.3	Some probability
V9_8	0.45 - 0	0.2	Probably not, very slight probability
V9_9	0.35 - 0	0.1	Not consider, we believe not unlikely

Table 4. Probability ranges for nine variables

Variable Notations	Probability Ranges	Average Values	Possible Verbal Expressions
V7_1	1 - 0.7	0.95	Certain, virtually certain
V7_2	1 - 0.55	0.8	Very probable
V7_3	0.9 - 0.4	0.65	Better than even chance, good possibility
V7_4	0.75 -0.25	0.5	Probable, even chance, about even chance
V7_5	0.6 - 0.1	0.35	Less than even chance, fair possible
V7_6	0.45 - 0	0.2	Probably not, very slight probability
V7_7	0.3 - 0	0.05	Not consider, almost no chance

Table 5. Probability ranges for seven variables

The probability ranges of the linguistic variables, shapes of membership functions, and their influence on analysis and output results are the main targets of our research. For the verification of three scenarios with four shapes of the membership functions, customers' responses were generated as the uniform and the Gamma distribution assuming 1000 responses in each sample. The equation for the Gamma distribution is as follows

$$f(x, \alpha, \beta) = \frac{1}{\beta^{\alpha} \Gamma(\alpha)} x^{\alpha-1} e^{x/\beta}$$

(11)

with $x \in [0,1]$, $\alpha = 4$, and $\beta = 0.1$.

Variable Notation	Range Name	Probability Range	Average Probability	Possible Verbal Expressions
V5_1	Positive extreme	1-0.65	0.9	Certain
V5_2	Positive anchor modification	0.95-0.45	0.7	Very probable
V5_3	Mid-range anchor	0.75-0.25	0.5	Probable
V5_4	Negative anchor modification	0.55-0.05	0.3	Some probability
V5_5	Negative extreme	0.35-0	0.1	Not consider

Table 6. Probability ranges for five variables

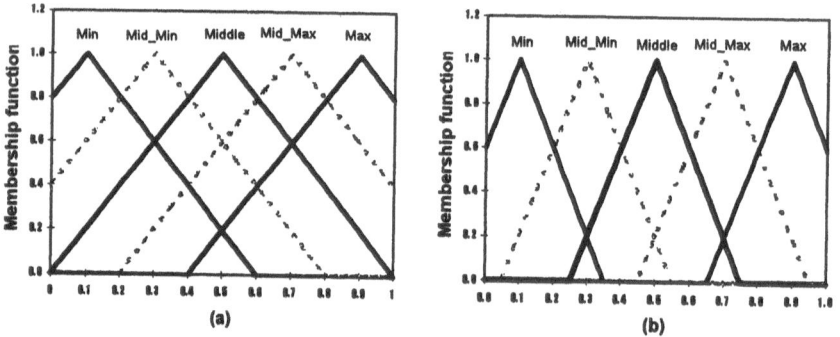

Figure 2. Delta-shaped membership functions for 5 verbal expressions.

The choice of the Gamma function is justified by the results of surveys in which most customers did not have strong purchasing intentions falling mostly in categories below 0.5 purchase probabilities [Juster, 1966]. Three Gamma-distribution options model the scenarios when the survey is carried out with 9 verbal expressions. During the post-survey analysis, the number of expressions is reduced by shifting customers from the expressions canceled to the nearest neighborhood. For example, moving from the 9th to 7th-variable scenario, the expression V9_6 on the level of 0.4 was canceled, and customers from this group were shifted to the group with the closest subjective probability in the 7th - variable scenario.

The membership functions of verbal expressions are standardized to represent the distribution function of the purchase probability for which a cumulative probability is equal to the average values presented in Tables 4, 5 and 6. The distributed probability for a particular variable is given as:

$$p_i^s(x) = w_i \frac{\mu_i(x)}{A_i(x)} \tag{12}$$

where w_i is the weighting coefficient equal to the average probability and μ_i is the membership function.

$$A_i(x) = \int_0^1 \mu_i(x)dx \text{ for } x \in [0,1] \tag{13}$$

This standardization can have the following physical interpretation. An average probability associated with a given variable represents the total customer intention, some kind of energy for purchase. Standardization does not change this energy which can be calculated as an integral of the standardized membership functions. When the integration is carried out from the right side, the probability of purchase intentions equals from 1 to 0 resulting in a fuzzy level of the purchase intentions for a given verbal expression, assuming the positive side for this direction.

$$FL_i(x_x) = \int_1^{x_x} p_i^s(x)dx \le w_i \tag{14}$$

where $x_x \in [0,1]$. The total level of purchasing intentions for a given sample can be computed as

$$N(x_c) = \sum_{i=1}^{n_i} N_i \int_1^x p_i^s(x)dx \tag{15}$$

where: n_i - number of verbal expressions;
N_i - number of customers in a given verbal group;

x_c - cut-off which represents a range of integration reaching the maximum for $x_c = 0$ if all verbal expressions are taken into account;
$N(x_o)$ - number of customers who will purchase a product for a given cut-off.

The Juster level of purchase intentions for the tree scenarios was calculated as

$$JusterLevel = \sum_{i=1}^{n_i} w_i N_i \Big/ \sum_{i=1}^{n_i} N_i \tag{16}$$

3.2. Calculation of purchase probability

Calculations of the cumulative probability of purchase for four membership functions and the uniform distribution of customer responses are shown in Figure 3(a). The cumulative probability depends more on the range of the membership function 'Wide' or 'Narrow' than on the delta or the trapezoidal shape. Differences between the cumulative probability obtained for the delta-wide shape, assumed to be the reference, and three others are shown in Figure 3b. Similar calculations were repeated for the Gamma distribution of customer responses showing small differences among cumulative probability functions obtained for four shapes of the membership functions, as shown in Figures 3c and 3d.

The composition of the cumulative probability of purchase allows to recognise how a particular verbal expression affects the total cumulative function (Figure 4). It also provides possibility to achieve a number of prospective customers when during a post survey analysis it occurs that the most realistic scenario requires the rejection of customer intentions below assumed values. For example, when the cut-off level is set up between the *Minimum* and *Mid_Min* variables (closer to the *Mid_Min* value), the number of prospective customers drops from nearly 400, calculated due to the Juster rule, to 340. Continuous adjustment of the cut-off level provides the new number of prospective customers. The second case, where the cut-off level was set up below the *Middle* expression, is shown in Figure 4b.

The calculations carried out for two scenarios: the 9th and 7th verbal expressions provide similar results that is the cumulative purchase function is an inverse S-shaped which gradually approaches the Juster level. The composition analysis allows the investigation of levels of customer purchases for various values of cut-offs, as shown in Figures 5 (a) and (b).

It is interesting how the reduction in the number of verbal expressions affects the cumulative purchase function. Simulations for the Gamma distribution of customer responses show very small differences (below 3%), as presented in Figure 5. These differences result partly from the step of integration that was equal to 0.01. The reduction of the integration step will diminish differences even more. Similar results have been obtained for the uniform distribution of responses.

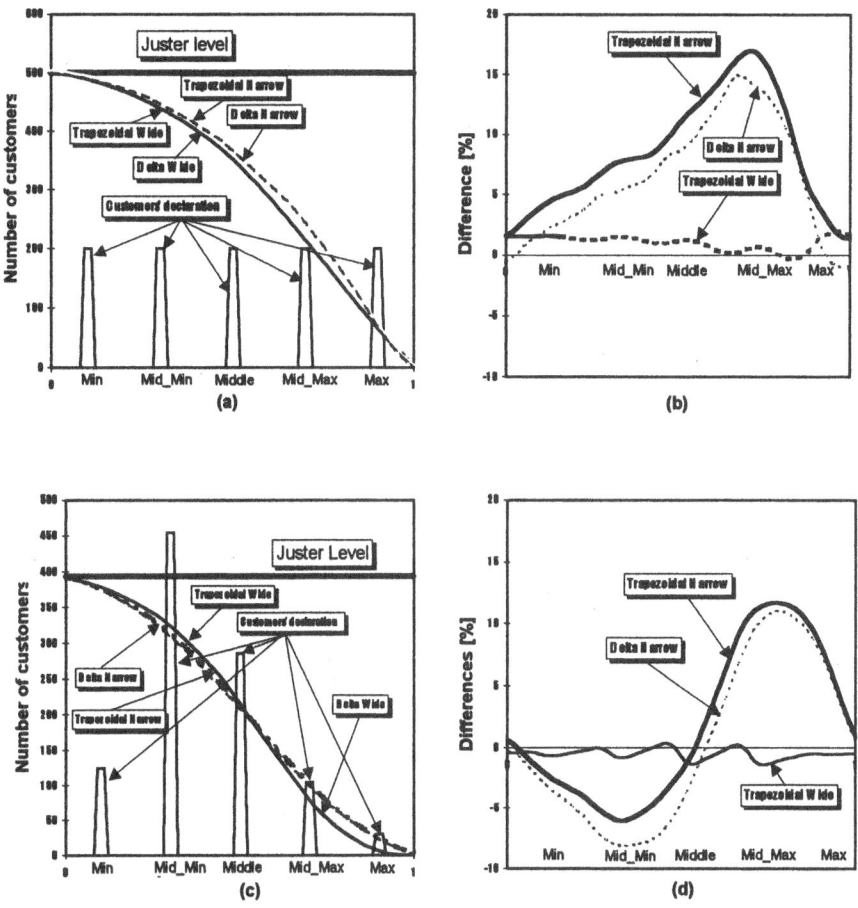

Figure 3. Cumulative probability of purchase. (a) and (b) - the uniform distribution, (c) and (d) - the Gamma distribution.

Figure 5 (b) presents the structure of cumulative purchase. It is visible how particular declarations construct the cumulative functions. The cut-off x_c set for any level of probability will provide the number of customers (N_{xc}) who will purchase a product when the verbal expressions with lower probabilities are rejected. It provides excellent opportunities for an analysis of prospective purchases for the various levels of subjective probabilities and verbal expressions.

Figure 6 shows the comparison of cumulative purchase functions for three scenarios with 9, 7, and 5 verbal expressions. The plots of these functions are very close to each other so that it is difficult to recognize them visually. The difference between cumulative purchase functions are calculated as:

$$\text{Difference}_1(x) = \frac{[CPF_1(x) - CPF_9(x)]}{CPF_9(x)}100\% \tag{17}$$

where: $CPF_1(x)$ - cumulative purchase function;
  1 - notation of a scenario ($1 = 5$ or $1 = 7$);
  $CPF_9(x)$ - cumulative purchase function for the scenario with 9 verbal expressions.

These differences are very small and they mostly result from the step of integration.

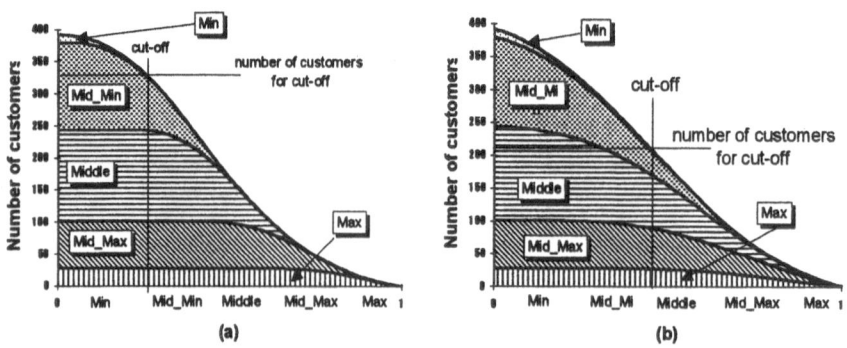

Figure 4. Composition of the cumulative probability.

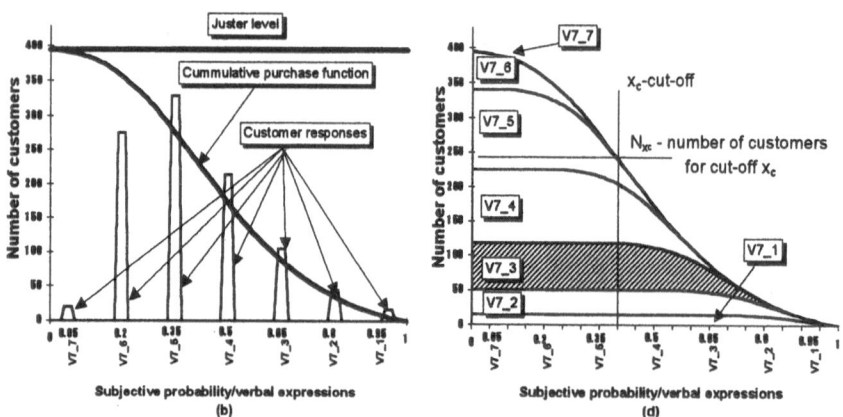

Figure 5. Cumulative probability of purchase for the Gamma distribution.

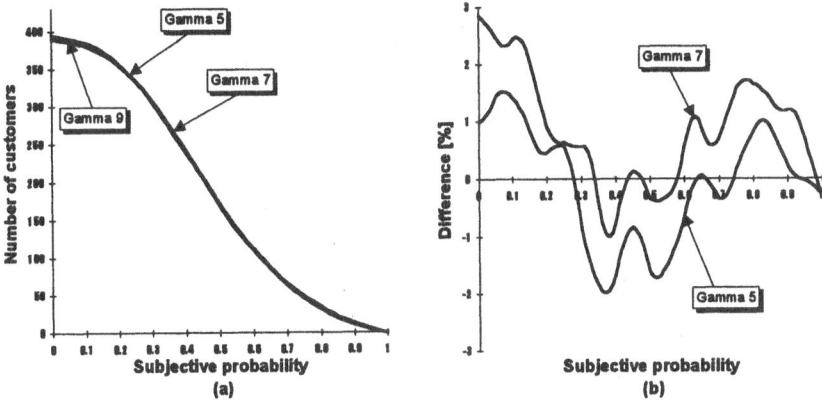

Figure 6. Differences between cumulative probability of purchase.

4. Conclusions

The study considering four options of fuzzy filters (delta and trapezoidal shaped with wide and narrow spans) with two distributions of customer responses (normal and Gamma distributions) and three scenarios of linguistic variables leads to the following results:

a) Applications of fuzzy filters to the analysis of survey responses result in cumulative probability of purchase with visible shares of each declaration to the cumulative probability of purchase determining the total number of prospective customers.
b) Fuzzy approach allows the application of cut-off levels indicating the number of prospective buyers for the assumed level of linguistic declarations or inversely, the determination of which declarations result in the purchase of equipment when the final sale is known.
c) The shape of filters does not significantly affect results. More influence was observed for a large span of filters over linguistic declarations.
d) The number of linguistic terms when applying fuzzy filters does not affect the results. Differences for the three considered options of linguistic variable sets: 9, 7, and 5 variables are below 3% for the cumulative probability of purchase.

An application of fuzzy filters to process data from surveys expressed in linguistic terms has two main advantages. First, it allows to consider, so characteristic for humans, uncertainty in customer responses and changing intentions between declarations expressed in surveys and real purchases. Customers who declared *very probable* purchase of equipment may change their mind and not follow their declarations. On the other hand, customers who declared *not consider* the purchase may in the end make decision on buying the equipment. Fuzzy filters seem to be the most appropriate tools to deal with human uncertainties.

Second, fuzzy filters reduce the impact of the number of linguistic variables and the levels of assigned subjective probabilities on the cumulative probability of purchase. Results obtained for two distributions of customer responses indicate that differences resulting from various options of linguistic terms are negligible.

Fuzzification of customer responses eliminates two significant problems in customer surveys: selection of the most appropriate number of linguistic terms and accurate assignment of subjective probability levels. This method allows to carry out studies for smaller number of linguistic variables with some freedom in selecting levels of subjective probabilities assigned to particular linguistic declarations. An example of subjective probability levels is shown in Table 6.

The method presented was derived for forecasting of new, energy efficient equipment penetration, but it is enough general to be applied to any type of studies involving peoples' declarations expressed in linguistic terms and the analysis of survey responses.

References

1. Bezdek, J., "Fuzziness vs. Probability Again (!,?)" *IEEE Transactions on Fuzzy Systems*, Vol. 2, pp. 1-3, (1994)
2. Budescu, D. V. and Wallsten T. S., "Consistency in Interpretation of Probabilistic Phrases," *Organizational Behavior and Human Decision Processes*, Vol. 36, pp. 391-405, (1985)
3. Budescu, D. V. and Wallsten T. S., "Subjective Estimation of Precise and Vague Uncertainties in Judgmental Forecasting", G. Wright and P. Ayton, J. Wiley and Sons, NY (1987) (1).
4. Budescu, D. V.and Wallsten, T. S. "Dyadic Decisions with Numerical and Verbal Probabilities," *Organizational Behavior and Human Decision Processes*, Vol. 46, pp. 240-263, (1990)
5. Comerford, R. B. and Gellings, C. W., "The Application of Classical Forecasting Techniques to Load Management," *IEEE Transactions on PAS*, Vol. 101, No 12, pp. 4656-4664, (1982).
6. Curtin, R. T., "Indicators of Consumer Behavior; the University of Michigan Surveys of Customers", *Public Opinion Quarterly*, Vol. 46, pp.340-352, (1982)
7. Dua, P. and Smyth, D. J. "Forecasting US Home Sales Using BVAR Models and Survey Data on Households' Buying Attitudes for Homes," *Journal of Forecasting*, Vol. 14, pp. 217-227, (1995)
8. Dubois, D. and Prade, H., "Fuzzy Sets and Statistical Data," *European Journal of Operational Research*, Vol. 25, pp. 345-356, (1986)

9. Dubois, D. and Prade, H., "Fuzzy Sets, Probability and Measurement," *European Journal of Operational Research*, Vol. 40, pp. 135-154, (1989)

10. Dubois, D. and Prade, H., "Fuzzy Sets - A Convenient Fiction for Modelling Vagueness and Possibility," *IEEE Transactions on Fuzzy Systems*, Vol. 2, No 1, pp. 16-21, (1994)

11. Gellings, P. E., (Editor), Demand Forecasting for Electric Utilities, The Fairmont Press, Inc. (1992)

12. Gendall, P., Esslemont D. and Day D. "A comparison of Two Versions of the Juster Scale Using Self-Completion Questionnaires," *Journal of the Market Research Society*, Vol. 33. No 3, pp. 257-263, (1991).

13. Hamm, R. M., "Selection of Verbal Probabilities: A Solution for Some Problems of Verbal Probability Expression," *Organizational Behavior and Human Decision Processes*, Vol. 48, pp. 193-223, (1991)

14. Hersh, H. M., Caramazza, A. and Brownell, H., "Effects of Context on Fuzzy Membership Functions," In Gupta, M. M., Ragade, R. L. and Yager R. R.(Eds) "Advances in Fuzzy Set Theory and Application", pp. 389-408, Amsterdam, The Netherlands North-Holland, (1979)

15. Hook, J. A. and Gendall, P. J., "A New Method of Predicting Voting Behavior," *Journal of the Market Research Society*, Vol. 35, No 4, pp. 361-371, (1993)

16. Irwin, J. R., "Buying/Selling Price Preference Reversals: Preference for Environmental Changes in Buying versus Selling Modes," *Organizational Behavior and Human Decision Processes*, Vol. 60, pp. 431-475, (1994)

17. Isherwood, B. C. and Pickering, J. B., "Purchase Probabilities and Consumer Durable Buying Behavior," *Journal of the Market Research Society*, Vol. 16, No 3, pp. 203-226, (1974).

18. Jonsson, B and Agren, A., "Forecasting car Expenditures Using Household Survey Data," *Journal of Forecasting*, Vol. 73, pp. 435-448, (1994)

19. Juster, F. T., "Consumer Buying Intentions and Purchase Probability", National Bureau of Economic Research, N.Y. (1966)

20. Kadane B. J., "Comment:Codifying Chance", Statistical Science, vol 5, (1), pp.18-20, (1990)

21. Kosko, B., "The Probability Monopoly," *IEEE Transactions on Fuzzy Systems*, Vol. 2, No 1, pp. 32-33, (1994)

22. Laviolette, M. and Seaman, J. W., "Evaluating Fuzzy Representations of Uncertainty," *The Mathematical Science*, Vol. 17, pp. 26-41, (1995)

23. Michalik, G., Khan, M. E. and Mielczarski, W., "Fuzzy Logic Approach to Forecasting Penetration of Energy Efficient Technologies," Proceedings of AUPEC'96, October, Melbourne, Vol. 1, pp. 13-18, (1996)

24. Mosteller, F. and Youtz, C., "Quantifying Probabilistic Expressions,*"* *Statistical Science*, Vol. 5, No. 1, pp. 1-12 (1990)

25. Oden, G. C., "Integration and Fuzzy Logic Information," *Journal of Experimental Psychology: Human Perception and Performance*, Vol. 3, pp. 565-576, (1977).

26. Regan, R. T., Mosteller, R. and Youtz C., "Quantitative meanings of verbal probability expressions," *Journal of Applied Psychology*, Vol. 74, No 3, pp. 433-442, (1989)

27. Savage, L. J., "The Foundation of Statistics", J. Wiley and Sons, Inc. NY, (1954).

28. Sullivan, W. G. and Claycombe, W. W., "Fundamentals of Forecasting", Preston Publishing Company, Inc. Peston, Virginia, (1977)

29. Turksen, I. B. and Willson, I. A., "A Fuzzy Set Preference Model for Consumer Choice," *Fuzzy Sets and Systems*, Vol. 68, pp. 253-266, (1994)

30. Wallsten, T. S., Buduscu, D. V., Rapoport, A., Zwich, R., and Forsyth B., "Measuring the Vague Meanings of Probability Terms," *Journal of Experimental Psychology: General*. Vol. 11, No 4, pp. 348-365, (1980)

31. Wright W. F., "Mood Effects on Subjective Probability Assessment," *Organizational Behavior and Human Decision Processes*, Vol. 52, pp. 276-291, (1992)

32. Zadeh, L. A., "Discussion: Probability Theory and Fuzzy Logic are Complimentary Rather than Competitive, " Technometrics, Vol. 17, No 3, pp. 271-275, (1995)

33. Zadeh L. A., "Fuzzy Logic Computing with Words," *IEEE Transactions on Fuzzy Systems*, Vol. 4, No 7, pp. 103-111, (1996)

34. Zysno, P., "Modelling Membership Functions," In Rieger R. R.(Ed) "Empirical Semantics", pp. 350-375, Bochum, Brodzmeyer, (1981)

A UNIVERSAL APPROACH TO ADAPTIVE FUZZY LOGIC CONTROLLER DESIGN WITH AN APPLICATION TO A POWER GENERATOR EXCITATION CONTROL

Omar Ghanayem and Leonid Reznik

Department of Electrical and Electronic Engineering
Victoria University of Technology
PO Box 14428 MCMC
Melbourne VIC8001
Australia
E-mail: {Omar,leonid}@cabsav.vut.edu.au

Abstract. The problem of computerisation of the adaptive fuzzy logic controller design has emerged nowadays as both practical and theoretical one. This paper proposes a universal adaptive fuzzy logic controller (FLC) structure with its application to the excitation control of a synchronous generator connected to an infinite bus through a transmission line. The proposed system features an automatic learning mechanism which enables on-line updating and tuning of the global input and output ranges of the FLC as well as an implicit tuning and updating of the controller classes and rules. The on-line tuning of the controller parameters is achieved via new concepts of fuzzy logic design. These novel concepts have been developed and implemented for the purpose of improving the efficiency of fuzzy learning based on the FLC status and the system performance. The system implementing this universal design approach, called SHAY, has been developed. The proposed system has been tested for an excitation control on a laboratory setup of a synchronous generator connected to an infinite bus where it has proved efficiency and robustness under different operating conditions.

Keywords: adaptive fuzzy logic controller, excitation control.

1. Introduction

Despite being comparatively new in the field fuzzy control technology has already been able to emerge as a new solution for many control problems. Nowadays fuzzy logic controller (FLC) domain of applications ranges from highly specialised military applications to simple domestic appliances. In fact, fuzzy logic algorithms are being used in some areas where other control algorithms have never been applied, areas such as business and data management and query systems. However, this bright side of the fuzzy logic algorithms is accompanied with many question marks. Anti-FLC critics claim that FLC lacks a structured design methodology as well as means of evaluating the FLC from stability and performance prospective. Such statements can be justified by an young age of the technology just partly. Other explanations are based on a quite specific and different from a classic approach nature of a FLC design methodology incorporating methods from both control engineering and artificial intelligence fields.

The design procedure of a FLC should cater for many factors and points all are open wide for investigation and improvement. The main problems in the FLC design are:

1. **the controller/plant interfacing**, which includes the input and output ranges as well as the membership functions for all the domains in the FLC.
2. **knowledge availability and processing**, this is mainly concerned with the rules base part of the FLC. Other details may be very important in this part, such as the inference mechanism, the aggregation and the operators.
3. **performance evaluation**, stability security and criteria of evaluation.
4. **practical implementation concerns**, hardware and software restrictions.
5. The great dependence of the above mentioned factors on each other makes the problem even more complex. Especially when dealing with very large and highly critical applications.

In the wake of these comments, FLC designers started to implement advanced FLC design methodologies and approaches trying to solve the main obstacles in the design and development of a FLC. Advanced FLC design structures such as adaptive [1-3], self organising [4,5], self learning [6,7] have been proposed by many authors.

However, none of these structures has emerged as a general or universal FLC design structure in spite of all the success reported in regard to the particular areas of an application. The reasons for this can be summarised in the following points:

1. Most of the proposed structures lack the global optimisation approach required to balance all of the FLC parameters into an optimal equilibrium in these parameters settings.
2. Adaptive and self organising FLCs require large amounts of time and processors memory when implemented in a microprocessor based controller.
3. Most of the proposed structures require extensive off-line parameter tuning and optimisation as these algorithms rely heavily on the accurate choice of some gain factors. Such systems demonstrate drastic and unpredictable behaviour under variations of the operational conditions.

This paper proposes a universal adaptive FLC design structure with an example of a practical implementation of the structure in real time systems. The proposed algorithm is called Stability Handling Algorithm, Transient and Steady-State, SHAY.

SHAY FLC design algorithm attacks the main FLC design obstacles in a novel approach of adaptive self learning mechanism that enables:

1. on line tuning for both the input and output ranges of the FLC,
2. adaptive tuning of the knowledge representation via altering the controller membership functions and rules in an implicit manner.

The learning mechanism is driven by the plant under consideration state as well as the FLC status. The structure also includes some parameters which when altered can help the designer to reach predefined control objectives, like maximum overshoot, rise time, etc. Some research of these parameters choice and their influence on the performance indicators has been performed.

The paper also fully describes the procedure of implementing the SHAY-FLC in a complex environment. The excitation control of a power generator is considered as an implementation study. The results are analysed and discussed.

2. Proposed FLC design structure- Shay

SHAY-FLC design concept is a structured set of the special modules connected in a logical chain. It consists of *main* and *supporting functions* as shown in Figure 1. The structure operation is based on some new concepts. This section starts with a description of these new concepts and tools which are considered as supporting functions for the main SHAY components explained later in the section.

2.1 New Concepts Applied in SHAY - *Supporting Functions*

The FLC operation is based on an initial evaluation of some parameters characterising the current state of the plant and the FLC. The calculation of these parameters is performed by supporting functions.

2.1.1. Operational angle concept

This concept represents a different look at a classical differential controller concept as it gives a measurement of the angle between two successive samples of its input (θ_{opr}), E_k and E_{k-1}, Figure 2. So it replaces the absolute deviation of the input signal (where the error signal is traditionally applied) with the ratio of the deviation to the signal value.

As it is clear from the Figure 2 a scaling factor A is used to scale E_{k-1}. The role of this factor in the performance evaluation of a SHAY-FLC is explained later.

2.1.2. Operational sector concept

E_k and E_{k-1} are both used in order to subdivide the operational domain into six operational sectors as shown in Figure 3. This space subdivision was initially proposed in [8]. The separation of the sectors is based on E_k, SE_{k-1} and ΔE_k which is E_k-SE_{k-1}.

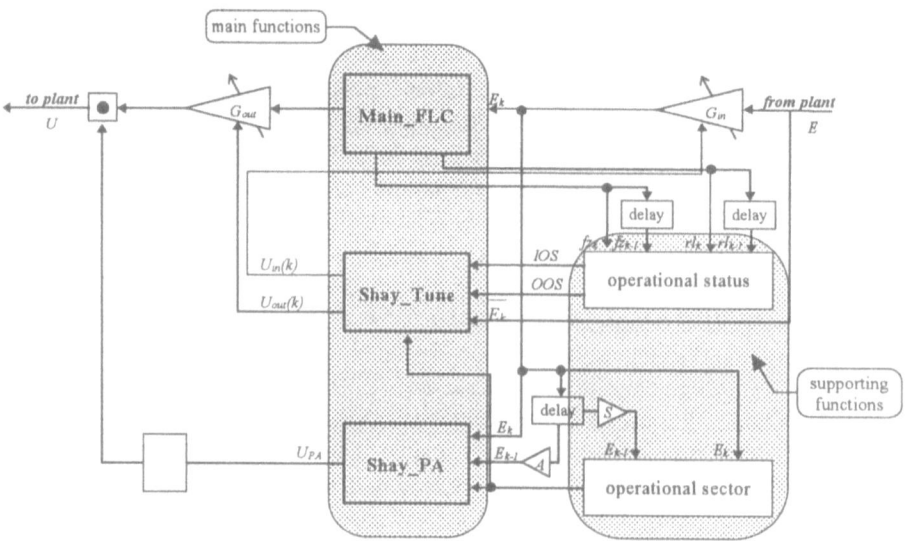

Figure 1. SHAY FLC design structure

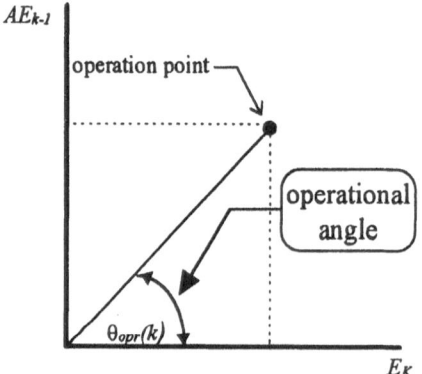

Figure 2. Operational angle

2.1.3. Operational status concept

This is a new tool introduced by the first author to evaluate the FLC. It is a fuzzy measurement of a suitability of the current FLC input and output ranges. This function keeps continuous supervision on the FLC under consideration. Its introduction has let a rough evaluation, if any adaptation is necessary at the current state, to be performed.

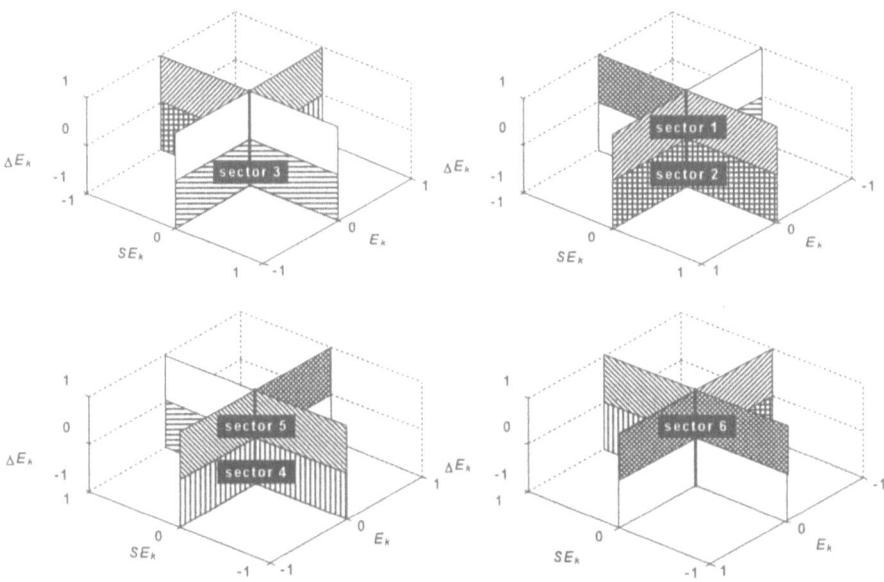

Figure 3. Operational sectors

The operational status is mainly concerned with the upper and lower boundaries of the FLC ranges, of whether inputs or outputs. Continuous monitoring of fuzzy inputs and outputs of a controller is performed in order to determine in which mode the current ranges are.

These modes are:

1. **under control mode** (R_1) which is the mode when only the lower classes of the FLC are active and the performance monitor is not satisfactory,
2. **normal control mode** (R_2) which is the case when the FLC is interactively responding to the system, with accepted performance monitor levels,
3. **over control mode** (R_3) which is the case when only the upper boundaries of the FLC are always active and the performance monitor is not giving satisfactory results.

The operational sector information is captured in an array called *sct*, where *sct*=[*sct1 sct2 sct3 sct4 sct5 sct6*]. *Sct* and θ_{opr} have enabled clear identification of the plant dynamics and proved to be very efficient in practical situations. The operational status information is kept in an array called *R*, where *R*=[R_1 R_2 R_3].

2.2. Main Functions

2.2.1. Main-FLC

This is a simple single input single output (SISO) FLC, where an initial knowledge about the system is stored in its rules and classes. The input to this FLC is E_k and it produces an unsigned output U_{main}. The contribution of U_{main} to the final control signal output from SHAY (*U*) is minor, but the role the Main-FLC plays in the overall structure is essential as it is the base for the whole learning mechanism used in SHAY.

2.2.2. SHAY-PA

This is an imitation of the proportional differential controller (PD). Two nonlinear functions *P* and *N*, shown in Figure 4, are used to produce the output of this function (U_{PA}) which is a signed signal calculated as either *P-N* or *N-P* according to the sector. The x-axis used in the figure is θ_{opr} and the angle at which *P*=*N* is called the switching angle (θ_{sw}). A different set of *P* and *N* functions is used for each sector. Two parameters in the figure, θ_{opr_min} and θ_{opr_max} are the angles that bound the sector.

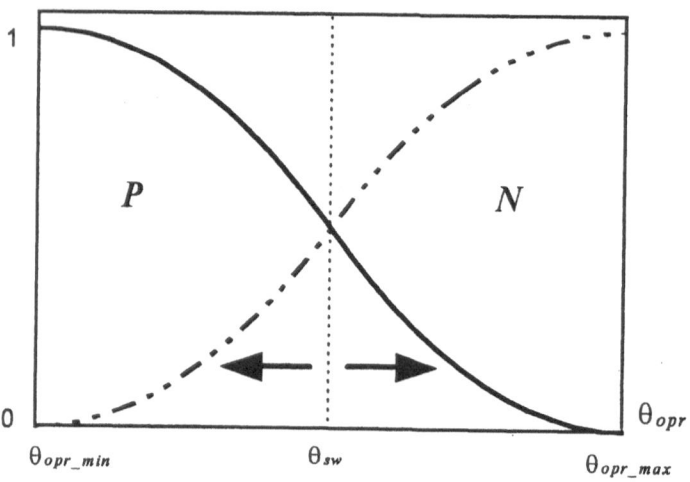

Figure 4. Nonlinear functions used in SHAY-PA

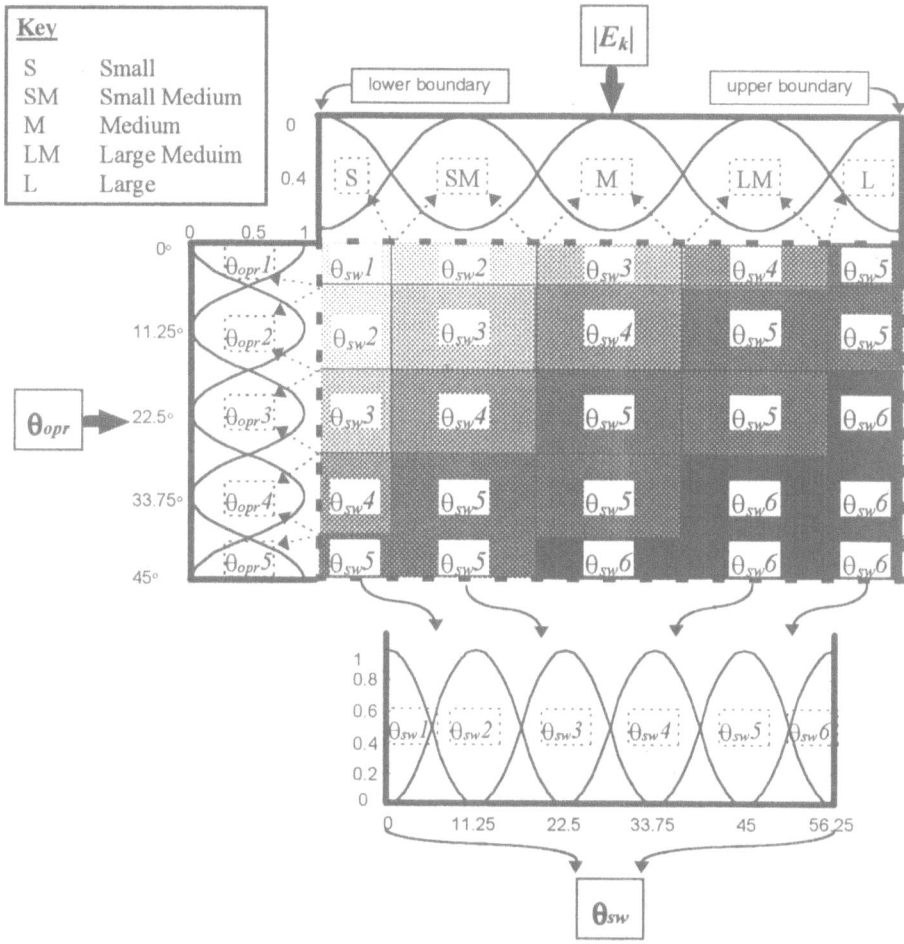

Figure 5. FLC used for sector 1

The arrows in Figure 4 indicate that θ_{sw} (the switching angle) is variable. θ_{sw} is being calculated on-line according to θ_{opr}, E_k and sct. An example of the fuzzy logic procedure used to derive θ_{sw} in sector 1 is shown in Figure 5 and the control surface of this FLC is shown in Figure 6. Figure 7 shows that different allocations of θ_{sw} yield significant differences in the values of U_{PA}.

2.2.3. SHAY-Tune

Values of R, Sct and E_k estimated at the previous stages are used to produce an updating signal for the input and output ranges of the Main-FLC (G_{in} and G_{out}

respectively), these signals are $U_{in}(k)$ and $U_{out}(k)$ used to update the global input and output ranges of the Main-FLC as

$$G_{in} = \sum_{k=0}^{M-1} U_{in}(k)$$

and

$$G_{out} = \sum_{k=0}^{M-1} U_{out}(k)$$

where M is the number of samples.

A learning period (L) is required in SHAY-Tune during which IOS_{av} and OOS_{av} (the operational status of the main FLC from input and output domains prepositives) are being calculated resulting in $IOS_{av} = \begin{bmatrix} R_{i1_{av}} & R_{i2_{av}} & R_{i3_{av}} \end{bmatrix}$ and $OOS_{av} = \begin{bmatrix} R_{o1_{av}} & R_{o2_{av}} & R_{o3_{av}} \end{bmatrix}$, M number of samples is collected during the L time. Then, both $R_{i_{av}}$ and $R_{o_{av}}$ are subject to fuzzy processing to produce nine flags used in later stages of the ranges updating process. The flags are produced according to the rule base in Table 1.

Table 1. Rule base used to produce the R flags

		$R_{i_{av}}$		
		$R_{i1_{av}}$	$R_{i2_{av}}$	$R_{i3_{av}}$
$R_{o_{av}}$	$R_{o1_{av}}$	R_{11}	R_{21}	R_{31}
	$R_{o2_{av}}$	R_{12}	R_{22}	R_{32}
	$R_{o3_{av}}$	R_{13}	R_{23}	R_{33}

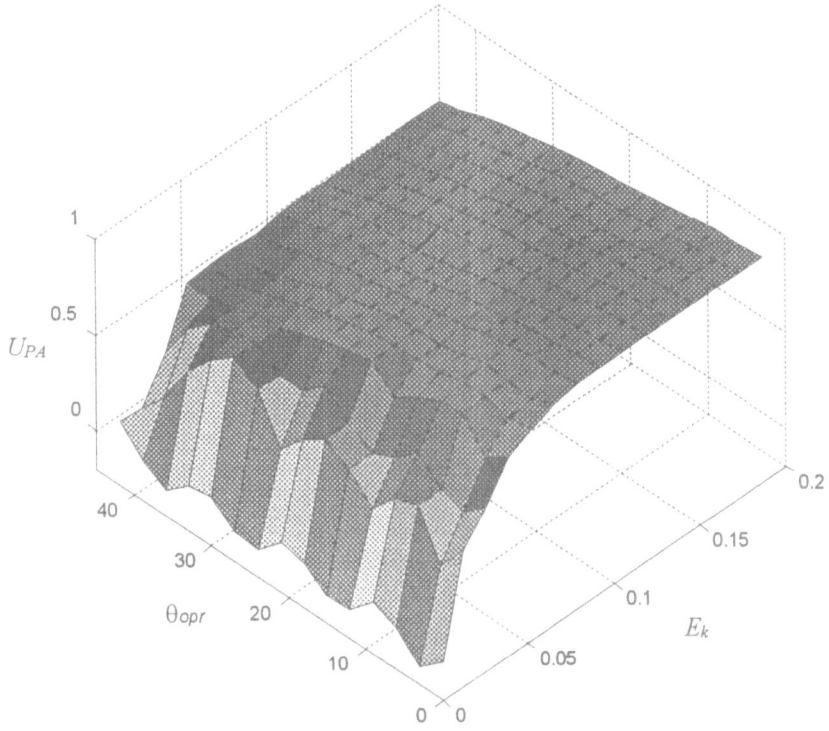

Figure 6. Control surface for sector 1

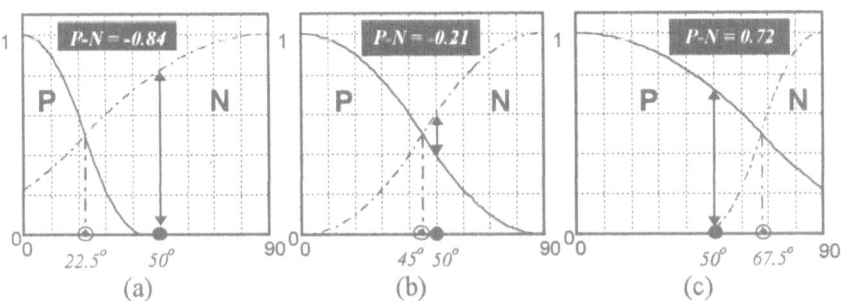

Figure 7. Effect of different θ_{sw} on P and N
(a) $\theta_{sw}=22.5°,\theta_{opr}=50°$, (b) $\theta_{sw}=45°$, $\theta_{opr}=50°$ (c) $\theta_{sw}=67.5°$, $\theta_{opr}=50°$

The truth value for each cell in the rule table is calculated as
$R_{ij} = \min(R_{ii_{av}}, R_{oj_{av}})$

The *sct* vector information is used to categorise the operation state into three different modes, **mode 1 (*md1*), mode 2 (*md2*)** and **mode 3 (*md3*)**, where:

$$md1 = \frac{\sum\limits_{k=0}^{M-1} sct1(k) + \sum\limits_{k=0}^{M-1} sct2(k)}{M}$$

$$md2 = \frac{\sum\limits_{k=0}^{M-1} sct4(k) + \sum\limits_{k=0}^{M-1} sct5(k)}{M}$$

The mode is determined by comparing *md1* to *md2*. If the greater one was greater than a threshold value (*T*) then the mode is the larger mode, either *md1* or *md2*. Otherwise the mode is set to be *md3* according to the following if-then rules

If (md1>md2) and (md1 > T)
*Then the mode is **mode 1***
Else   if (md2>md1) and (md2 > T)
        *Then the mode is **mode 2***
        Else
        *The mode is **mode 3***

The modes and the nine flags are applied in order to implement two updating strategies, Figure 8 shows these strategies. The final control signal (*U*) produced by a SHAY-FLC is calculated as: $U=(U_{main} \times U_{PA}) \times G_{out}$ (see [8-10] for further details about the SHAY FLC structure).

3. Research of the SHAY structure

The effect of the four factors, *S*: the sectors scale, *A*: the operational angle scale, *L*: the learning rate, and *T*: the learning threshold on a SHAY-FLC have been investigated. A comprehensive study was applied in an attempt to develop general patterns of how each factor effects the maximum overshoot, μ_p, one of the traditionally applied in control engineering performance indicators.

The study have been performed for all possible combinations of the *SALT* parameters according to the following:

- *S*: ranges from 0 to 2 with the incremental step size of 0.1, 20 cases
- *A*: ranges from 0 to 2 with the incremental step size of 0.1, 20 cases

- L: ranges from 0 to 2.5 with the incremental step size of 0.125, 20 cases
- T: ranges from 0 to 1 with the incremental step size of 0.05, 20 cases.

The investigation results are shown in Figure 9, and summarised in Table 2, the result shows the effect of the $SALT$ factors on μ_p .

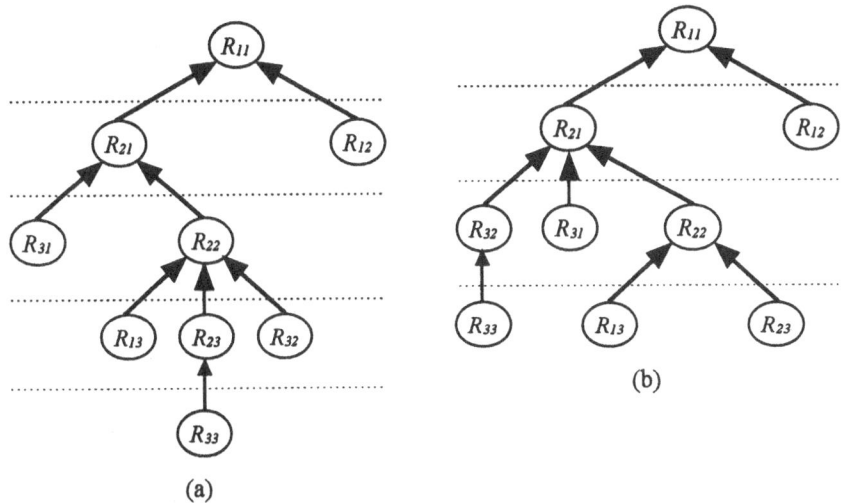

(a)

(b)

Figure 8. Modes updating strategy, (a) *md1* and *md2*, (b) *md3*

4. Implementation study

The proposed SHAY-FLC architecture was applied for the excitation control of a synchronous generator connected to an infinite bus through a transmission line. In order to cater for both voltage and rotor angle stabilities, the performance monitor used in the system considers both the rotor angle speed deviation ($\Delta\omega$) and the error in the generator terminal voltage (v_t) compared to a reference signal (v_{ref}). Generator parameters are given in Appendix.

The input to the Main-FLC as well as the inputs to SHAY-PA and SHAY-Tune were derived from the manipulation process of both signals in a Pre-Control (P-C) stage. This manipulation stage is explained in details in the following section.

4.1. P-C FLC Stage Description

4.1.1. Rotor angle stability

The P-C stage is applied to choose the corresponding set of the rules to provide three different control strategies:

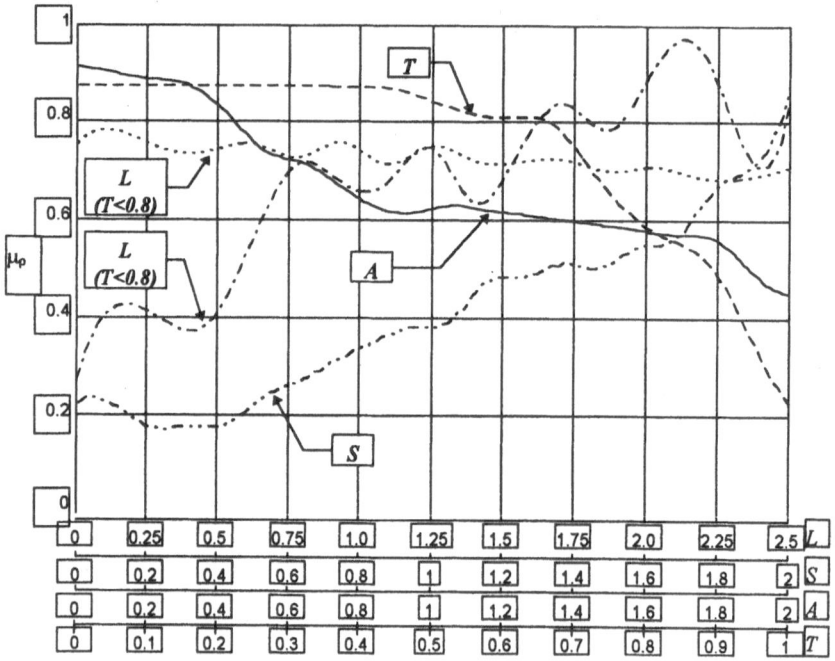

Figure 9. Influence of *SALT* factors on μ_p

Table 2. Effect of *SALT* factors on μ_p

Relation	Range	Conditions
$\mu_p \propto S$	$\forall S \in [0,2]$	
$\mu_p \propto \dfrac{1}{A}$	$\forall A \in [0,2]$	
$\mu_p \propto L$	$\forall L \in [0,2.5]$	*If T is large*
$\mu_p \propto \dfrac{1}{L}$	$\forall L \in [0,2.5]$	*If T is small*
$\mu_p \equiv constant$	$\forall T \in [0,0.5]$	
$\mu_p \propto \dfrac{1}{T}$	$\forall T \in [0.5,1]$	

1. **Acceleration control**: a set of rules that are responsible for giving the excitation system the signal to boost up the excitation in order to increase the generator electrical power output (P_e)

2. **Deceleration control**: rules that are responsible to indicate to the excitation system to bring down the excitation signal in order to reduce P_e

3. **Breaking control**: rules that are responsible of producing the precautionary signals that will help the system to over come any anticipated future undesired behaviour.

The P-C FLC is shown in Figure 10 with the three control strategies marked with different shading cells. Figure 11 shows the control surface of the P-C FLC.

4.1.2. Voltage stability

The generator terminal voltage stability is accounted for by considering the error in the generator terminal voltage (e_{vt}) compared to a reference voltage (v_{ref}), e_{vt} is calculated as:

$$e_{vt} = v_{ref} - v_t$$

The performance monitor used as an input to SHAY-FLC (E) is calculated as

$$E = \Delta\omega` + e_{vt}$$

where $\Delta\omega`$ is the output of the P-C FLC.

The final control signal out of the SHAY-exciter (U_{fd}) which is used as the input to the field of the synchronous generator is calculated as:

$$U_{fd} = \sum_{k=0}^{M-1} U(k)$$

where k is the number of samples and U is the output of SHAY-FLC. A fully detailed block diagram of the SHAY exciter is shown in Figure 12.

The implementation study was performed on a laboratory setup of a synchronous generator connected to an infinite bus through a transmission line. The SHAY excitation system was implemented in a C-code downloaded to the digital signal processing board (DSP), DPC/C40 board mounted with two TMS320C40 processors[1]. Figure 13 shows a general view of the laboratory setup used. A 5 KVA synchronous machine was used as a generator and a dc motor was used as a prime mover. The adjustable reactors as well as variable load banks were also applied in order to imitate the most of the real time components of a power system network. A general layout of the laboratory setup employed is shown in Figure 14.

---

[1] TEXAS INSTRUMENTS C40 PROCESSORS

The implementation study required the development of both hardware and software tools to facilitate an implementation of the digital form of SHAY-FLC as well as to allow safe interfacing between the synchronous generator and the DSP board in addition to providing flexible man/machine interface through software. Figure 15 shows a flowchart of performing the SHAY-FLC algorithm. Table 3 describes the tests and operational conditions used in the laboratory experiments. The test results are shown in Figures 16-21.

5. Conclusion

The tests results highlight the superiority of the proposed FLC structure. The superiority of the proposed SHAY scheme can be derived from the following facts:

1. it gives a good chance for better tuning of the main FLC parameters in the design stage, considering that the input/output ranges problem is solved using the Shay-Tune.
2. the structure of the SHAY is system independent.
3. the on-line operation of SHAY supports the overall control loop robustness under variations of operating conditions.

The tests were performed under a very wide range of the operational conditions. Operation points P1 and P2 represent the normal and an up-normal operational conditions. The SHAY exciter was still able to adapt to the new environment and secure the stability of both the rotor angle and the terminal voltage. The tests undergone have confirmed the perspectiveness of the proposed universal approach and the SHAY structure in design of complex engineering systems.

References

[1] Han J. Y. and McMurray, V., " Adaptive Fuzzy Logic Controller Based on Parameter Estimation", Proceedings of the 35th Midwest Symposium on Circuits and Systems, New York, USA, *vol 3*, pp. 808-811, 1992.
[2] Fei, J. and Isik, C., " Adaptive Fuzzy Control Via Modification of Linguistic Variables", IEEE international conference on fuzzy systems, New York, USA, pp. 399-406, 1992.
[3] Young-M. P.; Un-Chul, M.; Lee, K.Y., " A Selforganising Fuzzy Logic Controller for Dynamic Systems Using a Fuzzy Auto-regressive Moving Average (FARMA) Model", IEEE Transactions on Fuzzy Systems *vol*: 3 Iss: 1 pp 7, 1995.

[4] Pave, L. and Chelaru M., " Neural Fuzzy Architecture for Adaptive Control", IEEE International Conference on Fuzzy Systems, pp 1115-1122, New York, USA, 1992.

[5] Isomursu, P. and Rauma, T., " A Self Tunning Fuzzy Logic Controller for Temperature Control of Superheated Steam", Proceedings of the Third IEEE Conference on Fuzzy Systems, IEEE Word Congress on Computational Intelligence, New York, USA, *vol* 3, pp. 1560-1563, 1994.

[6] Moudgal, V.G., Kwong, W.A., Passino, K.M. and Yurkovich, S. ,"Fuzzy Learning Control for A Flexible-Link Robot", IEEE Transactions on Fuzzy Systems, *vol*: 3 Iss: 2, pp 199-210 ,May 1995.

[7] Sutton R. and Jess, I. M., " Real-Time Application of A Self Organising Autopilot to Warship Taw Control", IEEE International Conference in Control'91, pp. 827-832, Edinburgh, UK, March 1991.

[8] Ghanayem, O. and Berangi, R., " On Line Identification of Input/Output Ranges of A FLC Scheme", January 1996, pp. 150-154, FLAMOC'96 Conference, *vol.* 2, Sydney, Australia.

[9] Ghanayem O. and Reznik, L., "A Hybrid AVR-PSS Controller Based on Fuzzy Logic Technique", April 1995, pp. 347-351, CONTROL'95 Conference, *vol.* 2, Melbourne, Australia.

[10] Ghanayem and L. Reznik, "A New Reasoning Approach and its Application in Power System Stability", August 29-September 1 1995, pp.1527 - 1532, Proceedings of the Third European Congress on Intelligent Techniques and Soft Computing (EUFIT'95), Aachen, Germany, vol.3.

302

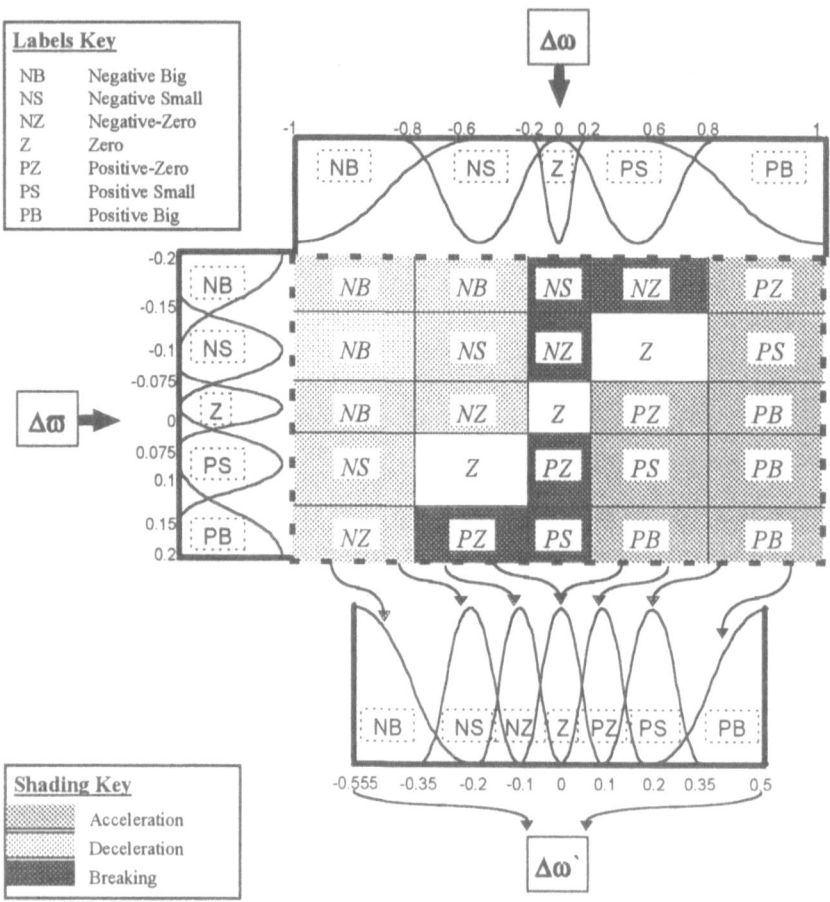

Figure 10. P-C FLC

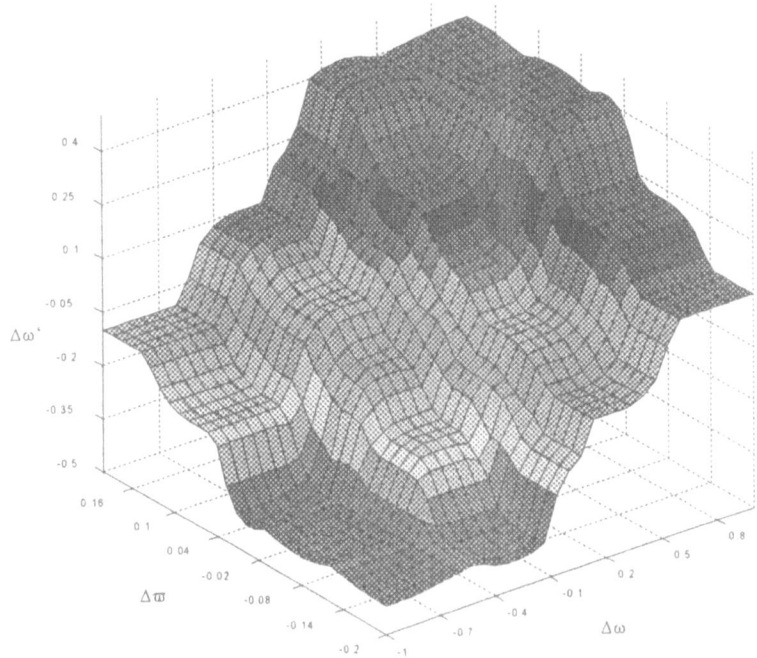

Figure 11. P-C FLC control surface

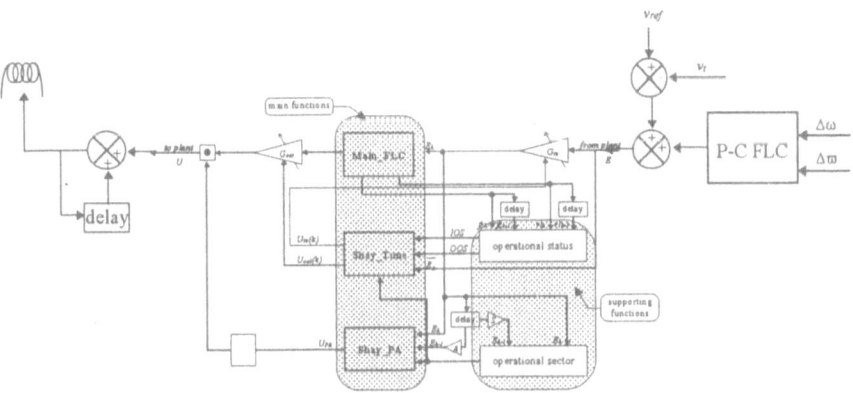

Figure 12. SHAY exciter block diagram

304

Figure 13. General view of laboratory setup

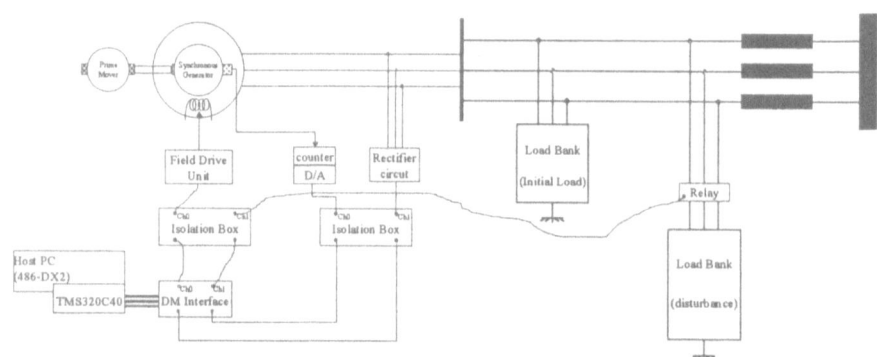

Figure 14. Laboratory setup layout

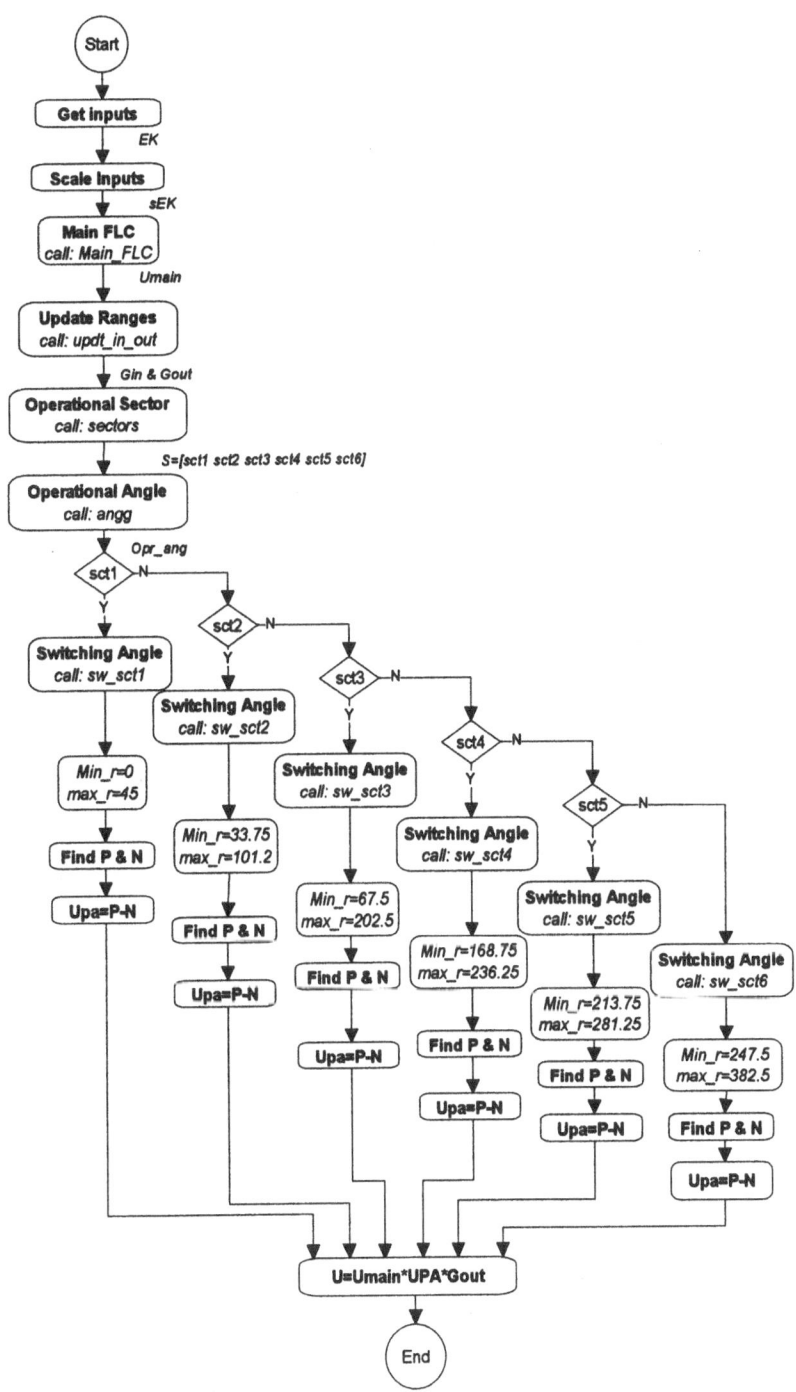

Figure 15. Flowchart for implementing a SHAY-FLC

Table 3. Small load changes testing conditions

Active Power (pu)	Operating conditions Test	Power Factor	Disturbance	Duration
0.9	TEST-SL1	0.9 (lag)	30% resistive load	5 sec
0.5	TEST-VS2	0.6 (lag)	10% change in v_{ref}	5 sec
0.9	TEST-SC2	0.9 (lag)	L-l-g short circuit	100 msec

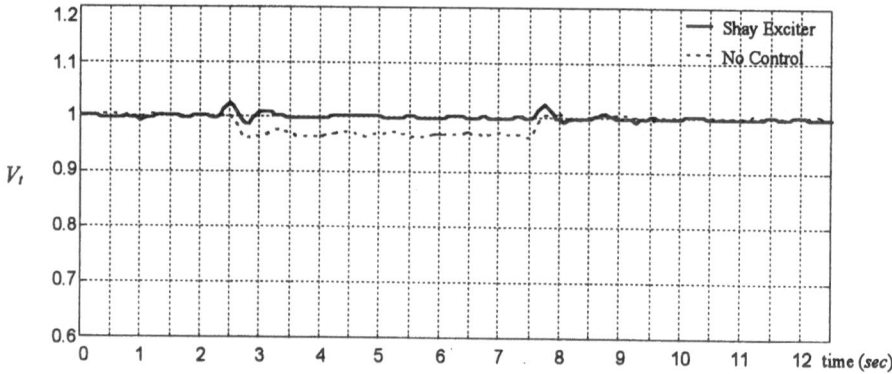

Figure 16. Terminal voltage (*pu*), TEST-SL1

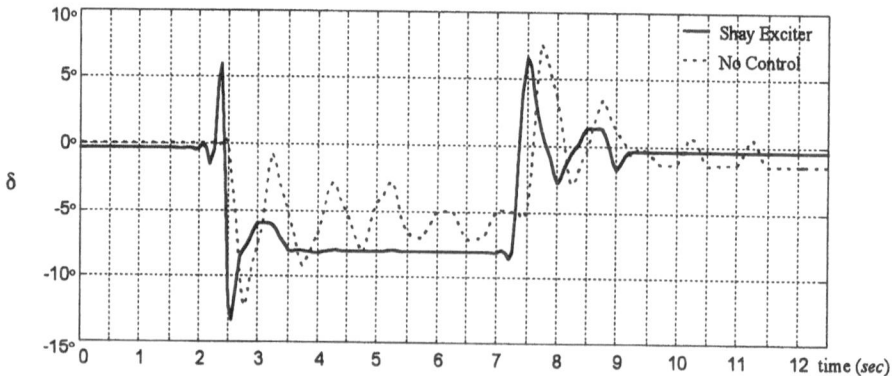

Figure 17. Rotor angle (electrical degrees), TEST-SL1

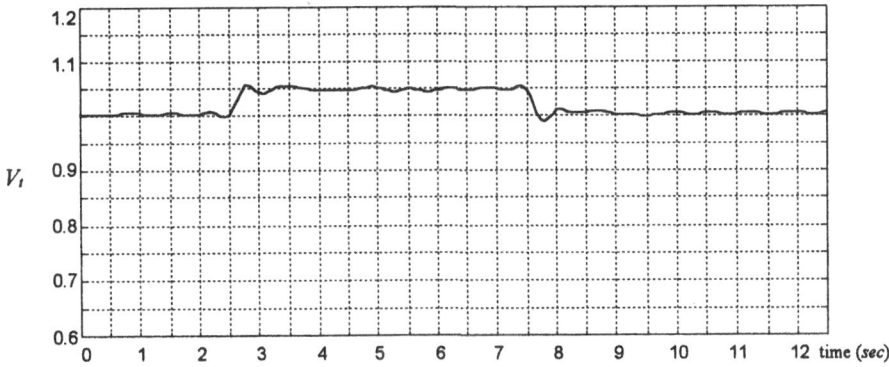

Figure 18. Terminal voltage (*pu*), TEST-VS2

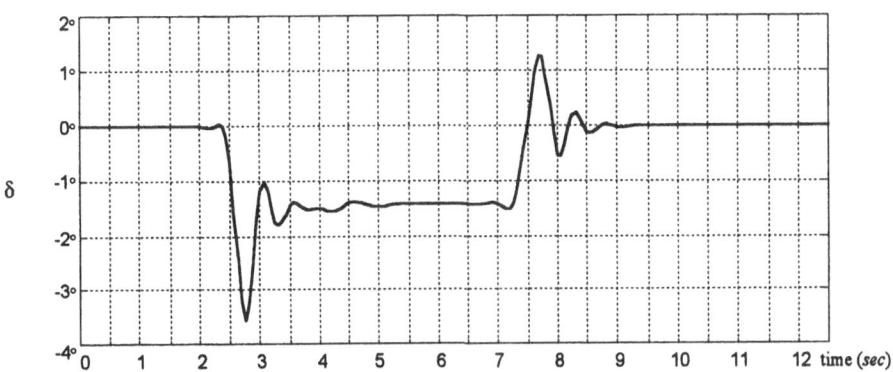

Figure 19. Rotor angle (electrical degrees), TEST- VS2

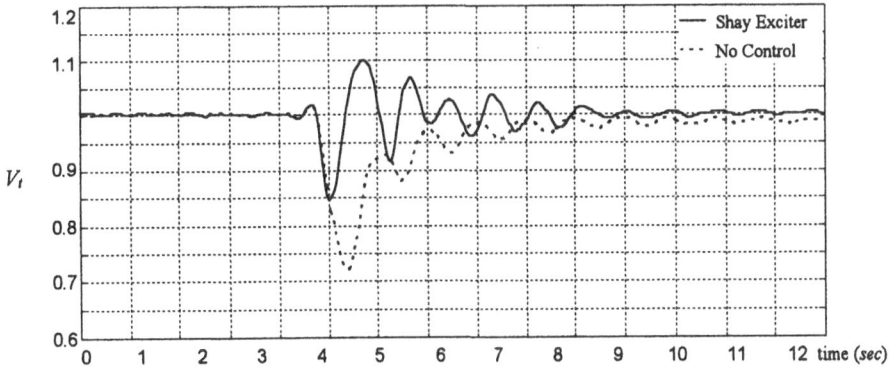

Figure 20. Terminal voltage (*pu*), TEST-SC2

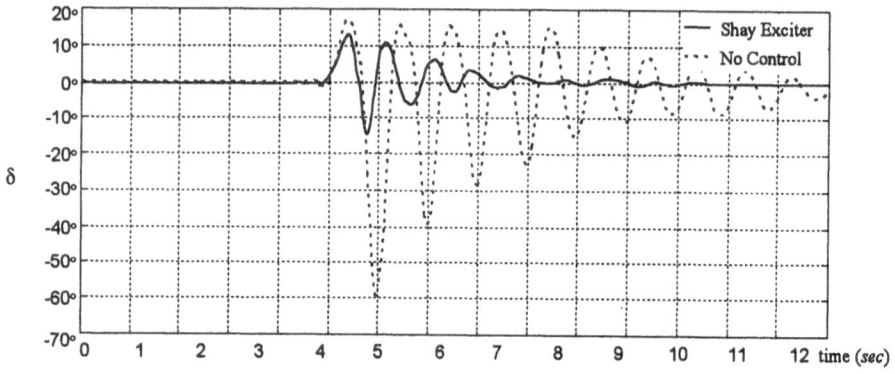

Figure 21. Rotor angle (electrical degrees), TEST-SC2

APPENDIX

Synchronous Generator Parameters	Transmission Line Parameters
X_d=1.027 pu	R_e=0.02 pu
$X_{d'}$=0.479 pu	X_e=0.4 pu.
X_q=0.489 pu	
T'_{do}=0.345 sec.	
H=0.764 sec.	
ω_B=314 rad/sec.	

FUZZY LOGIC APPLICATIONS IN DIAGNOSING MECHATRONIC SYSTEMS

Tapio Rauma and Matti Kurki

VTT (Technical Research Centre of Finland) Electronics
P.O.Box 1100
FIN-90571 Oulu
Finland
E-mail: Tapio.Rauma@vtt.fi

Abstract. In this paper, possibilities of using fuzzy system models as a part of model-based fault diagnosis systems are discussed. Methods and experiences of using fuzzy system models at different stages of fault diagnosis systems are presented. In this study, fuzzy system models are used in fault detection and inverse fuzzy system models in fault localization. Inverse fuzzy models are discussed in the light of real-world applications. The mathematical theory is skipped entirely. Additionally we present some examples of using a fuzzy system model in an interface between a fault diagnosis system and a (human) operator. We also briefly discuss a total management of the life-cycle of mechatronic systems. We divide the life-cycle of a mechatronic system into four stages: design, development, installation (integration), and operation and maintenance. When a mechatronic device is being designed and built, it is difficult to analyze the behavior of the device before testing it in practice. It is common that many prototypes of the device are tested at different stages of the development process. The main idea of our approach is that a fuzzy model of a system is built as early as possible, and the model is used, for example, in analyzing the validity of design work done so far. The model is updated and used for different purposes throughout the development and installation process. Finally, the model is used as a part of fault diagnosis and control systems.

Keywords: Fault Diagnosis, Knowledge Based Systems, Fuzzy Modeling

1. Introduction

One of the most significant application areas for AI and other intelligent techniques is fault diagnosis. Fault diagnosis has appeared to be a very demanding task for so called traditional methods, for example because of dynamically changing environments and unpredictable phenomena. In recent years, the most used methods have been based on knowledge based systems. [1]

Fuzzy logic has proved to be a useful technique when reasoning in uncertain circumstances or with an inexact information, which is very often the case with fault diagnosis. In this paper we concentrate on possibilities of using fuzzy logic in model-based fault diagnosis systems. The common problem in model-based systems is that the model is always inaccurate. In general, this makes it difficult to define by reasoning whether a fault exists or not.

In this study we present some experience on using fuzzy system models in model-based fault diagnosis. As a fuzzy system model we mean a behavioral model of the system to be analyzed. Experiments of using fuzzy system models in fault detection and localization are presented and discussed. In fault localization tasks we used inverse fuzzy system models.

The original principles of fuzzy modelling were presented in 1973 by Dr. L.A. Zadeh [2]. The principles are based on the assumption that the possibility to make more accurate models with more detailed information becomes difficult with so called traditional methods, and therefore other methods are needed to fulfill the needs. Compared to commonly used fuzzy logic controllers, the basic difference of fuzzy system models is that they produce the process output for given inputs rather than the process input (a control value) based on given input variables. Methods of developing and using different kinds of fuzzy models have been studied by several authors and are presented, for example, in [3 - 6]. A common feature of these studies is that in all of them fuzzy system models are constructed based on expert knowledge rather than mathematical knowledge. Fuzzy system models clearly have properties of both quantitative and qualitative models, but are considered to be more qualitative than quantitative [7].

In the end of the paper we briefly discuss the possibilities of using fuzzy system models at different stages of the life-cycle of mechatronic systems. In our approach, the life-cycle of mechatronic systems is divided into four stages: design, development, installation, and operation and maintenance. Traditionally, knowledge and experience that are acquired at the first stages of the life-cycle have not been used effectively at later stages. In this study the purpose is to form links between the different stages with fuzzy system models. A model is constructed at the early stages of the development of a mechatronic system and is updated at every stage of the development cycle. At first, models are used to assist the development of the system. Behind the scene, this fairly easily leads to a monitoring and fault diagnosis system to be used during normal operation of the device.

2. Model-based fault diagnosis with fuzzy logic

2.1. About fault diagnosis

A fault diagnosis system can be divided into four stages [1]:

1. Preprocessing the measurements.

2. Fault detection.
3. Fault localization.
4. Recovery.

Preprocessing is often needed, for example, to filter measurements or to discard useless information. In the fault detection stage, the operation of a device is monitored and symptoms of possible faults are produced. The task of fault localization is to find out an exact source of the fault. Recovery is generally understood as the procedure of correcting the faulty part of the device and returning the device back to normal operation.

In general, there are two possible knowledge representation ways to handle the fault diagnosis problem: shallow and deep knowledge. In systems based on shallow knowledge, the diagnosis is based on previous fault cases. This approach is useful if the required diagnosis knowledge is available and the number of possible fault types is limited. The latter approach is typically used in cases where shallow knowledge cannot be used.

There are several reasons for using fuzzy logic in diagnosing the operation of a device. Numerous statements encouraging the use of fuzzy logic rather than other AI or statistical techniques are presented in [8]. All of these emphasize the overall difficulty of acquiring the essential diagnostic knowledge, for example, choosing the most reliable knowledge sources and combining information from many inexact sources. The difficulty of knowledge acquisition is a well-known problem in knowledge based systems. An excellent overview of knowledge-based systems can be found, for example, in [9].

In fault diagnosis, there are typically phenomena that are difficult to handle with traditional methods. These include, for example:

- The general difficulty to infer current and especially a near future availability of a system. For example, a small leakage in a hydraulic component may not cause any problems in a week or two.
- The knowledge of appearing faults is often inexact, incomplete, or unreliable. Depending on the case, the existing knowledge can be difficult to use effectively in the fault diagnosis system.
- A fault does not always need quick service. It seems to be sensible to wait for several faults before fixing a device. However, if faults are severe, the device must be serviced quickly.
- In the model-based approach, the model often differs from the measured data. The difficulty is to make the decision whether there really exists a fault or not.
- Referring to previous statements, the advantages of using fuzzy logic are:
- The possibility to also use other than ON/OFF information about the condition of the device.
- The ability to predict incipient faults. With fuzzy indicators, weak symptoms can also be submitted to operators. This also reflects on the overall confidence of the diagnosis system as faults are not declared to be absolutely true.

- The seriousness of symptoms can often be analyzed with fuzzy logic to find out the right time to start the appropriate procedure to fix the problem.
- Inadequate and even inconsistent knowledge can be utilized in fuzzy logic based systems easier than in traditional expert systems.
- Typical of fault diagnosis is that faults do not appear in the same way every time. This causes difficulties to exact fault localization and encourages the use of more vague methods.
- Uncertain results produced by the decision making process can be presented to operators with the help of fuzzy membership functions and fuzzy rules.

Great effort is usually not needed in developing fuzzy logic based systems. Fuzzy logic makes it possible to utilize multiple knowledge sources, and versatile development tools are available. In addition to this, the way of representing knowledge in fuzzy systems supports using the same knowledge in different tasks: the same fuzzy model can, for example, be utilized in both fault detection and localization [10].

2.2. Fuzzy interface between a human and a computer

The model used in our studies describes the normal behavior of a process or a device. With that kind of a model, the fault detection task is usually carried out by comparing the output of the model with the data that is measured from the device: a fault exists if the output of the model differs considerably from the measured information. In complex applications the difference is sometimes not the best way to express the results of the fault detection stage. For example, if an existing fault only produces weak symptons (i.e. the difference is small) or the constructed model is not accurate, additional information may be needed even at the fault detection stage.

In [11] there is presented an approach of using fuzzy reasoning in a case where the behavior of a fuzzy system model (FSM) differs considerably from the measured behavior of a device. A specific fuzzy reasoning module is applied in analyzing the difference (Figure 1).

The methods were developed to monitor the operation and condition of a drilling machine produced by Tamrock Ltd., Finland. The general frame for a model-based fault diagnosis system was built and the behavior of a relatively small hydraulic subsystem was modeled with fuzzy logic. The behavior of the modeled subsystem was unstable and changed from time to time without any apparent reason (fault). In this case, it was impossible to use the difference between the model and the measurements in fault detection. Instead of using the absolute value of the difference in fault detection we calculated a fuzzy estimate to describe how well the model and the measured data corresponded to each other.

To make the approach clear, let us write it in the form of a very simple example of a rulebase. Among others, the next three rules could be included:

IF model_output IS *falling* AND measured output IS *falling* THEN fault_state IS *no_fault*

IF model_output IS *falling* AND measured output IS *stable* THEN fault_state IS *some_fault*

IF model_output IS *falling* AND measured output IS *raising* THEN fault_state IS *full_fault*

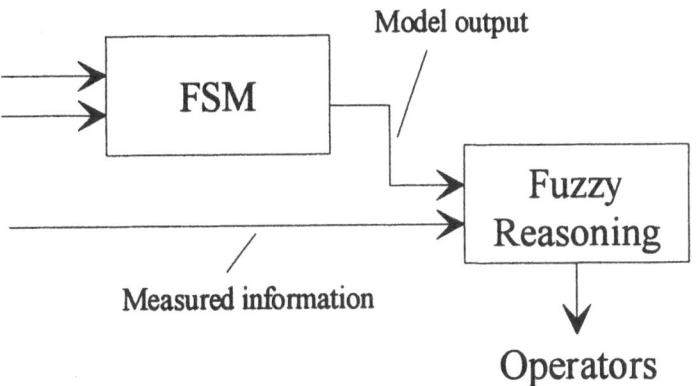

Figure 1. Using fuzzy reasoning to submit information to operators

The approach proved to be useful in a case where the model hardly ever produces similar results compared to the measured data. In Tamrock's case the condition of the modeled subsystem could easily be monitored and, based on our experience, proper monitoring is essential, for example, in deciding whether there is a real need for service in the near future or only much later.

Figure 2 presents the user interface of the system. In the user interface numeric information is presented with graphs that help in examining long term trends, for example.

2.3. Inverse fuzzy models in fault detection and localization

Inversion of a (traditional) process model is familiar to control engineering, but in recent years some research has been carried out on using inverse fuzzy models in control systems [5], [12]. In these studies it has been stated that one advantage of using fuzzy system models is the possibility to use the same model for many purposes: controller design, monitoring, and fault detection.

We have used inverse fuzzy models in fault localization [10]. Figure 3 describes the main parts of the developed fuzzy fault diagnosis system framework.

The methods are based on a fuzzy system model which describes the normal behavior of a device (or a subdevice). The output of the fuzzy system model is compared to the measured data in fault detection to find out if a fault exists or not. Inverse fuzzy system models are used to find out the exact location of the fault.

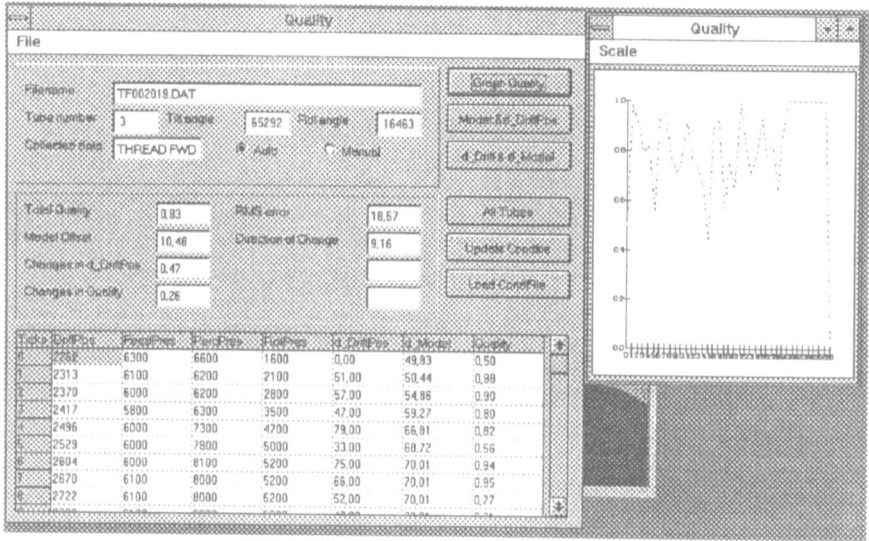

Figure 2. The user interface of the monitoring system

The first task in developing a fault diagnosis system like the one described in Figure 3 is to construct a fuzzy system model. As usual when working with fuzzy systems, there are many possibilities in constructing the model. Probably the most common approach nowadays is to use domain (expert) knowledge in building the rulebase and in setting the membership functions. Data is usually needed in fine tuning the membership functions. The possibility to use data can also be of great help when the validity of the model is analyzed. Our approach was to analyze the model with a tuning method (see [13]). If the model can be tuned with a set of data, the validity of the model is in all probability good. If tuning does not improve the accuracy of the model, redesign and further development are needed.

Complex fuzzy mathematics is usually not necessary in fuzzy system models. For example in [10] the applied methods were the max-dot method in inference and the center of gravity method in defuzzification. Because the normal behavior of a device is modelled, it is possible to use specific learning or adaptation methods in updating the model. Choosing a certain learning method may affect the choice of methods for inference and defuzzification.

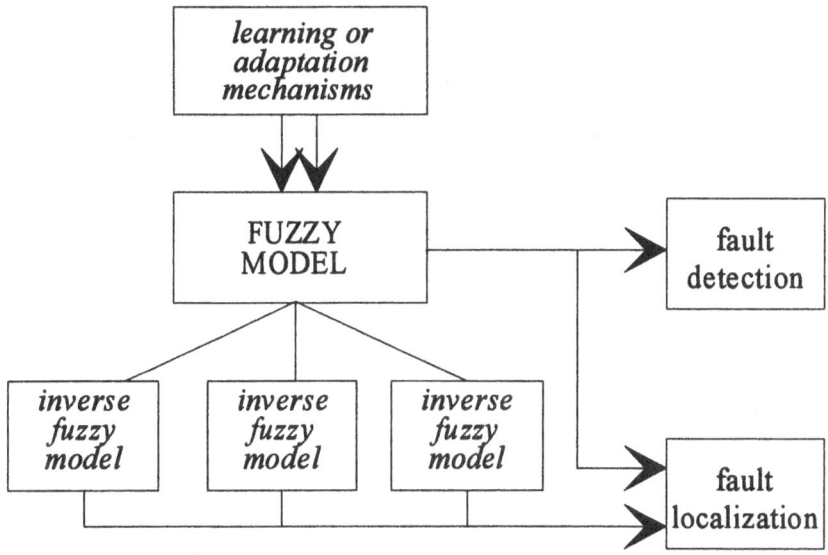

Figure 3. The framework for using a fuzzy system model in fault detection and localization

The inversion of the rulebase of the Mamdani type fuzzy model can be carried out swapping the premise and the consequence. In transferring input membership functions into output membership functions, the position of the peaks of the triangles remains the same. In transferring output membership functions into input membership functions, the functions are set to overlap each other.

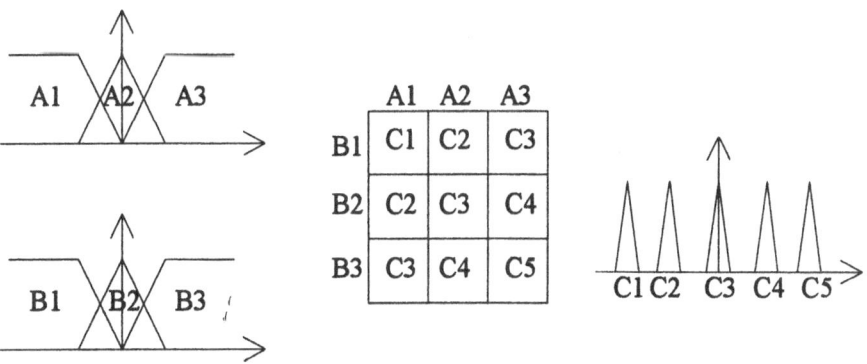

Figure 4a. An example of a fuzzy system model

In Figures 4a-b, an example of the inversion of a fuzzy system model is presented. Figure 4a illustrates a MISO (two inputs and one output) fuzzy system model and Figure 4b presents the inversion of the fuzzy model in Figure 4a. The rulebases are presented in matrices: for example in Figure 4a the rules are of the form *IF Var_A is A_i AND Var_B is B_j then Var_C IS C_k*. When the inversion is carried out, the *Var_C* is swapped for either *Var_A* or *Var_B* (in Figure 5, *Var_C* is swapped for *Var_B*).

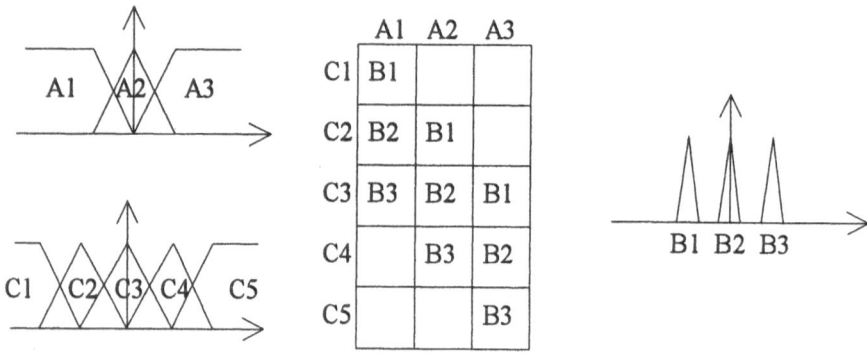

Figure 4b. An inverse fuzzy system model

The empty cells in the matrix of Figure 4b may seem surprising. However, the empty cells can easily be explained in terms of possible cases. Because only the normal behavior of a device or a process is modeled, the impossible combinations of different membership functions do not exist. For example, in Figure 4a it is impossible to have a situation where the *Var_A* is *A1* and *Var_C* is else than *C1*, *C2*, or *C3*. This is why combinations (*A1,C4*) and (*A1,C5*) do not exist in Figure 4b. In developing a fault diagnosis system this must be taken into account: for example, a fault situation may suddenly produce impossible combinations.

The maximum number of inverse fuzzy models is the number of input variables in the model. For example, in the case of Figure 4 the number of possible inverse fuzzy models is two.

In [10] there are described both the methods of using inverse fuzzy system models in model-based fault diagnosis systems and testing them. The methods are based on the assumption that if there is only one faulty variable in the modeled system, it can be found using inverse fuzzy models. In the case of Figure 4a, the fault is detected by comparing the output variable *Var_C* to the corresponding measurement. The fault is localized by comparing the outputs of the inverse fuzzy models to the corresponding measurements. The source of the fault is the variable that acts as an output variable in an inverse fuzzy model that produces a faulty output. If all of the inverse fuzzy models tend to produce a faulty output, then the fault source is the original output variable.

The methods were tested in analyzing the accuracy of the temperature measuring devices of a steel roller. A fuzzy system model based merely on expert knowledge was built. The fault localization methods based on the inversion of the fuzzy model were used together with three different measuring devices. Testing was carried out with manipulated data (faulty information was manually set into some of the measurements). The results of the testing showed that the original source of a fault was found with inverse fuzzy models. It is worth noticing that the results are clear although in this study the model was constructed quickly and without any automatic tuning methods.

Based on these examples, the method seems very promising. However, the reliability of the method should be carefully analyzed before putting it into practice in different applications. Based on our experience, methods to be used in different fault diagnosis applications can only be chosen after identifying the problems. In this case, difficulties may occur if many faults exist at the same time. However, such cases are difficult to handle anyway, regardless of techniques and methods.

2.4 Synthesis of fault diagnosis and control

Combining a fault diagnosis system and a control system appears to be an interesting area of research, although references to real-world applications are difficult to find. In our point of view, there seems to be at least two arguments that encourage using fault diagnosis information in a control system:

1. In some cases a faulty device/process can still be operated if the control system can adapt to the changing circumstances.
2. If the condition of a controlled (sub)system is not taken into account, this may lead to the control system causing more faults or making existing faults more severe.

In other words, the aim is to take into account parameters that change during the operation of a device and to adapt the control system to the changing environment. If this is possible, the device can be operated more or less in the normal way and the required maintenance can be arranged later.

Thinking of the possibilities of applying this into practise, the fault diagnosis information could be used for example in changing the intensity of an existing controller(s) or in changing timing parameters of the controlling system. However, the efforts needed for putting these methods into practise largely depend on the complexity of the system. The information of a fault diagnosis system can only be used in a control system if we can be certain that the fault diagnosis system produces reliable information. This cannot always be guaranteed, especially in complex systems.

3. About the life-cycle of a mechatronic system

The life-cycle of a mechatronic system can be described in several ways depending on the viewpoint (e.g. software engineering, or mechanic system design). It is common that different parts of a mechatronic system are designed and developed quite separately without any veritable links before they are integrated into one total system. [14]

In this study we describe the life-cycle on an extremely general level. Figure 5 presents the stages that are used: design, development, installation, and operation and maintenance.

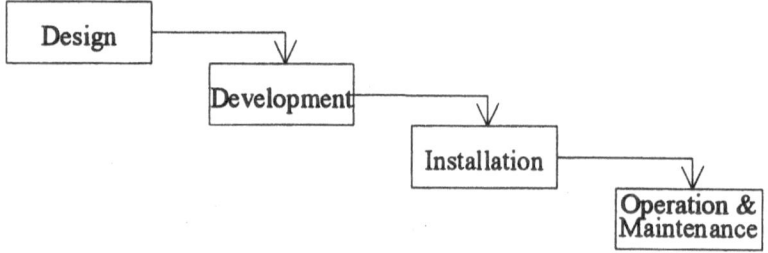

Figure 5. Stages of the life-cycle of a mechatronic system

Considering the stages in Figure 5, the following main tasks are assigned:

1. Design

   - Specifying the demands of the device.
   - Formulating the general structure of the device.

2. Development

   - Dimensioning of different subdevices.
   - Choosing the hardware (microprocessor, operating system, etc.).
   - Coding the software.

3. Installation

   - Integration of mechanical (sub)devices, hardware, and software.
   - Integration tests.

4. Operation

   - Normal operation in real environment.

- Monitoring the operation of the device.
- Diagnosing the condition of the device.
- Services (scheduled or as needed).
- Improving the efficiency of the device.

When a mechatronic device is being designed and built, it is difficult to analyze the behavior of the device before testing it. It is common that many prototypes of the devices are tested at different stages of the development process. Multiple prototyping requires more effort from the software development as even the smallest changes can cause considerable new demands to the software. One way to reduce costs is to ensure the validity and quality of the design and development work before constructing the actual device.

To manage the total life-cycle from the design stage to the operation stage, it is obviously useful to clear out the interfaces between different stages carefully. In other words, knowledge and arguments concerning decisions made at previous stages should be submitted to the following stages to avoid discrepancies in the different stages of the life-cycle of the device.

4. Applying fuzzy system models in different stages of the life-cycle

As discussed in Chapter 2, there are several factors that support using fuzzy logic in different tasks of fault diagnosis systems. Our methods are based on model-based techniques which use especially fuzzy system models. Considering the life-cycle of a mechatronic system (presented in Chapter 3), the main use of fuzzy logic seems to be in the latter part of the life-cycle (Operation and Maintenance).

Possibilities of using fuzzy logic at different stages of the life-cycle are presented in Figure 6. The approach is based on model-based techniques, and the model is built with fuzzy logic. The main idea of the approach is that the model of the system is built as early as possible. The same model is updated and used for different purposes throughout the development and installation stages. Finally, the model is used as a part of fault diagnosis and control systems. The development of a model is at first based on expert knowledge. Later, when knowledge from other sources (for example data) is available, parameters of the model are adjusted to produce similar results as a real system. The similar idea has been used for example in software engineering, where the final software is the collection of several prototypes produced during the life-cycle [15].

Knowledge based modeling is chosen because it has been proven to be useful, especially in the operation stage of the life-cycle [8]. Additionally, there seems to exist a large amount of possibilities for using especially fuzzy logic based models in different tasks [10].

5. Summary and conclusions

In this paper, the possibilities of fuzzy logic in fault diagnosis systems are discussed. In fault diagnosis, uncertainty management is essential. Because of dynamically changing parameters of a device/process, a reliable fault diagnosis system is difficult to develop. With fuzzy logic it is more natural to use uncertain and incomplete information than with AI or statistical methods.

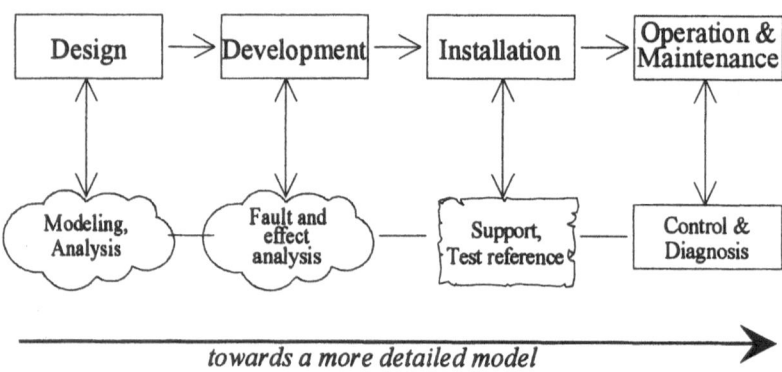

Figure 6. Fuzzy logic in different stages of the life-cycle

In model-based fault diagnosis it has been typical that several types of models are needed to carry out different diagnosis tasks. This can make the development and maintenance of the system complicated. This problem can be reduced by using fuzzy system models. In Chapter 2 a framework of using a fuzzy model in fault detection and inverse fuzzy models in fault localization is presented. An inverse fuzzy model can be derived from a normal fuzzy model without any additional information. The test results of the methods seem promising but as far as we are aware of, the method has not yet been used in real-world applications.

Presenting information to an operator can be a pitfall in fault diagnosis systems. The basic problem is that all the necessary information should be given to an operator and meaningless information should be rejected. The difference between meaningful and meaningless can be difficult to set. In this paper we have presented methods to filter the information with a fuzzy logic module. Our approach was designed for and used in an extremely complex environment, where the behavior of the diagnosed device changes a lot from time to time. In such a case it proved to be impossible to construct a model with acceptable performance. Our methods were tested in diagnosing a hydralic subsystem of a drilling machine and, according to our experience, the long-term condition of the modeled subsystem could be monitored fairly easily.

One prospective use of fault diagnosis systems is to use the information of fault diagnosis systems in controlling devices or processes. With this approach, a device could be operated even if faults exist. To operate a faulty device, an adaptive control system has to be used.

Typical for the development of a mechatronic device is that different design tasks are carried out separately and the results of the design are integrated and tested in the end of the design and development stage. To reduce the need for unwanted changes in mechanics or software we presented an approach, in which a fuzzy model of the behavior of a system is built at the early stages of the design and development procedure. With the model, the behavior can be analyzed without actually constructing the device. The model is updated (tuned) when additional information is available along the progress of the system development. At the operation and maintenance stage the model is used in model-based control systems and in model-based fault diagnosis systems.

References

[1] Kurki, M. (1995): "Model-based Fault Diagnosis for Mechatronic Devices", Espoo: Technical Research Centre of Finland, VTT Publications 223. 116 p. ISBN 951-38-4761-6

[2] Zadeh, L.A. (1973): "Outline of a New Approach to the Analysis of Complex Systems and Decision Processes", IEEE Transactions on Systems, Man, and Cybernetics, SMC-3. pp. 28-44.

[3] Driankov, D., Hellendoorn, H., and Reinfrank, M. (1993): "An Introduction to Fuzzy Control", Springer-Verlag, Berlin, 316 p.

[4] Yager, R.R., and Filev, P.D. (1994): "Essentials of Fuzzy Modeling and Control". John Wiley & Sons, N.Y., 388 p. ISBN 0-471-01761-2

[5] Babuška, R., Sousa, J., & Verbruggen, H.B. (1995): "Model-Based Design of Fuzzy Control Systems", Proceedings of EUFIT'95. Aachen, Germany, August 28-31. Aachen: Verlag Mainz. Vol. II, pp. 837-841. ISBN 3-930911-67-1

[6] Rauma, T. & Alahuhta, P. (1996): "Experiences of Developing a Model Based Fuzzy Logic Controller", Proceedings of FLAMOC'96, Sydney, Australia, January 15-18. Vol. II, pp. 135-139. ISBN 0-646-27132-6

[7] Sugeno, M. & Yasukawa, S. (1993): "A Fuzzy-Logic-Based Approach to Qualitative Modeling", IEEE Transactions on Fuzzy Systems. Vol. 1, no. 1, February 1993.

[8] Ulieru, M. & Isermann, R. (1993): "Design of a Fuzzy-Logic Based Diagnostic Model for Technical Processes", Fuzzy Sets and Systems 58, pp. 249-271.

[9] Johannsen, G. & Alty, J.L. (1991): "Knowledge Engineering for Industrial Expert Systems", Automatica, Vol. 27, no. 1, pp. 97 - 114. Pergamon Press plc. Oxford. Great Britain.

[10] Rauma, T., Kurki, M., & Alahuhta, P. (1996): "An Approach of Using Fuzzy Logic in Fault Diagnosis", To be published in Proceedings of EUFIT'96, Aachen, Germany, September 2-5.

[11] Alahuhta, P. (1996): "An Approach of Using Fuzzy Modelling in Fault Diagnosis of a Mechatronic Device" (in Finnish), University of Oulu, Department of Electrical Engineering, Oulu, Finland. Diploma Thesis, 57 p.

[12] Sousa, J., Babuška, R., & Verbruggen, H.B. (1995): "Adaptive Fuzzy Model-Based Control", Proceedings of EUFIT'95, Aachen, Germany, August 28-31. Aachen: Verlag Mainz, 1995. Vol. II, pp. 865-869. ISBN 3-930911-67-1

[13] Rauma, T. (1996): "Knowledge Acquisition with Fuzzy Modeling", To be published in Proceedings of FUZZ-IEEE'96. New Orleans, LA, September 9-11.

[14] Preston, M.E. (1993): "Mechatronic Product Development Survey and Examples in Mechatronics", CISM Courses and Lectures, no. 338, pp.19-89.

[15] Agresti, W. (1986): "What Are the New Paradigms?", In: Agresti, W. (ed.) New Paradigms for Software Development, Washington D.C., IEEE Computer Society Press, pp. 6-10.

APPLICATION OF FUZZY LOGIC TO HANDOFF CONTROL IN CELLULAR MOBILE COMMUNICATION NETWORKS

Patrick S.K. Leung

Department of Electrical and Electronic Engineering
Victoria University of Technology
P.O. Box 14428
Melbourne, VIC 8001
Australia

1. Introduction

Mobile communications area [4,6] in the past two decades was marked by an extraordinary growth rate and rapid technological advances. In the continental US alone, there are currently over 8 million cellular mobile telephone users that is forecasted to grow to more than 20 million by a turn of the century. On top of this, there are some 65 million cordless telephones and 8 million pagers in operation today. All these radio services are imposing a heavy demand on the already congested frequency spectrum. The situation is further aggravated by the emerging multimedia PCS (Personal Communications System) and wireless mobile computer networks that strive to provide a simultaneous real-time mix of interactive audio, video and data services to every person without an interruption irrespective of time and location. Such exciting developments have triggered an explosive demand for the radio capacity unprecedented in the history of telecommunications. As a result, there is a likewise upsurge of research activities in finding new techniques to improve the wireless communication system capacity in recent times.

In this paper, we look at the application of fuzzy logic to improve the handoff characteristics of cellular wireless communication systems. The effect of different membership functions and decision rules on the performance of a fuzzy-logic-aided handoff procedure is investigated in a typical mobile radio environment. Sugeno inference method is used and the results are compared with the conventional approach.

2. Cellular concept and handoff mechanism

2.1. Cellular concept

Capacity of a wireless network is fundamentally limited by the number of radio channels available to it. A network with N radio channels is capable of serving only N mobile subscribers at any one moment. This limitation is worrying to designers of many public mobile communication networks. Due to the scarcity of radio frequency resources worldwide, it is unrealistic to expect any sizable allocation of radio channels for public radio network usage. But all the same, network designers are charged with the responsibility to provide services to as many mobile subscribers as possible. For instance, the popular GSM (Global System for Mobile) [5] digital mobile communication system is allocated with only 125 full-duplex radio channels and yet is expected to serve many tens of thousands of subscribers simultaneously. Obviously, special frequency management techniques are needed before such public mobile radio system becomes practical.

Traditionally, the traffic capacity of a wireless network is improved by adopting the cellular design concept and cell cite engineering. In this approach, the entire coverage area of the network is divided into smaller regions called "cells" which are then grouped into clusters of K cells each. The available frequency channels are likewise divided into K subsets to be assigned one per cell within the cluster. In this way, the same frequency spectrum can be reused repeatedly in other cell clusters without causing undue intercellular cochannel interference (CCI). As a result, even a small number of radio channels can be made to service a mobile subscriber population size many times over. This design approach forms the basis of many modern cellular digital mobile communication systems such as GSM in Europe and many parts of Asia, IS54 in USA, IS95 in Korea, PHS in Japan and the evolving PCS systems for the future.

As an illustrative example, Figure 1 depicts the so-called "K=4" cellular scheme [6] where the network's entire service area is divided into smaller regions called "cells". These cells are grouped into clusters of four cells each. In this scheme, the available radio channels are also divided into four subsets so that each subset of radio channel is assigned to a different cell in the cluster (see Figure 1). The numerical value (1,2,3 or 4) associated with the cell identifies the frequency subset assigned to it. With careful planning, cells with the same frequency subset are placed as far apart as possible to keep the undesirable intercellular cochannel interference below the acceptable level. It is interesting to note that this "K=4" frequency reuse scheme is adopted in many GSM digital cellular mobile communication systems around the world.

2.2. Handoff mechanisms

In a cellular network, each cell within a cluster is assigned a basestation that operates on its own set of radio frequencies. While a mobile unit is inside a cell's boundary, all its communication needs are handled by the basestation associated with that cell. But when the unit moves towards the cell edge and cross the cell boundary to the adjacent cell, the control of it must be handed over to the basestation associated with the adjacent cell. This process of handing over the control of a mobile unit from one basestation to another constitutes the "handoff"

[6,1,3] operation. Properly designed handoff algorithms are essential for the efficient operation of any cellular radio system.

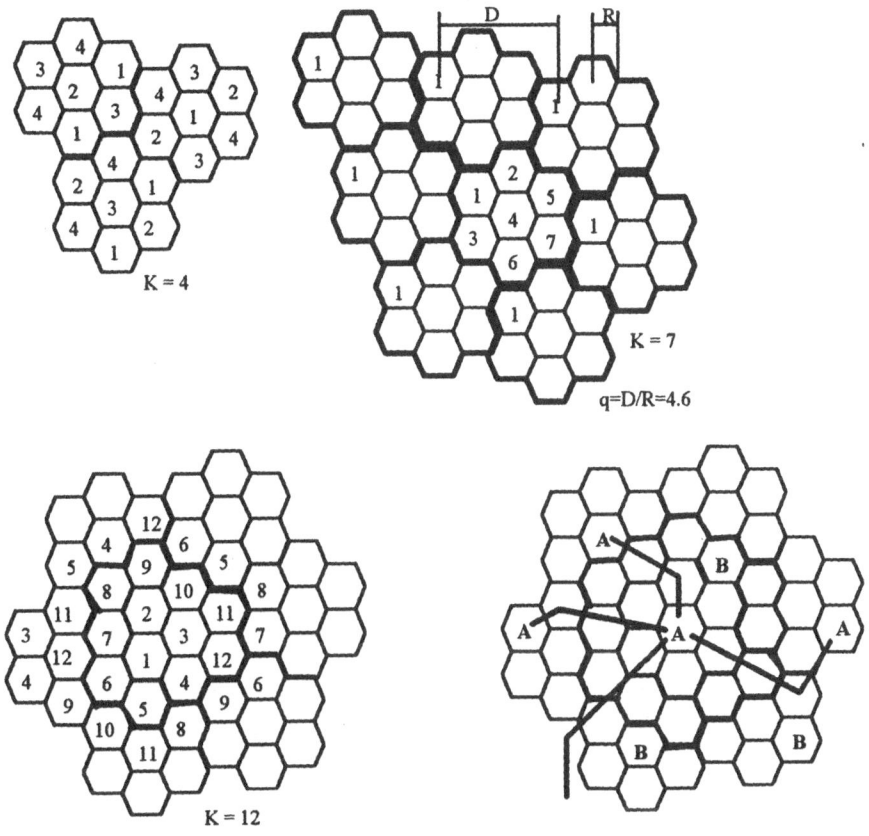

Figure 1. Cellular network structure with K=4, K=7 and K=12 [6]

In principle, the decision to initiate a handoff can be made based on the measurements of (i) the received signal strengths from the communicating and neighboring basestations, (ii) the distance from basestations, (iii) the prevailing signal-to-noise (SNR) or carrier-to-interference (CIR) ratios, and (iv) the bit error rate etc. However for simplicity, received signal strength is often used as the sole indicator for the need of handoff between cells in many practical mobile networks.

In GSM for example, the mobile unit continuously monitors the receive signal strength from several of its neighboring basestations. These measurements are periodically sent back to the controlling basestation and form the basis in deciding whether a handoff between cells should be made or not. This is the so-called MAHO (Mobile Assisted Handoff) scheme. It has the advantage of enabling the network to make more intelligent handoff decisions as the basestation now has information on signal strengths in *both* the forward and reverse radio channels.

To deliver good quality services to the mobile user, the handoff algorithm must be fast acting to avoid any audible interruption on the call. In addition, it must also minimize the number of any unnecessary handoffs. Unnecessary handoffs are the ones made in situations where the previous link would have continued to give satisfactory performance. They affect the call quality (even causing possible loss of the call) and increase the switching and signalling load.

From the switching and signalling point of view, only one handoff should ideally be performed when a mobile crosses a cell boundary. Furthermore, handoff should occur at location as close to the planned cell boundary as possible. This is to avoid the undesirable expansion of a cell's apparent coverage area and hence forcing it to carry more than its planned traffic load. However, good and reliable handoff algorithm is difficult to design due to the stochastic nature of the mobile radio channel as discussed in Section 3.2.

3. Conventional handoff strategy

3.1. Ideal environment

In conventional handoff strategy, the handoff decision is based on the difference d_i of the receive signal strengths from the two competing basestations. In the scenario depicted in Figure 2, the mobile unit is moving from basestation A to basestation B in a constant speed V. A and B are D meters apart and the mobile is currently d meters away from the basestation A.

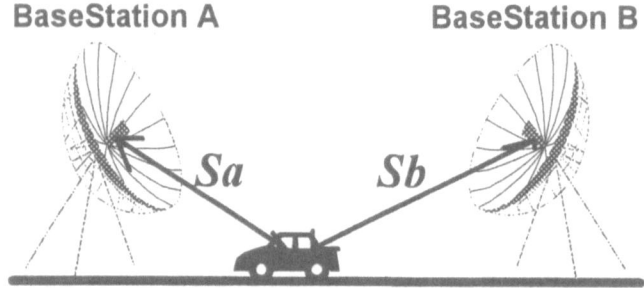

Figure 2. Mobile communications with two basestations A and B

In an ideal environment without shadow fading, the receive signal strengths $S_A(d)$ and $S_B(d)$ from the basestations A and B are given by

$$S_A(d) = K_1 - K_2 \log(d) \qquad (1)$$
$$S_B(d) = K_1 - K_2 \log(D - d) \qquad (2)$$

The constant K_1 relates to the transmit power from the basestation, K_2 characterizes the radio path loss (with $K_2 \approx 30$ being typical in urban environment). Figure 3 plots these as function of d. The curves are seen to be smooth with a clear point of intercept. In this case, the logical place for handoff to occur is obviously the midway point between the two basestations where $S_A(d) = S_B(d)$. This desired result can be achieved by adopting the simple handoff rule that (i) If $di < 0$, no handoff to basestation B occurs where $di = S_B(d) - S_A(d)$, and (ii) if $d_i \geq 0$, then handoff to basestation B.

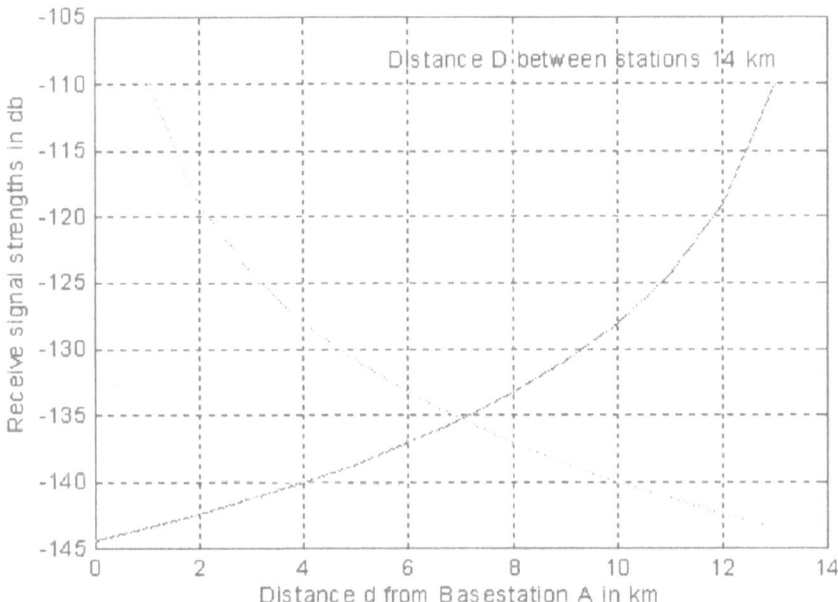

Figure 3. Receive signal strengths as a function of a distance from the basestation A. Ideal environment without shadow fading

3.2. Non-ideal fading environment

Unfortunately, shadowing in the real-world environment makes the receive signal strength unpredictable (see Figure 4). In this environment, the mobile's received signal strengths (in dB) from the basestations A and B are respectively:

$$S_A(d) = K_1 - K_2 \log(d) + u(d) \qquad (3)$$

and

$$S_B(d) = K_1 - K_2 \log(D - d) + v(d) \qquad (4)$$

The random variables $u(d)$ and $v(d)$ account for the random signal fluctuations due to shadowing effect of the surrounding terrain. Result from many propagation measurements shows them to follow the well-known log-normal distribution [6]. Hence mathematically, both $u(d)$ and $v(d)$ are modelled as independent identically distributed zero-mean Gaussian random variates with standard deviation σ_s. Recent propagation measurements [2,3] further show them to be self-correlated with negative exponential auto-correlation functions:

$$E\{u(d_1)u(d_2)\} = \sigma_s^2 a^{-|d_1 - d_2|} \qquad (5)$$

$$E\{v(d_1)v(d_2)\} = \sigma_s^2 a^{-|d_1 - d_2|} \qquad (6)$$

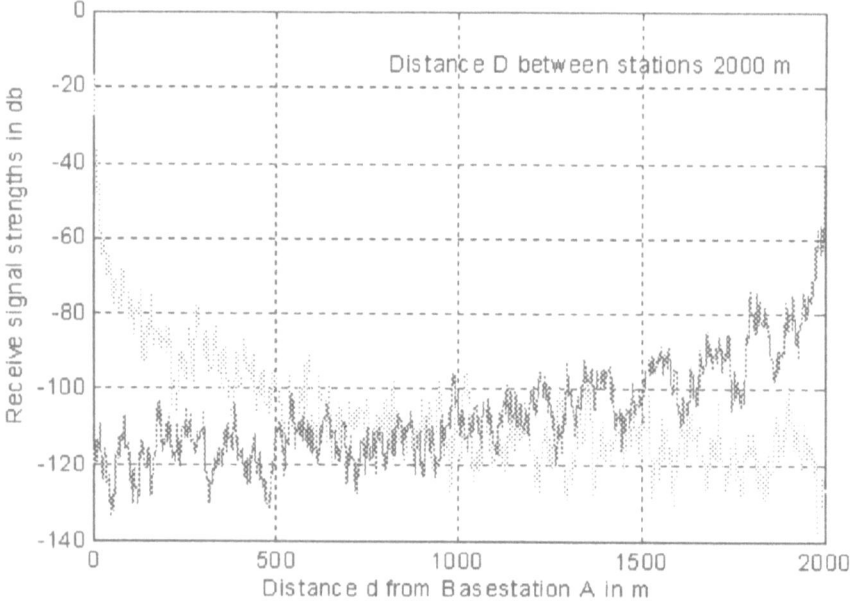

Figure 4. Receive signal strengths as a function of a distance from the basestation A. Real-world environment with log-normal shadow fading

The degree of auto-correlation in these random processes is given by the constant:

$$a = \varepsilon^{\frac{VT}{k}}$$

where ε is the shadow fading's correlation coefficient at two sample points k meters apart. V is the speed of the mobile and T is the time interval between taking samples. Typical shadowing in suburban environments at 900 MHz is generally found to have σ_s between 6 to 8 dB and ε=0.82 over a distance of k=100 m [2].

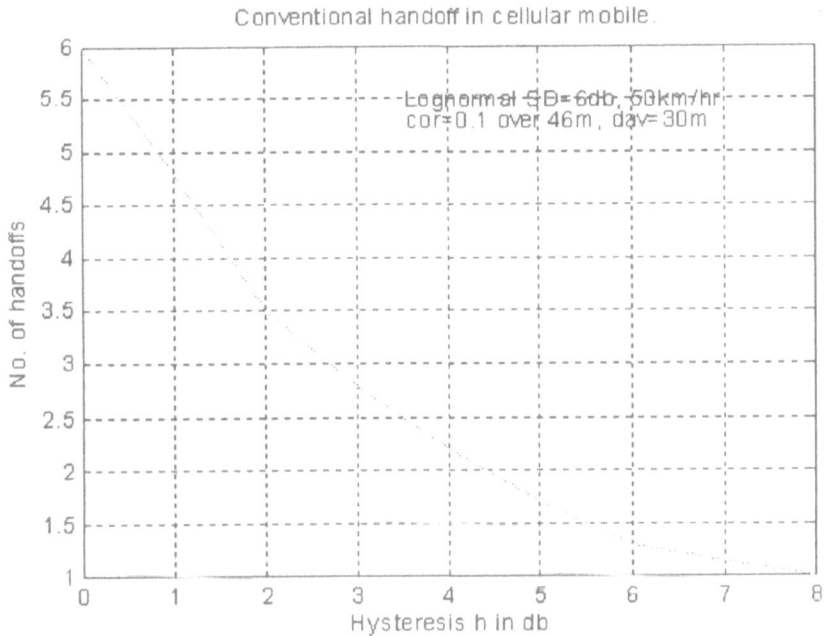

Figure 5. Number of handoffs as a function of hysteresis level with conventional handoff algorithm

This random fluctuation in the receive signal strength causes spurious handoffs which is very annoying to the mobile users. In general, spurious handoffs can be partially avoided by two techniques. They are (i) smooth out the signal variation by averaging, and (ii) introducing a hysteresis h in the decision rule.

Under this circumstance, an improved handoff decision rule is: (i) If $d_i < h$, no handoff from the current station occurs, and (ii) if $di \geq h$, then handoff to the alternative basestation. Here, $d_i=(S_a - S_c)$ is the difference in receive signal strengths. Current basestation is the one currently controlling the mobile unit, and alternative station is the one to which control of the mobile unit can potentially be handed over.

Figure 5 plots the characteristics of a conventional handoff scheme for different hysteresis levels h setting in dB. In this study, the shadow fading follows log-normal distribution with standard deviation $\sigma_s= 6$dB and an autocorrelation factor of 0.1 over a distance of 46m. The mobilenunit is travelling at a speed of 50Km/Hr.

As expected, the number of handoff decreases when the magnitude of the hysteresis increases. For instance, Figure 5 shows that a hysteresis of $h=4$ dB reduces the number of handoff from 6 to 2.2. When h is increased to 8dB, the number of total handoffs is reduced to one. This means a total absence of any spurious handoff with this h value.

4. Fuzzy logic aided handoff

In the conventional handoff algorithm with hysteresis, the handoff decision is based on a hard decision on the receive signal strength difference value d_i and the hysteresis level h. But because of the stochastic nature of the radio channel, d_i is a random variable. As a result, there appears to be an inherent conceptual mismatch between this and the hard-and-fast decision process based on level crossing.

With fuzzy logic [7,8], we can avoid this conceptual incompatibility by allowing the hysteresis level h to be an integral part of the fuzzy logic reasoning process. In this way, both d_i and h can go through the same operations of (i) fuzzification, (ii) rule evaluation, and (iii) defuzzification together. The benefit is a closer match between the random receive signal strength difference d_i and the handoff decision making process. The outcome should be a better handoff decision with a reduction of (i) unnecessary handoffs and (ii) decision delay. In the following, we propose two fuzzy-logic-aided handoff algorithms.

4.1. Algorithm I

This new algorithm for handoff control is based on fuzzy logic with Sugeno inference method. Membership functions of the difference d_i in receive signal strength, as shown in Figure 6, incorporate the hysteresis level. In this algorithm, we define only one rule, viz: If the difference in receive signal strength between the alternative and current basestations $d_i = S_a - S_c$ is positive large (u_{21}), then handoff to the alternative basestation.

In terms of the actual operation, we compute the decision variable $\Gamma = u_{12} - u_{21}$ from the value of d_i according to the membership functions depicted in Figure

6. The decision rule in this case is: (i) if $\Gamma \geq 0$, no handoff occurs, and (ii) $\Gamma < 0$, then handoff to the alternative basestation.

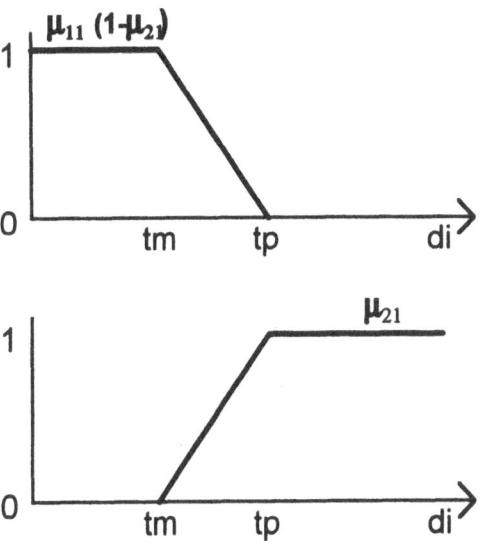

Figure 6. Membership functions for the fuzzy-logic-aided handoff algorithm I

The performance of this fuzzy-logic-aided handoff algorithm is plotted in Figure 7. It shows that the proposed algorithm outperforms the conventional handoff scheme at any hysteresis levels. For instance, with a hysteresis level of h = 4dB, our proposed scheme generates 1.5 handoffs that compares favorably with the conventional scheme that produces 2.2 handoffs.

4.2. Algorithm II

This algorithm improves upon the previous one by having two rules rather than one as before. The membership functions as shown in Figure 8 include S_a as well as d_i . The two rules specified in this algorithm are as follows:

- **Rule 1:** If the signal strength S_a from the alternative basestation is weak (u_{12}), then no handoff occurs
- **Rule 2:** If the signal strength S_a from the alternative basestation is strong ($1 - u_{12}$), and the difference $d_i = S_a - S_c$ is positive large (u_{21}), then handoff to the alternative basestation.

With this algorithm, we compute the decision variable $\Gamma = u_{11}*u_{12} - u_{21}*u_{22}$ from the prevailing d_i and S_a values and the membership functions shown in

Figure 8. The decision rule remains: (i) if $\Gamma \geq 0$, no handoff occurs, and (ii) if $\Gamma <$ 0, then, handoff to the alternative basestation.

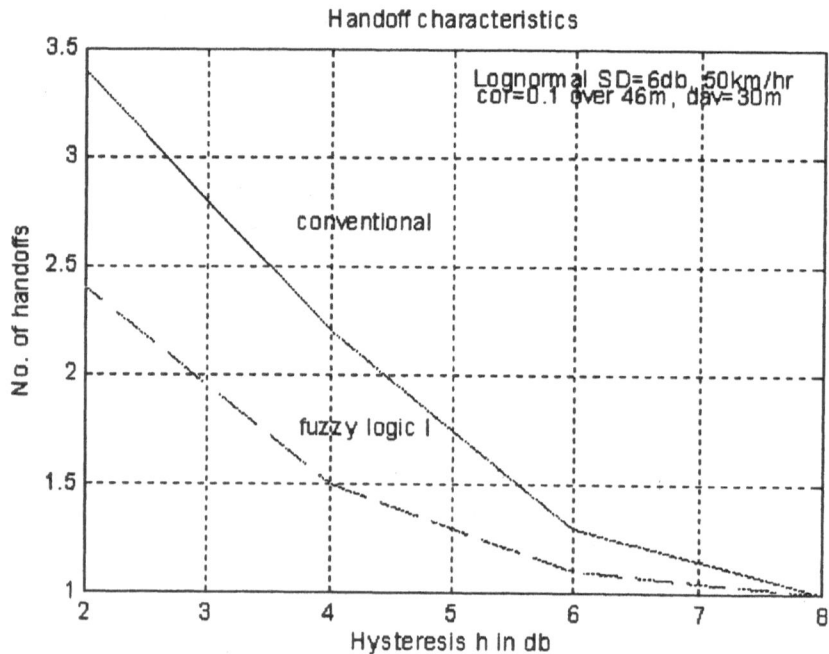

Figure 7. Number of handoffs as a function of hysteresis level with the fuzzy-logic-aided handoff algorithm I

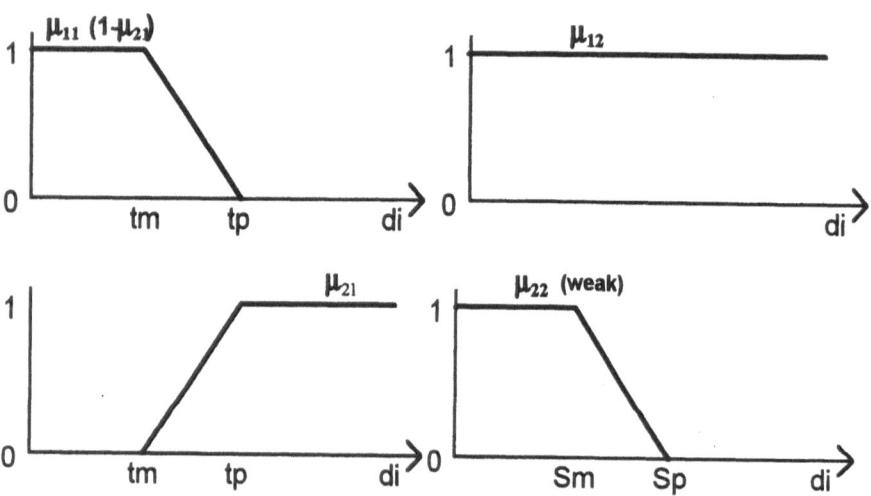

Figure 8. Membership functions for the fuzzy-logic-aided handoff algorithm II.

Performance of this improved handoff control algorithm is plotted in Figure 9. It is seen to outperform significantly the conventional method. For example, with a hysteresis of h=4dB, this algorithm generates an average of 1.1 handoff which compares well with the 1.5 handoffs with the previous algorithm and 2.2 handoffs with conventional method.

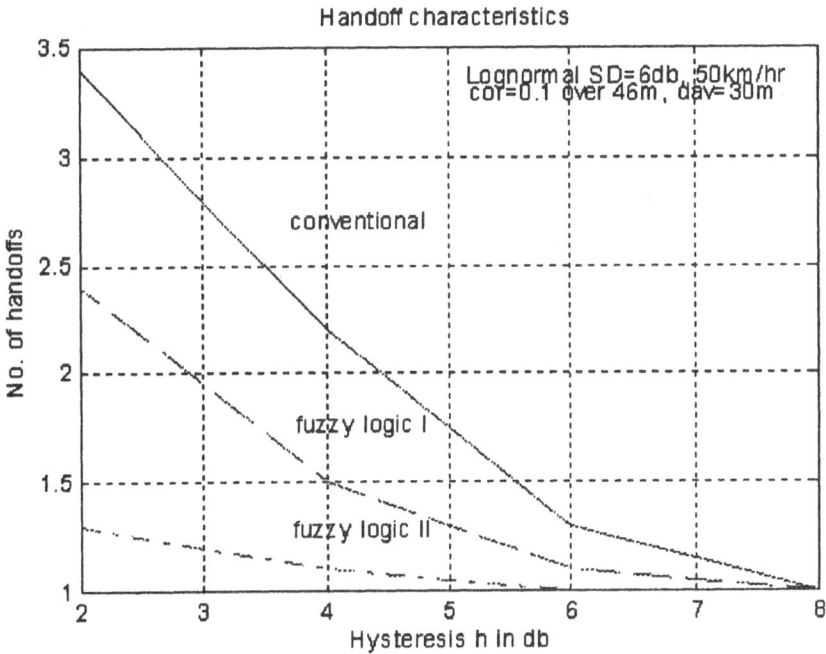

Figure 9. Number of handoffs as a function of hysteresis level with the fuzzy-logic-aided handoff algorithm II

5. Conclusion

In this investigation, the handoff strategies in cellular mobile communication systems are reviewed. A new handoff control mechanism using fuzzy logic is proposed. Preliminary investigation result shows that fuzzy logic is capable of reducing the number of spurious handoffs experienced in cellular mobile communication environments. We also illustrate that further performance improvement can be obtained by introducing more elaborate fuzzy logic rules. We thus conclude that fuzzy logic is a promising technique for improving the handoff characteristics of cellular mobile communication systems.

References

[1] R. Vijayan, J. Holtzman, "A Model for analyzing handoff algorithms", IEEE Trans. on Vehicular Technology, Vol.42, No.3, pp.351-356, August 1993.

[2] M. Gudmundson, "Correlation model for shadow fading in mobile radio systems", IEE Electronic Letters, Vol.27, No.23, pp.2145-2146, Nov. 1991.

[3] M. Gudmundson, "Analysis of handover algorithms", IEEE Proceedings of Vehicular Technology Conference, pp.537-542, May 1991.

[4] D. Goodman, "Second generation wireless information network", IEEE Trans. Veh. Tech. Vol.35, pp.366-374, 1991.

[5] M. Rahnema, "Overview of the GSM system and protocol architeture", IEEE Communications Magazine, pp.92-100, April, 1993.

[6] W. Lee, "Mobile Cellular Telecommunications Systems", 2nd edition, McGraw-Hill Book Company 1995.

[7] L.A. Zadeh, "Outline of newapproach to the analysis of complex systems and decisionprocesses", IEEE Trans. Systems, Man and Cybemetrics, Vol. SMC-3, No.1, pp28-44, 1974.

[8] K.S. Ray, "Analysis design implementation and critical appreciation of fuzzy logic controller, Fuzzy reasoning in information, decision and control systems", pp.199-275, Kluwer Academoic Publishers, Netherlands, 1994.